T0337872

ULTRASONIC TECHNOLOGY FOR DESICCANT REGENERATION

ULTRASONIC TECHNOLOGY FOR DESICCANT REGENERATION

Ye Yao

Shanghai Jiao Tong University, Shanghai, China

Shiqing Liu

Zhejiang Normal University, Jinhua, China

Library of Congress Cataloging-in-Publication Data

Yao, Ye.
  Ultrasonic technology for desiccant regeneration / Ye Yao, Shiqing Liu.
      pages cm
  Includes bibliographical references and index.
  ISBN 978-1-118-92160-9 (cloth)
  1. Drying agents–Drying. 2. Ultrasonic waves–Industrial applications. 3. Air conditioning–Equipment and supplies. 4. Ultrasonic cleaning. I. Liu, Shiqing (Physicist) II. Title.
  TP159.D7Y36 2014
  660'.28426–dc23

                                                                2014016564

A catalogue record for this book is available from the British Library.

ISBN: 9781118921609

Typeset in 10/12pt TimesLTStd by Laserwords Private Limited, Chennai, India
Printed and bound in Singapore by Markono Print Media Pte Ltd

1   2014

Contents

About the Authors

Dr Ye Yao is an Associate Professor at the School of Mechanical Engineering, Shanghai Jiao Tong University, China. He received his PhD from Shanghai Jiao Tong University (SJTU), China. He was promoted as Associate Professor of SJTU in December 2008. From September 1, 2009 to September 1, 2010, he performed his research work in the Ray W. Herrick Laboratory at Purdue University (PU), USA. He was awarded as Excellent Reserve Youth Talent (First Class) and SMC Excellent Young Faculty by SJTU, respectively, in the year 2009 and 2010, and got the Shanghai Pujiang Scholars Talent Program in the year 2012. His current interests of research mainly include: (1) heat and mass transfer enhancement assisted by ultrasound; and (2) HVAC modeling and optimal control for energy conservation. He has successfully published about 100 academic publications and 30 patents and one academic monograph (sole author). He is now the peer reviewer of many international academic journals, such as the *International Journal of Heat and Mass Transfer*, *International Journal of Thermal Sciences*, *International Journal of Refrigeration*, *Energy*, *Building and Environment*, *Energy and Buildings*, and *Applied Energy*.

Dr Shiqing Liu is a Professor at the School of Mathematical and Information Engineering, Zhejiang Normal University, China. He received his PhD from Shanxi Normal University, China. His current interests of research are mainly applied acoustic and ultrasound transducers. He has published about 40 academic publications and over 10 patents in his research domains.

Preface

With global warming and the rapid improvement of people's living standards, energy consumption by air conditioning (AC) systems in buildings is on the rise. It has been noted that the dehumidification process accounts for a large proportion of energy consumption by an AC system. In southern areas of China where the climate is very hot and humid, the percentage of energy to be consumed by the dehumidification process in an AC system will be more than 40%. By using adsorption/absorption dehumidifying technology, the heat and moisture load of air can be processed separately, and a higher energy efficiency will be achieved compared with the conventional cooling dehumidification method. In addition, no condensation of water happens during the air dehumidification process with the adsorption/absorption method, which effectively prevents virus and mold from breeding, and hence improves indoor air quality (IAQ). Therefore, people are paying more attention to the adsorption/absorption dehumidifying method as the key technology for developing high-performance of AC systems.

Regeneration of desiccant is a crucial process during the air dehumidification cycle with the adsorption/absorption method. It will produce great influence on the energy efficiency of desiccant AC systems. The conventional regeneration method by heating is found to be energy-wasting due to the relatively higher regeneration temperature of some desiccant materials. So, we have put forward the ultrasound-assisted regeneration method in this book. The fundamental theory of the novel regeneration method is summarized as follows: ① The mechanical effect of ultrasound causes a series of rapid and successive compressions. This can reduce the thickness of boundary layer near the surface of solid desiccants and bring about the enhancement of mass transfer during regeneration. Meanwhile, the ultrasonic heating effect causes a temperature rise in solid desiccants and enhances internal moisture diffusivity known as "rectified diffusion." ② For liquid desiccants, the cavitation effect induced by power ultrasound sprays the solution into numerous tiny droplets with a size range of 40–80 μm, which improves the regeneration rate of liquid desiccants through enlarging the contact area between the air and the desiccant solute instead of increasing the solution temperature.

The study in this book demonstrates that ultrasound-assisted regeneration can significantly increase energy efficiency of regeneration, shorten regeneration time and hence improve performance of the desiccant AC system. In addition, the temperature for regeneration can be reduced by introducing power ultrasound, which provides favorable conditions for the utilization of low-grade thermal energy (e.g., solar energy and waste heat) in the desiccant regeneration.

This book is edited based on recent studies on ultrasound-assisted regeneration. It consists of six chapters as below:

Chapter 1 introduces the background of the topic to be illustrated in this book; it includes a literature review on up-to-date technologies related to desiccant materials, desiccant dryer systems and regeneration methods, and gives basic knowledge about ultrasound and methods for producing ultrasound.

Chapter 2 deals with models for ultrasound-assisted regeneration for silica gel, presenting experimental and theoretical results and including a parametric study of the new regeneration method.

Chapter 3 investigates the effect of ultrasound on the regeneration of a new honeycomb-type desiccant and includes a parametric study on ultrasound-assisted regeneration.

Chapter 4 introduces the mechanism of the ultrasound-assisted regeneration for the liquid desiccants, and studies the effects of the ultrasonic atomization on the liquid desiccant regeneration.

Chapter 5 deals with the working principle and design calculation method for longitudinal and radial vibration ultrasonic transducers that have potential applications in ultrasound-assisted regeneration.

Chapter 6 presents several desiccant air-conditioning systems in which ultrasound-assisted regeneration is employed.

The book is written by Dr Ye Yao (Associate Professor at the Institute of Refrigeration and Cryogenics, Shanghai Jiao Tong University, China) and Dr Shiqing Liu (Professor at the Institute of Mathematics and Physics, Zhejiang Normal University, China). Chapters 1, 2, 3, 4 and 6 as well as the appendix have been written by Dr Ye Yao, and Chapter 5 has been written by Dr Shiqing Liu and Dr Ye Yao.

Acknowledgements

The study work related to the book has been financially supported by Shanghai Pujiang Program (2012) and several National Nature Science Foundations (No.50708057; No.11274279; No.11074222) as well as Shanghai Jiaotong University Academic Publishing Fund (2013). Meanwhile, this book has been successfully chosen as China Classics International Academic Publishing Project (2014). The publication of the book will be an important reference for related research fields.

I would like to express my appreciation to those who have educated, aided and supported me: my mentors Prof. Ruzhu Wang (Shanghai Jiao Tong University), Prof. Guoliang Ding (Shanghai Jiao Tong University), Prof. Xiaosong Zhang (Southeast University, China); my collaborators Mr Beixing He (senior engineer at the Institute of Acoustics, Chinese Academy of Sciences) and Prof. Houqing Zhu (Institute of Acoustics, Chinese Academy of Sciences); and my students, including my PhD Candidate Yang Kun (who designed most of the computer programs), Dr Weijiang Zhang (who carried a large number of experimental studies related to this book), my Master Candidate Godwin Okotch (who revised the language errors), Weiwei Wang and Zhengyuan Zhu (who participated in some experimental tests and measurements).

Finally, I offer my heartfelt gratitude to the editorial director Dr Fangzhen Qian and Mrs Yingchun Yang at Shanghai Jiao Tong University Press for their help, cooperation, advice and guidance in preparing this edition of the book.

Ye Yao
Shanghai Jiao Tong University
December 30, 2013

Nomenclature

a_u Ultrasonic absorptivity by medium

A_o Pre-exponential factor of Arrhenius equation, m^2/s

A_w Activity of water

A_ϕ Debye-Huckel constant for the osmotic coefficient

AEE   Average energy efficiency, %

AMR   Additional moisture removal brought about by ultrasound, kg

AMRC   Additional moisture removal capacity brought about by ultrasound, kg/s

ASEC   Adiabatic specific energy consumption, J/(kg moisture desorption)

B Standard atmosphere pressure, Pa

c Specific heat, J/(kg.°C)

c_Y Specific heat ratio

cos   Cosine function

C_m Mechanical quality factor of transducer

C_w Moisture concentration in the mainstream air, kg/m^3

$C_w{}^*$ Concentration on the surface of liquid droplet, kg/m^3

COP   coefficient of performance

CR   Contribution ratio of ultrasonic effect to the total enhancement of regeneration

CRT   Conditioned regeneration time, s

d Diameter, m

D Diffusion coefficient, m^2/s

DCOP   Dehumidification coefficient of performance

E   Energy consumption, J; or NRTL binary interaction energy parameter; or Young's modulus of the material, Pa; or electric field

E_a Activation energy, kJ/mol

EP   Enhancement percentage of regeneration, %

cosh	Hyperbolic cosine function	ER	Enhanced ratio of regeneration brought by ultrasound
cot	Cotangent	ERE	Experimental relative error
c_o	Adiabatic sound velocity in the air, m/s	ERARR	Enhancement ratio of average regeneration rate
$c_{e,cp}$	Equivalent vibration velocity, m/s	ERERR	Experimental relative error of regeneration rate, %
C	Heat capacity rate, W/°C	ESR	Energy-saving ratio
C_o	One-dimensional cutoff capacitance of the piezoelectric ceramic, F	ESEC	Excess specific energy consumption, J/(kg moisture desorption)
C_f	Drag coefficient	f	Acoustic frequency, Hz
f_c	Activity coefficient	MAMR	Maximum additional moisture removal, kg
F	Force, N	MEEU	maximum energy efficiency of ultrasound, %
g	Gibbs energy of molecules; or acceleration of gravity, m/s^2	MMD	Mass mean diameter, m
g'	Derivative of equilibrium isotherm	MR	Dimensionless moisture ratio
g_{33}	Voltage constant of piezoelectric ceramic	MRC	Moisture removal capacity, kg/s
G	Mass flow rate, kg/(m^2.s)	MRS	Mean regeneration speed, kg/s
h	Enthalpy, J/kg; or height, m	MRE	Mean Relative Error, %
H	Adsorption (desorption) heat of desiccant, kJ/(kg water)	n	Electromechanical conversion factor of piezoelectric ceramic
H_m	Coefficient of heat transfer, W/(m^2.C)	N	Molar flux, mol/(m^2.s); or number of droplets or piezoelectric ceramic wafers
i	Unit of the imaginary number	Nu	Nusselt number
I	Sound intensity, W/m^2; or electric current, A	NRTL	Nonrandom two-liquid theory
I_x	Ionic strength in mole fraction scale	p	Pressure or tensile stress, Pa
$J_0(x)$	The zero-order Bessel function of the first kind	P	Power, W
$J_1(x)$	The first-order Bessel function of the first kind	PE	Prediction error, %
k	Wave number, 1/m	q	Moisture ratio in medium, kg water/(kg dry medium)
k'	Modified complex wave number	Q	Heat trtansfer rate, W
K_m	Coefficient of mass transfer or mass transfer flux, kg/(m^2.s)	r	Radius, m
l	Length, m	r_0	Latent heat of vaporization of water at 0°C, J/kg

L	Height of the packed bed or thickness of particle surface layer or mean free path, m	R	Dynamic flow resistance, kg/(m^2.s); or gas constant, kJ/(mol.K)
m	Mass, kg	RD	Regeneration degree
M	M-type honeycomb desiccant; or molecular weight, kg/kmol	R_m	Mass transfer resistance, m$^2 \cdot$ s/kg
R_V	Vibration speed ratio of the front surface to the rear surface of the transducer	SPL	Sound pressure level
RR	Regeneration rate, kg/s	t	Temperature
Re	Reynolds number	tan	Tangent function
RE	Regeneration enhancement, kg/s; or regeneration effectiveness	tanh	Hyperbolic tangent function
s	Strain, m/m	T[t]	Temperature, K [°C]
sc	Strain constant	TSEC	Total specific energy consumption, J/(kg moisture desorption)
ss$_r$	Elastic flexibility coefficient, m^2/N	u	Velocity or sound wave propagation speed, m/s
sin	Sine function	u_s	Induced velocity of air due to ultrasonic oscillation, m/s
s^D_{33}	Elastic flexibility coefficient under constant axial electric displacement, m^2/N	U	Overall heat transfer coefficient of heat exchanger, W/(m^2.°C)
s^E_{33}	Elastic flexibility coefficient under constant axial electric field, m^2/N	UF	Ultrasonic frequency, Hz
		UP	Ultrasonic power, W
sinh	Hyperbolic sine function	V	Volume, m^3; or voltage, V
S	Area, m^2	w^*	Humidity of air on the surface of solid or liquid, kg/(kg dryair)
SEC	Specific energy consumption, J/(kg moisture desorption)	w	Humidity of air in the main stream, kg/(kg dryair)
SMD	Sauter mean diameter	x	Distance or spatial space, m; or mole fraction in the mixture; or concentration by mass
Sh	Sherwood number	$Y_0(x)$	The zero-order Bessel function of the second kind
Sc	Schmidt number	$Y_1(x)$	The first-order Bessel function of the second kind
S_V	Volumetric surface area of solid desiccant, m^2/ m^3	z (or Z)	Acoustic impedance in medium, Pa.s/m^3
SS	ss-type honeycomb desiccant		

Greek Letters

δ	Thickness, m	φ	Relative humidity, %
λ	Coefficient of thermal conductivity, W/m·°C; or wave length, m	ρ	Density, kg/m^3
v	Kinematic viscosity, m^2/s; or Poisson's ratio	η	Working efficiency
α	Sound attenuation coefficient in medium; or NRTL non-randomness factor	ϕ	Electric displacement
β_{33}^T	Voltage constant of the piezoelectric ceramic	τ	Time, s
ε	Void fraction of desiccant bed; or porosity of particle; or effectiveness of a heat exchanger	Υ	Shear modulus of material
ε_{33}^T	Dielectric constant	ψ	Structure factor of the packed bed
ε_{m-e}	Transverse electro-mechanical coupling coefficient	κ	Slope of the straight line
ε_e	Effective electro-mechanical coupling coefficient	μ	Dynamic viscosity, Pa.s
ω	Acoustic angular frequency or resonance angular frequency of transducer, rad/s	γ	Mass flow rate ratio of liquid desiccant to air; or the ratio of the outer radius to the inner radius of the metal cylindrical shell
σ	Surface tension, N/m; or stress, Pa	ς	Tortuosity factor
χ	Closest approach parameter of the Pitzer-Debye-Huckel equation	ϑ	Extension factor of the conical rod
ξ	Vibration displacement, m	Δ	Increment or absolute error
$\dot{\xi}$	Vibration velocity, m/s		

Subscripts

a	Air	e	Equivalent size of pore in the packed bed; or equilibrium state; or effective
ads	Adsorption	env	Environmental or ambient
ave	Average	f	Falling
c	Cool fluid	fc	Front cover of the transducer
dry	Dry sample	g	gas
des	Desorption	h	Hot fluid

deh	Air dehumidifier	hx	Heat exchanger
in	Inlet	*qb*	Saturated sate
i	Inner surface	SRC	Short-range contribution to the activity of molecules
ini	Initial state	reg	Regenerator or regeneration
K	Knudsen	rc	Rear cover of the transducer
LRC	Long-range contribution to the activity of molecules	*s*	Solid desiccant or liquid droplet
m	Mechanical; or mean value	syn	Synergistic effect
mcs	Metal thin-walled cylindrical shell	ta	At the regeneration air temperature
mole	Molar	tar	Target (or terminal) value
min	Minimum	teff	Ultrasonic heating effect
max	Maximum		
mcs	Metal cylindrical shell	ts	At the solution temperature
NU	Without ultrasonic radiation	T	Ultrasonic transducer
o	On the radiation surface of ultrasonic transducer; or outer surface	U	In the presence of an ultrasonic field
ord	Ordinary	v	At constant volume
out	Outlet	veff	Ultrasonic mechanical effect
p	At constant pressure	vap	Vapor or vaporization
pc	Piezoelectric ceramic	w	Moisture or water

1

Introduction

1.1 Background

Dehumidification is an important air-handling process in the air-conditioning system, which aims at reducing the level of humidity in the air. This is usually for health reasons, as humid air can easily result in mildew growing inside a residence and causing various health risks [1]. It is also necessary in many industrial or agricultural situations where a certain low level of air humidity ought to be required to be maintained. Traditionally, the moist air is commonly dehumidified through the refrigerant cooling method, that is, the air is first cooled below the dew-point temperature (DPT) to condense moisture out, and then reheated to a desired temperature before it is delivered to the occupied spaces. This method not only results in additional energy dissipation due to the cooling–heating process, but also works against the energy performance of the chiller system because of the lower refrigerant evaporating temperature required. To improve the energy efficiency of the air-conditioning system, an independent humidity control system that integrates liquid/solid desiccant devices with a conventional cooling system has been developed to separate the treatment of sensible and latent load of moist air [2–4]. This system can bring about high chances of energy conservation, for example, avoiding excess cooling and heating, utilizing waste heat rejected by machines [5], and solar energy [6] to accomplish the dehumidification. Furthermore, dehumidification with dehumidizers has been proved to be beneficial for improving IAQ (Indoor Air Quality) [7].

As shown in Figure 1.1, the working cycle of the desiccant dehumidification method consists of the following three stages: adsorption (from A to B), regeneration or dehydration (from B to C), and cooling (from C to A). During the repeated adsorption–regeneration–cooling cycle, the regeneration conditions will produce great influence on the performance of water vapor adsorption on dehumidizer [8]. Although the higher regeneration temperature will contribute to increasing the desiccant volume of dehumidizer, it may be disadvantageous to the energy efficiency of the desiccant system because high-temperature regeneration will not only consume a large amount of thermal energy for heating, but also result in much energy dissipation during the following process of cooling. In addition, a higher regeneration temperature is not beneficial for utilizing the low-grade thermal energy which can be freely available. From this point of view, new types of dehumidizers [9] with lower regeneration temperatures are attractive in this field of application. In fact, the limitation of high regeneration temperature for the traditional

Ultrasonic Technology for Desiccant Regeneration, First Edition. Ye Yao and Shiqing Liu.

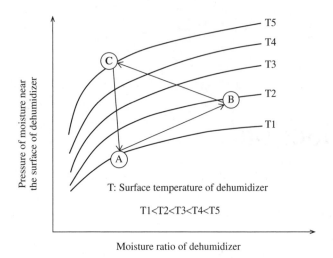

Figure 1.1   Working cycle of dehumidification using dehumidifier

desiccants may be overcome by some nonheating dehydration methods, such as centrifugal forces [10], pulsed vacuum [11], pulsed corona plasma [12, 13], and electro-osmotic [14, 15]. These nonheating methods help enhance heat and mass transfer in media through some kind of physical force, and this makes it possible to reduce the regeneration temperature of desiccants.

The high-intensity ultrasound is another nonheating technology that can improve the dehydration process of moist material, and hence becomes a promising regeneration method for desiccants [16]. Some researchers [17–21] have studied the regeneration of some adsorbents with ultrasound, such as the activated carbon and resins of different sorts. They found that the special effects (e.g., cavitation and micro-oscillation) induced by high-intensity ultrasound could overcome the affinity of the adsorbed species with the adsorbent surface and accelerate the molecular transport toward and from the adsorbent surface, and hence, the regeneration rate of these adsorbents would be greatly improved under the impact of ultrasonic radiation. The regeneration of the adsorbents for water treatment is made in the solid–liquid system where the mass transfer occurs on the interface between solid and liquid, while the desiccant regeneration is often performed in either solid–gas or liquid–gas environment.

In recent years, we have made a series of studies on the new regeneration method with ultrasound [22–58], based on which this book is edited. The primary objectives of this book are to illustrate clearly the impact of ultrasound on the regeneration of solid and liquid desiccants used in the air-conditioning systems, and manage to reveal the mechanism of regeneration enhancement brought by ultrasound. Meanwhile, the design theories of the ultrasonic transducers for the desiccant regeneration are discussed, and the desiccant air-conditioning systems based on ultrasound-assisted regeneration are proposed.

1.2   Literature Reviews

1.2.1   Desiccant Materials

Desiccants are a class of adsorbents/absorbents that have a high affinity for water vapor. They can be either liquids or solids. Examples of liquid desiccants are salt solutions, such as lithium chloride or calcium chloride, and some organic liquids, such as triethylene glycol. Since the 1930s, liquid desiccants have been used in industrial dehumidifiers. The liquid desiccants used in these systems commonly are very strong solutions of ionic salts of lithium chloride and calcium chloride. These ionic salts have the attractive characteristic that the salt themselves have essentially zero vapor pressure, and so vapors of the desiccant will not appear in the air supplied by the liquid desiccant air-conditioning system (LDACS). However, the liquid desiccants (chemically related salt solution, for example, solutions of lithium and calcium chloride) are normally corrosive. This corrosiveness requires that all wetted parts within the LDACS be protected and that no droplets of desiccant are entrained in the supplied air.

Solid desiccants are highly porous materials that adsorb water by mechanisms of chemical adsorption of water molecules onto sites on the walls of the pores, physical adsorption of successive layers of water molecules, and capillary condensation within the pores. Examples of solid desiccants are silica gel, molecular sieves, and natural zeolites. Silica gel, which is made of highly porous amorphous silicon oxide binding water molecules in random inter-section channels of various diameters, has been widely used for the air dehumidification and cooling system [59–63] due to its relatively lower regeneration temperature compared with other desiccants like molecular sieves. The silica gel's adsorption capacity is relatively small at low humidity levels but increases as humidity rises. Figure 1.2 compares linear isotherm with typical isotherms for molecular sieve and silica gel [64]. It shows that the silica gel has a larger moisture adsorption capacity than the molecular sieve when the air humidity is above 40%. At room temperature in saturated air, silica gel will pick up 35–40% of its weight in moisture.

To develop the new desiccant materials with improved sorption capacity and low regeneration temperature for the air-conditioning applications, the hybrid desiccant materials impregnating a host porous material (silica gel, vermiculite) with hygroscopic salt (calcium

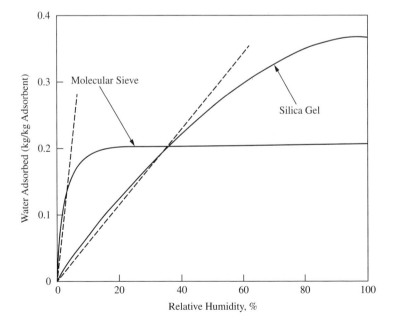

Figure 1.2 Linear approximation to properties of silica gel and molecular sieve

chloride, lithium chloride) have been put forward and studied by Aristov *et al.* [65–69]. Thoruwa *et al.* [70] developed low-cost solar regenerative solid clay–$CaCl_2$-based desiccants to continue the drying process at night. The compositions of the desiccants were given by percentage of mass. The bentonite–$CaCl_2$ (type 1: 60% bentonite, 10% $CaCl_2$, 20% vermiculite, and 10% cement) desiccant had a maximum moisture sorption of 45% (dwb (dry weight basis)). The moisture sorption of Bentonite $CaCl_2$ (type 2: 65% bentonite, 5% $CaCl_2$, 20% vermiculite, and 10% cement) and kaolinite–$CaCl_2$ (type 3: 65% kaolinite, 5% $CaCl_2$, 20% vermiculite, and 10% cement) desiccants were the same with values of 30% (dwb). Liu and Wang [71] obtained a composite adsorbent used for air dehumidification by impregnating silica gel with calcium chloride. And Jia *et al.* [72], developed a new composite desiccant material for the high-performance cooling system. The composite desiccant was a two-layered material that consists of a host matrix with open pores (silica gel) and a hygroscopic substance (lithium chloride) impregnated into its pores. The pore surface area of the composite desiccant was $194\,m^2/g$ and the pore diameter was 3.98 nm.

1.2.2   Types of Desiccant Dryer

Each type of desiccant materials has its advantages and disadvantages. The best desiccant material has a high adsorption capacity for all ranges of relative humidity (humidification process) and can be regenerated at low temperature. The type of desiccant selected will depend on the intended applications.

1.2.2.1   Solid Desiccant Drying System

Solid desiccants cause a pressure drop in the processed air when it passes through the desiccant material. Solid desiccant systems are normally in the form of stationary or rotary wheel beds for packing the desiccant materials [73].

A system with a pre-cooler, double-stage systems (two desiccant wheels and a four-partition desiccant wheel type), and a batch system with an internal heat exchanger is presented in Figure 1.3. The batch system with the internal heat exchanger was found capable of operating at the lowest heated air temperature around 33 °C and at a cooled air temperature of 18 °C [74].

A solar dryer system integrated with desiccant material to dry fresh maize was constructed by Thoruwa *et al.* [75]. As shown in Figure 1.4, a flat plate solar air heater is connected to the drying chamber and the solid desiccant material is mounted above the maize bed. Bentonite clay and $CaCl_2$ materials are selected as desiccant materials due to their low cost and high moisture sorption. The desiccant has a moisture sorption of 45% and can be regenerated at 45 °C. The saturated desiccant bed is regenerated by solar energy during daytime. The dryer can dry 90 kg of fresh maize from 38 to 15% within 24 h.

Thoruwa *et al.* [76] built and tested a prototype dryer that provides dehumidified air at night using solid bentonite $CaCl_2$ as the desiccant material (as shown in Figure 1.5). A photovoltaic panel and a 12 V battery were used to drive the electric fan and produce constant air flow. The collector had an area of $0.921\,m^2$ containing 32.5 kg of desiccant, which could produce an airflow of $2\,m^3/min$ throughout the night. The relative humidity of the dehumidified air was about 40% below the ambient level and the temperature increased by 4 °C. The desiccant was regenerated by solar radiation during the daytime. The system could capture and utilize more than 50% of the incident solar energy.

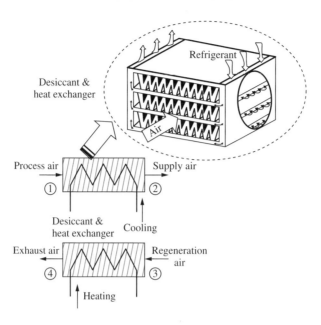

Figure 1.3 Batch-type system with internal heat exchanger [74]

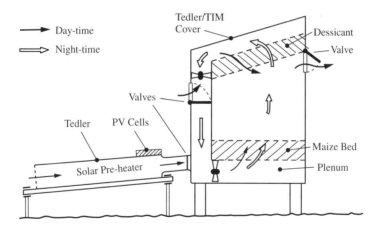

Figure 1.4 Operation of solar-desiccant dryer [75]

Shanmugam and Natarajan [77] developed forced convection and desiccant integrated solar dryer (as shown in Figure 1.6) to investigate its performance. Bentonite–$CaCl_2$ (type 1) was used as the desiccant to continue the drying operation during the off-sunshine hours, during which air inside the drying chamber was circulated through the desiccant bed by a two-way fan. During the hot weather, a flat plate collector heated the air, and a blower forced the hot air into the drying chamber. At the same time, solar radiation regenerated the desiccant bed. Results showed that the desiccant drying system produced more uniform drying and improved

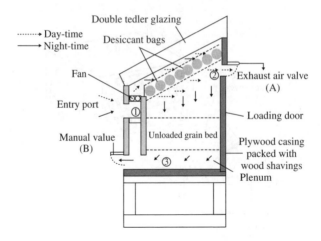

Figure 1.5   Integrated desiccant/collector dehumidifier [76]

1—Blower; 2—flat plate collector;
3—drying chamber; 4—insulator;
5—absorber plate; 6—bottom plate;
7—transparent cover; 8—desiccant bed;
9—plywood; 10—air inlet; 11—duct
for air exit; 12—trays; 13—two-way
fan; 14—valve; 15—plywood.

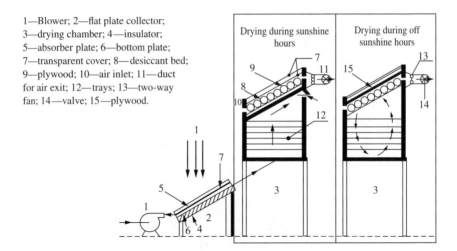

Figure 1.6   Desiccant integrated solar dryer [77]

the quality of the dried product. The characteristic and structural integrity of the desiccant remained stable even after a year. A reflective mirror was used to concentrate solar radiation on the desiccant bed. The reflective mirror improved drying performance of the desiccant by 20% and decreased drying time by 4 and 2 hours for pineapple and green peas, respectively.

Nagaya *et al.* [78] designed and developed a desiccant-based low temperature drying system to dry vegetables (as shown in Figure 1.7). The drying system was equipped with heating and air circulation control to maintain a constant temperature of 49 °C. Experimental results showed that drying vegetables using this technique produced good product uniformity and maintained their fresh color, original texture and shape, and high vitamin content. The desiccant-rotor dehumidifier was divided into three zones: an operating zone in which the

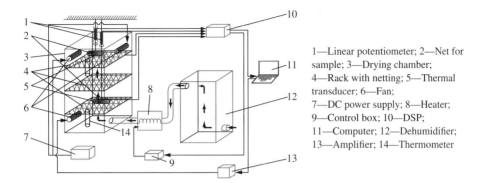

1—Linear potentiometer; 2—Net for sample; 3—Drying chamber; 4—Rack with netting; 5—Thermal transducer; 6—Fan; 7—DC power supply; 8—Heater; 9—Control box; 10—DSP; 11—Computer; 12—Dehumidifier; 13—Amplifier; 14—Thermometer

Figure 1.7 Desiccant drying system with controlled temperature and airflow [78]

silica gel captured moisture from the air (supplying dry air to the drying chamber), a recovery zone in which wet air was heated and blown out and a heat collection zone in which the silica gel released heat to external fresh air.

Madhiyanon *et al.* [79] developed a hot-air drying system integrated with a rotary desiccant wheel (as shown in Figure 1.8) to dry coarsely chopped coconut pieces. The dryer consisted of two air circuits. The first air circuit dried the product and operated in a closed or partially open system. The second air circuit regenerated the desiccant. Ambient air was dehumidified through the adsorption section of the desiccant wheel, while heated air from the second air circuit regenerated the saturated desiccant to remove moisture. A blower was used to supply air for regenerating the silica gel and drying the product, respectively. Two separate 1 kW electrical heaters heated the air for drying and regeneration.

1.2.2.2 Liquid Desiccant Drying System

Generally, using liquid desiccant to construct a dryer is more complicated than using solid desiccant. However, a liquid desiccant system is flexible and can position the regeneration area far away from the dehumidification zone, allowing localized dehumidification. The advantage of the liquid desiccant is that regeneration can be done at a lower temperature with high moisture removal capacity. Liquid desiccant can also absorb organic and inorganic contaminants from the air [80].

Rane *et al.* [81] developed liquid desiccant-based dryer (LDBD) with higher energy efficiency. A $CaCl_2$ solution was used as liquid desiccant. The contacting device was used to transfer the moisture in the absorber and regenerator. Compared to conventional packing, the surface density of the contacting device was about 120–185% higher. The generation process is divided into two stages (as shown in Figure 1.9). First, the dilute liquid desiccant is heated by an external heat source, boiled in the high temperature regenerator (HTR), and then the steam and liquid desiccant mixture are separated in the separator. Second, the hot liquid desiccant flows through the tube to the low temperature generator (LTR) and condenses. The water from the dilute liquid desiccant at contacting disks is transferred to the air due to the vapor pressure difference between the liquid desiccant and the air. There are 70 units of contacting disks rotated at 3–5 rpm. A chimney helps circulate the air by the buoyancy force for moisture removal.

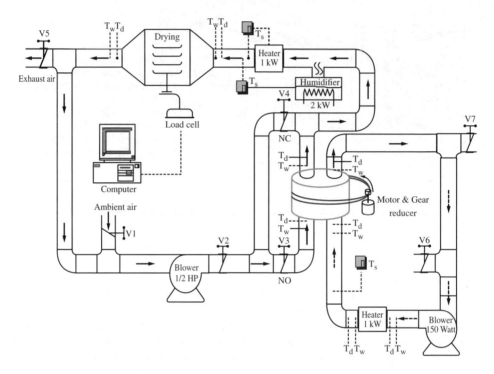

Figure 1.8 Hot-air dryer integrated with rotary desiccant wheel [79]

The liquid desiccant drying method has been used in drying green gel cast ceramic parts to shorten the drying time and to avoid defects due to the release of residual stresses. Barati *et al.* [82] studied the kinetics of one-dimensional drying of green gel cast ceramic parts using the Fickian model. Results showed that higher ceramic loading, higher sample thickness, and lower concentration of the liquid desiccant solution decreased the drying rate. Experiments have been done involving the immersion of green gel cast parts in an aqueous or nonaqueous solution of PEG1000 as liquid desiccant [83]. Cracking, bending, and warping, which are the common defects during conventional drying methods, were eliminated by using this method, and drying time was reduced by about 10 times. An aqueous solution of the liquid desiccant could achieve more homogeneous drying. However, drying rate in an aqueous solution of PEG1000 was lower than that of a nonaqueous solution.

Zheng *et al.* [84] used the liquid desiccant method for drying $BaTiO_3$-based semi-conducting ceramic gel cast parts. The gel cast parts were immersed in the liquid desiccant. The removal of water from the gel cast parts was due to the osmotic difference between the liquid desiccant and the gelled polymer in the part. Results showed that increasing the loading of green gel cast parts to more than 45% (in volume) would reduce the stresses developed during drying, and a higher concentration of the liquid desiccant would not induce any defects and would produce a smooth surface ceramic. However, the part with lower thickness and higher solid content in the gel would increase the ceramic density. The gel cast parts could be dried safely at room conditions or in an oven just after the critical stage of drying process of liquid desiccant.

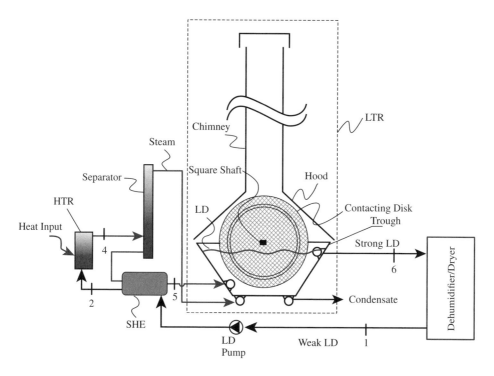

Figure 1.9   LDBD with two-stage regenerator [81]

Trunec [85] conducted osmotic drying of gel cast alumina in water solutions of polyethylene glycol (PEG) with different molecular weights in the range of 1000–80 000 g/mol. Results showed that the PEG solution with highest molecular weight was the most efficient liquid desiccant. Up to 30% water content from gel cast bodies immersed in a 43% (by mass) solution of PEG 80 000 could be removed. Uniform and crack-free drying could be achieved by osmotic drying in the PEG solution with a high molecular weight.

1.2.2.3   Solid vs. Liquid Desiccant Drying System

Figure 1.10 presents the classification of the desiccant-based dehumidification system. The solid desiccant is more widely used in drying applications compared to liquid desiccant. This is because the solid desiccant requires simple construction of drying system. Most of the solid desiccant is designed in the form of rotary wheel beds. In contrast to the solid desiccant, the liquid desiccant has a lower regeneration temperature and a higher moisture removal capacity. Other advantages of using the liquid desiccants in a drying system also include continuous drying even during off-sunshine hours, more uniform drying, and easier humidity control. However, the liquid entrainment in the processed air and the corrosion of the desiccant salt are the main challenges for the practical use of the liquid desiccants. The hybrid-based system is the combination of solid or liquid desiccant materials, which may have the advantages of both.

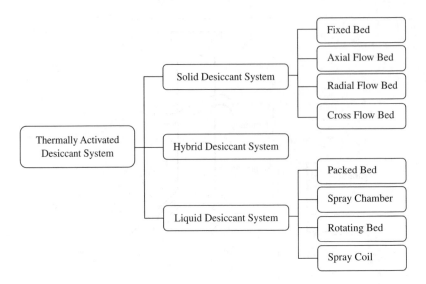

Figure 1.10   Classifications desiccant dehumidification system

1.2.3   Regeneration Methods

Desiccants work based on the principle of moisture transfer due to the difference of vapor pressure between the air and the desiccant. The cool desiccant with low moisture content will adsorb moisture from air until its vapor pressure is in equilibrium with the air. The regeneration of desiccant is actually a process of moisture removal from the desiccant (i.e., a drying or dehydrating process), which happens when the vapor pressure on the desiccant surface is higher than the surrounding air. Overall, the regeneration methods can be categorized into two groups: the heating method and the nonheating method. The heating method mainly uses electrical heater, waste heat, solar energy, heat pump or microwave radiation, and the nonheating method may employ membrane osmosis, electro osmosis, pulsed corona plasma, or ultrasound. The nonheating regeneration methods can help to reduce the regeneration temperature of desiccants and improve the energy performance of the dehumidification system. Therefore, they have been noted by people in recent years.

1.2.3.1   Heating Method

Electrical Heater
An electrical heater is a simple application that exhibits regeneration of desiccant material and is also a consistent heat source. However, the main drawback of electrical heaters is their high energy consumption. Due to its high operating cost, sometimes the electrical heater is only used as a back-up energy source if solar energy or waste heat is not available or is not enough.

Mandegari and Pahlavanzadeh [86] studied the efficiency of a desiccant wheel system in which the heater was used for the regeneration (as shown in Figure 1.11). The study showed that the adiabatic efficiency of the desiccant wheel mainly depended on dehumidification and

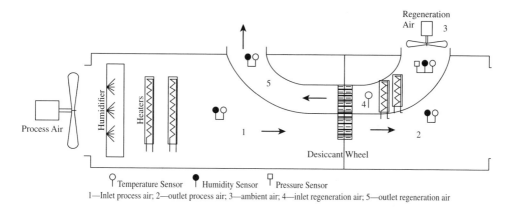

1—Inlet process air; 2—outlet process air; 3—ambient air; 4—inlet regeneration air; 5—outlet regeneration air

Figure 1.11   Desiccant wheel experimental setup [86]

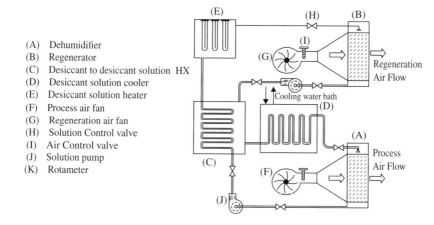

(A)   Dehumidifier
(B)   Regenerator
(C)   Desiccant to desiccant solution HX
(D)   Desiccant solution cooler
(E)   Desiccant solution heater
(F)   Process air fan
(G)   Regeneration air fan
(H)   Solution Control valve
(I)   Air Control valve
(J)   Solution pump
(K)   Rotameter

Figure 1.12   Cross flow liquid desiccant dehumidification system [87]

regeneration efficiency, and it increased when the desiccant wheel speed was increased from 4 to 24 rev/h.

Bassuoni [87] designed a cross flow liquid desiccant dehumidification system (as shown in Figure 1.12) and investigated the performance of the structured packing cross flow air–liquid desiccant contacting surfaces on the dehumidification and regeneration of the system. In the dehumidification system, an electric heater was used to heat the liquid desiccant for the regeneration.

Waste Heat

The utilization of waste heat for the regeneration in a desiccant dehumidifier system is one of the best alternatives because it can reduce the cost for the regeneration. However, it is only suitable for occasions where the exhaust waste heat at temperatures between 60 and 140 °C can be available.

The US Department of Energy (DOE) developed the Integrated Energy System (IES), which aimed at improving the overall energy efficiency of distributed generation (DG) systems. The system was integrated with the waste heat recovery and thermally activated (TA) technologies [88]. The TA technology used the hot exhaust gas of the DG system to heat, cool, and regenerate the desiccants in the dehumidification systems. The exhaust gas from the DG system could be used directly or routed to a heat recovery unit (HRU) through an air-to-water heat exchanger. The micro turbine generator (MTG) could be operated individually or integrated with various waste heat recovery. The overall efficiency of IES increased from 5 to 7% when used with an exhaust-fired desiccant dehumidification unit.

Solar Energy

The use of solar energy for the process of regenerating desiccant material has been studied extensively since it is a free energy source. The initial cost of a solar collector system is not very cheap. The payback period must be fully considered during the utilization of solar energy. In addition, solar radiation is weather-dependent; therefore, back up energy or energy storage is required to continue the drying process when solar energy is not available.

Lu *et al.* [89] developed two solar desiccant dehumidification regeneration systems known as SDERC (Solar Dehumidification and Enhanced Radiative Cooling) and SRAD (Solar Regeneration and Dehumidification). The SDERC system mainly consisted of a glazed metal chamber, a solid-desiccant bed, three separate axial flow fans, a brass radiation cooling duct, a three-way valve mechanism, and an evaporative cooler. The assembly of the SDERC system was depicted in Figure 1.13. During the night, indoor air flowed through the solid desiccant bed and then through an evaporative cooler to decrease the temperature before it returned to the house. During the daytime, solar energy heated the glazed chamber and air passed through the saturated desiccant for the regeneration process.

Xiong *et al.* [90] studied a two-stage solar-powered liquid desiccant dehumidification system with two types of desiccant solutions (as shown in Figure 1.14). In this system, the air was

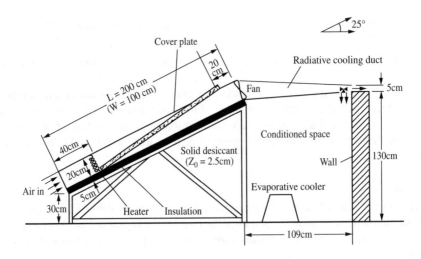

Figure 1.13   Installation of SDERC system [88]

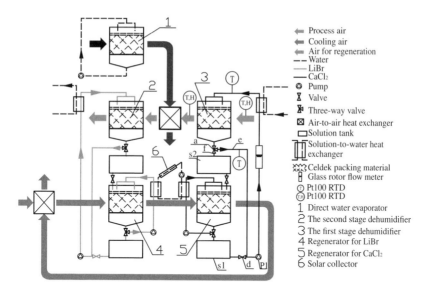

Figure 1.14 Two stage liquid desiccant [89]

dehumidified by a pre-dehumidifier and a main dehumidifier using $CaCl_2$ and lithium bromide (LiBr). The inter-cooling effect occurred between the two dehumidification stages through an air-to-air heat exchanger.

Alosaimy and Hamed [91] used a flat plate solar water heater to regenerate the liquid desiccants. The system mainly included a solar water heater with a storage tank, a water-to-air heat exchanger, and a packing of a honeycomb type (as shown in Figure 1.15). The water was heated by solar energy through the solar water collector. Then, the hot water in the tank was circulated in a heat exchanger by a pump. Hot air from the heat exchanger was blown to the packing for the regeneration of $CaCl_2$ solution. Experimental results showed that solar energy could regenerate up to 50% of the solution at 30% solution concentration.

Heat Pump

Heat pump is considered as an energy-efficient dryer system due to its low energy consumption. The combination of heat pump system and desiccant system in the drying applications improves energy efficiency and produces lower humidity of the processed air. This system is also called the hybrid desiccant system. Heat released by the heat pump through the condenser can be used to regenerate the desiccant materials. An evaporator and a desiccant material can carry out the dehumidification process to produce processed air with better conditions at low energy consumption.

Wang et al. [92] designed and developed a hybrid system combining heat pump and desiccant wheel (as shown in Figure 1.16) to produce low-cost drying and supply low DPTs of air. This system was used for rapid surface drying to avoid re-condensation at low DPT and low dry-ball temperatures (DBTs) in the range of 10–20 and 20–30 °C, respectively, after the product was dried. The heat rejected by the condenser was used to regenerate the desiccant wheel. Moisture from the ambient air was removed in a dehumidification process by condensation from the evaporator and adsorption using the solid desiccants.

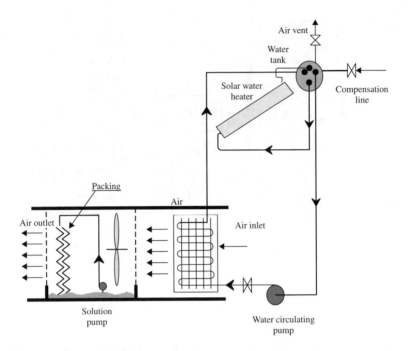

Figure 1.15   Regeneration of liquid desiccant by solar energy [90]

Microwave Radiation

In the conventional heating method, the hot air is usually employed as a heating medium for the desiccant regeneration. However, hot-air heating requires a long time to heat the entire desiccant rotor, because thermal energy is indirectly transferred from hot air to the rotor. Moreover, it is well known that lowering the regeneration temperature, especially below 80 °C, leads to a significant decrease in the humidity control performance due to insufficient water desorption. Microwave irradiation supplies energy to the whole material body, allowing the materials to produce thermal energy. As a consequence, the temperature of the materials rises rapidly. Based on the merits of microwave irradiation, a novel hybrid regeneration process combining microwave and conventional hot air (HA) heating has been proposed by some researchers [93–98]. Combination of both heating methods is expected to achieve higher regeneration rate due to the direct and rapid heating by microwave irradiation in addition to indirect heating by hot-air flow, and it will promote the utilization of the low-temperature thermal energy.

A desiccant wheel system with the microwave-assisted regeneration was designed and investigated by Mitsuhiro *et al.* [97, 98]. The system mainly consisted of a microwave irradiator, circular wave guide, microwave dummy load, desiccant rotor, and electric heater (as shown in Figure 1.17). The desiccant rotor coated with synthesized zeolite was used and installed inside the aluminum waveguide. Microwave energy was supplied from an irradiator through a horizontal circular waveguide, and was finally absorbed by circulating water flow. The regeneration characteristics of the desiccant rotor were experimentally investigated under conditions of microwave heating, hot-air heating, and combined heating at various microwave powers

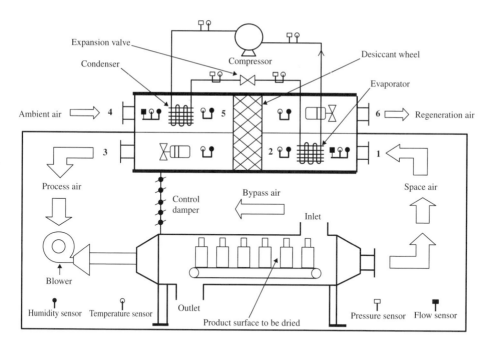

Figure 1.16   Hybrid desiccant system [91]

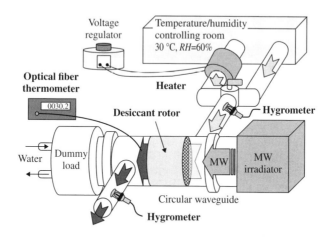

Figure 1.17   Desiccant wheel with microwave-assisted regeneration [92]

and hot-air temperatures. The study showed that the combined heating was effective for leveling nonuniform temperature distribution in the rotor, and the combined heating could acquire a larger regeneration rate and regeneration degree compared with the either microwave or hot-air heating. By applying the combined and microwave heating method to the dehumidification systems could reduce the switch time between the adsorption and the regeneration of the desiccant.

1.2.3.2 Nonheating Method

Membrane Osmosis

Osmosis is a process in which the solvent is transported through the membrane as a result of a difference in trans-membrane concentration. If the system is not subjected to any external influence, such removal of excess solvent results in the establishment of a hydrostatic pressure difference. The reverse osmosis process is characterized by the use of pressure in excess of the osmotic pressure to force the solution of salt at the same temperature through a selective membrane capable of rejecting the dissolved salts. The process name is derived from the phenomenon whereby the water under an applied pressure driving force flows in the opposite direction to that normally observed in an osmotic process where the driving force is the concentration gradient [99].

Reverse osmosis (RO) has been successfully applied to desalination of seawater in which saline water with some concentration of dissolved salts is distilled into pure water when passed through a membrane. In a similar manner, weak desiccants (e.g., calcium chloride and lithium chloride) may be distilled by removing the water from the solution with a suitable membrane [100].

Al-Sulaiman *et al.* [101] studied the energy performance of a cooling system with two-stage evaporative coolers using liquid desiccant dehumidifier between the stages (as shown in Figure 1.18). The reverse osmosis process was used for regeneration by mechanical energy, and an MFI zeolite membrane was proposed for separation of water from the weak desiccant solution. The osmotic pressure that separated product water from the weak calcium chloride solution under equilibrium conditions was found to be 24.4 MPa. Obviously, the major energy requirement associated with this cooling system is the energy for regenerating the weak liquid desiccant.

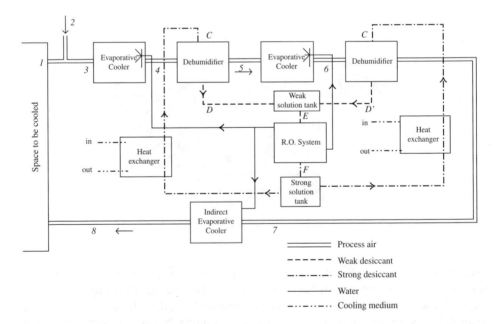

Figure 1.18 Schematic for liquid desiccant cooling system using RO regeneration [98]

A thermally driven flat plate air gap membrane distillation liquid desiccant regenerator for lithium chloride in dehumidification applications has been theoretically modeled by Alexander *et al.* [102]. According to the model results, the regenerator with membrane materials removed 11.4 g/(min·cm^3) of moisture with a COP (coefficient of performance) of 0.372 for the inlet solution concentration of 0.38, the solution flow rate of 50 ml/min, and the heated solution temperature of 135 °C.

Electro Osmosis

EO (electro-osmosis) flow is a flow in porous media or micro-channel structure, driven by the so-called EO force with an applied electrical field. The EOF (electro-osmosis force) dynamics in porous media have been well studied both theoretically and numerically. For most types of EO flow, an electric double layer (EDL) will be formed spontaneously at the liquid/solid interface due to electrochemical reaction. As depicted in Figure 1.19 [14], when contacting with an electrolyte (water in the regeneration system), the Si–O on the internal surfaces of the porous solid desiccant dissociates to act on regions of net charge. A layer of ions is firmly absorbed to the surface of the solid and does not move in the electric field applied, which is called a compact layer. And an equal number of opposite ions, which is called the diffuse layer, is in a liquid water state and less attracted to the surface. These two layers form the EDL [103]. The density of ions reduces as the distance from the surface increases, and so does the potential. The potential in shear plane of the two layers is called zeta potential [104], and it is negative for the solid desiccants. Once an electric field is supplied to the EDL, cations of the diffuse layer are driven by the Coulomb force from the anode to cathode so that the liquid in this area flows with the ions [105]. The fluid outside the EDL travels completely dependent on the viscous force. It therefore forms the EO flow which travels from the anode to cathode. The solid desiccant, which has both electro-osmosis and adsorption characteristics, adsorbs moisture from the moist air, and the water vapor turns into liquid water on the surfaces of

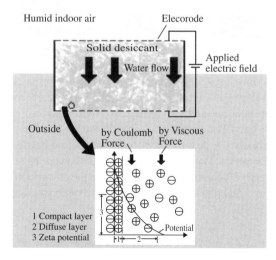

Figure 1.19   Schematic of the principle of the EO regeneration method [101]

the material. Then, with the effect of an electric field, the electro-osmosis transports the water away and the material could continue adsorbing the water vapor in the air.

Possible improvement methods for EO regeneration have been investigated by Qi and Tan et al. [15]. These include changing the anode material, changing the cathode layout, applying interrupted power, and optimizing electrical field strength. Through detailed experiments and analysis, it was found that applying platinum-plated titanium mesh as an anode could improve the working lifetime from 6 to over 120 hours and effectively reduce the Joule heating effect simultaneously; laying a piece of filter cloth under the cathode could enhance the EO regeneration rate up to 0.0021 g/s; the application of interrupted power could increase the regeneration rate by up to 1.5 times; the optimal on-off-time was found at 30 seconds : 1.3 seconds with 17 V/cm electric field strength and 30 seconds : 0.8 seconds with 11 V/cm; and the most suitable value of electric field strength was observed as ranging from 8.5 to 13 V/cm in the EO regeneration system.

Li et al. [106] also investigated the method of electro-osmosis based regeneration for a solid desiccant and its potential application in the HVAC (Heating, Ventilation and Air Conditioning) field, particularly for the dehumidification process in an air-conditioning system. The energy consumption in an EO integrated air-conditioning system was found to be averagely 23.3% lower than the conventional air-conditioning system with respect to different configurations in the air handling process.

Although the electro-osmotic regeneration for the solid desiccant has been proved to have many merits, such as regeneration without the heat source, energy-saving, and simple structure [14], it is still at the experimental stage.

Pulsed Corona Plasma

The concept of nonthermal plasma desorption or regeneration was first investigated by using methyl ethyl ketone (MEK, $CH_3CH_2COCH_3$) [12], and then the concept of plasma desorption was further demonstrated with benzene [107] and NO_x [108]. The mechanism of plasma desorption is thought to be due to the impact of high-energy electrons and excited molecules, or possibly electrostatic attraction among the ionized molecules adsorbed on the adsorbent and ionized background molecules. This mechanism might occur because the gas molecules are not chemically but physically bonded to the adsorbent [109]. The use of the nonthermal plasma may have several advantages; for example, the system can be operated at room temperature and atmospheric pressure, and it does not generate excessive heat.

Yamamoto et al. [13] investigated the characteristics of nonthermal plasma desorption with the experimental setup shown in Figure 1.20. The moist air passed through the reactor, where water vapor adsorption material and the plasma desorption unit were placed. The moisture concentration was measured both inside and outside the reactor by using a photo-acoustic, single-gas monitor analyzer. The adsorption material was placed between the multiple needles and perforated plate electrodes. The experimental results showed that the water vapor desorption per unit power (energy efficiency for desorption) and the desorption rate (or the gradient of desorption) for the nonthermal plasma was superior to that of conventional thermal desorption.

Ultrasonic Radiation

In recent years, a lot of research has been done on the nonheating method by using ultrasonic technology. A series of studies have proved that the ultrasound-assisted regeneration could significantly improve the dehydration kinetics of desiccants and energy efficiency [22–42].

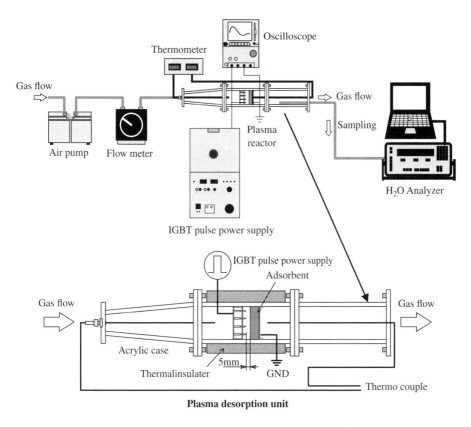

Figure 1.20 Experimental setup for water vapor adsorption and desorption [14]

Meanwhile, several Chinese invention patents have been proposed for the applications of the new regeneration technology in practical engineering [31–37]. These study results will be presented in detail in the following chapters.

1.3   The Proposed Method

1.3.1   Basic Knowledge about Ultrasound

Sound is a special form of energy transmitted through pressure fluctuations in air, water, or other elastic media. Any displacement of a particle of this elastic medium from its mean position results in an instantaneous increase in pressure. When leveling, this pressure peak not only restores the particle to its original position but also passes on the disturbance to the next particle. The cycles of pressure increase (compression) and decrease (rarefaction) propagate through the medium as a sound wave. Ultrasound is an oscillating sound pressure wave with a frequency greater than the upper limit of the human hearing range. Ultrasound is thus not separated from "normal" (audible) sound based on differences in physical properties, only the fact that humans cannot hear it. Although this limit varies from person to person, it is

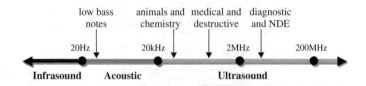

Figure 1.21   Approximate frequency ranges corresponding to ultrasound [13]

approximately 20 kHz in healthy, young adults. Ultrasound devices operate with frequencies from 20 kHz up to several gigahertz (as shown in Figure 1.21) [110].

On a micro scale, sound is characterized by pressure (p_{sound}) and particle velocity (u_{sound}). The product of these two parameters is called sound intensity (I_{sound}) – a vector normal to the direction of sound propagation:

$$I_{sound} = p_{sound} \times u_{sound} = \frac{Force}{Area} \times \frac{Distance}{Time} = \frac{Energy}{Area \times Time} = \frac{Power}{Area} \qquad (1.1)$$

As shown in Figure 1.22, sound generated by a point source with power W propagates as a spherical wave. Therefore, sound intensity is inversely proportional to the square of the distance from the sound source. The variations of both pressure and velocity follow a sinusoid; if they are in phase, the peak pressure occurs at the same time as the peak in the particle velocity, and the product of these two gives the intensity, which is not only the maximum instantaneous intensity but also the maximum time-averaged intensity.

On a macro scale, sound is primarily characterized by the frequency (f), which relates the speed of wave propagation (u) (sound velocity) to the wavelength (λ):

$$f = \frac{u}{\lambda} \qquad (1.2)$$

The second main quantity used to characterize sound on a macro scale is the amplitude of sound pressure level (SPL) on the decibel (dB) scale which takes 20 μPa as the reference level.

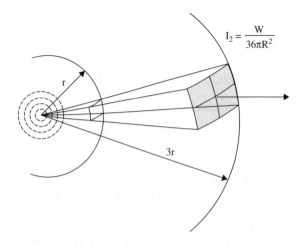

Figure 1.22   Sound intensity around the point source of acoustic energy

In a free acoustic field such as in an open air or anechoic chamber, the pressure and intensity levels in the direction of propagation are many times more than which travels in all directions with equal magnitude and probability (reverberation chamber). The pressure and intensity levels are different, and this difference is known as the pressure-intensity index. Aside from the frequency and sound intensity, another key point for the acoustic applications (e.g., sound-assisted dryers) is the mode of energy propagation. Sound energy can be propagated as longitudinal waves (also called compression waves) or transverse waves in which vibration of the particle in the material occurs perpendicularly to the direction of wave motion. Since the latter cannot be propagated in gases and liquids except for highly viscous liquids over very short distances (fraction of a millimeter), sound-assisted dryers are normally designed to accommodate longitudinal waves [111].

In a free space, the sound source can be considered as a point source. In practical industrial applications, however, sound is either shapes such as horn, paraboloid, and ellipsoid. In both cases, such sound radiation can be regarded as coming from a plane source. This results in a constant sound intensity (the Fresnel zone), whereas outside this zone (the Fraunhofer zone), the sound intensity decreases inversely with the square of the distance from the plane source, that is, in the same way as for a point source (as shown in Figure 1.23). The length of zones and the sound intensity distribution may be crucial for the configurations of sound-assisted dryers. The Fresnel zone for the plane source 10 cm in diameter is negligible (a couple of millimeters) for sound at 100 Hz (cf., frequency of pulse combustion), but extends for 15.6 cm in the range of ultrasound at 20 kHz and 31.2 cm at 40 kHz. According to the frequency of pressure pulsation, sound can be classified as infrasound ($f < 20$ Hz), sound (audible) (20 Hz $< f < 20$ kHz), and ultrasound ($f > 20$ kHz).

Ultrasonic applications are rigidly classified into low- and high-intensity applications. Low-intensity applications are made typically in the megahertz frequencies and acoustic power up to tens of milliwatts and usually do not alter material properties under operation. In contrast, high-intensity ultrasound is generally used for changing the properties of the material through which it is passed or altering the physical–chemical processes. High-intensity applications are made at low frequencies (i.e., from 20 to 40 kHz), and these are usually used in drying and dewatering.

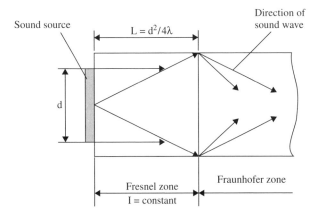

Figure 1.23 Propagation pattern from a plane sound source [111]

In the case of a plane progressive wave penetrating the material being dried, the sound energy is attenuated due to various mechanisms including relaxation processes, viscous shearing effects, and molecular absorption. Neglecting the scattering of sound energy, the amplitude of acoustic pressure at a distance x from the material surface at which the incident sound pressure is p_o may be expressed as:

$$p(x) = p_o \exp(-\alpha x) \tag{1.3}$$

where, α is the attenuation coefficient, which is a function of material properties and sound frequency, and it is often determined experimentally.

Equation (1.3) can be written conveniently in terms of sound intensity, which reflects the energy flux per unit surface area:

$$I = \frac{p^2}{u\rho} \tag{1.4}$$

where ρ is the material density and u the sound velocity. Hence,

$$I(x) = I_o \exp(-2\alpha x) \tag{1.5}$$

Assuming no heat loss from the absorbing volume, the rate of temperature rise is given by

$$\frac{dt}{d\tau} = \frac{2\alpha I}{\rho c} \tag{1.6}$$

where, t is temperature of medium, °C; τ is time, s; c is the specific heat of material, J/(kg·°C).

1.3.2   Sound Generation

In industrial applications, sound waves are generated by a transducer, which converts the original form of energy to the energy of oscillatory motion. Such transducers are classified into five main groups [112]:

1. Piezoelectric. The periodic changes in the physical size of certain crystals (such as quartz, tourmaline, and zinc oxide) due to applied electric potential generate mechanical vibrations that are propagated as sound waves. Used in the range from 20 kHz to 10 GHz.
2. Magnetostrictive. Mechanical vibrations are caused by changes in the physical size of certain metals such as nickel, cobalt, and iron, or certain nonmetals known as ferrites due to an external magnetic field. Used in the range of 40–100 kHz.
3. Electromagnetic. The vibration of a solid armature (e.g., membrane in loudspeakers and microphones) is due to coupled electric and magnetic fields. Used at $f < 50$ kHz.
4. Electrostatic. The periodic variation of charges in an electrical capacitor of a special design induces mechanical vibration. Used at $f < 100$ kHz.
5. Mechanical. Sound waves are generated due to the action of a truly mechanical device such as rotating counterbalanced weights or a mechanical device energized by the kinetic energy of the working fluid (sirens and whistles). Used at $f < 50$ kHz.

For high-intensity generation and propagation of sound in gases at frequencies up to about 25 kHz, mechanical generators are used almost exclusively because of their design and operational simplicity, energy capability, and low cost as compared to the other types of

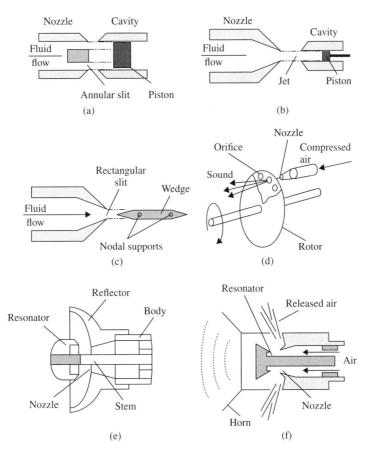

Figure 1.24 Principles of sound generation in mechanical sound generators [111]. (a) Galton whistle, (b) Hartman whistle, (c) wedge resonator, (d) dynamic siren, (e) modified Hartman whistle, and (f) Branson sound generator

transducers. These generators that produce the so-called *airborne sound* are used in sound and sound-assisted convective drying. Figure 1.24 presents the most frequently used types of mechanical generators from the group of cavity resonators (Galton and Hartman whistles), wedge resonators, and sirens. Usually, in a modified Hartman whistle, a rod is centrally positioned along the axis of the gas jet. Such a design allows for a more compact generator of increased efficiency. In the Branson pneumatic sound generator, the exhaust air jet is separated from the sound field, which does not only result in a higher sound intensity (no air turbulence effect) but also the generator to be used in processing materials for which contact with air is not acceptable. One of the possible designs of industrial sound generators is shown in Figure 1.25. When rotated at 4000–9000 rpm and fed with 0.138 m^3/s of compressed air at 0.3–0.5 MPa, this dynamic siren ($d = 0.2$ m) emits sound up to 180 dB and 8 kW of acoustic power [113].

In processing of liquid systems such as dewatering of slurries and pasty materials, piezo-electric, or magnetostrictive generators are used. The active part of such a generator (called a driver) generates and transmits mechanical vibrations through a solid rod (a booster) to

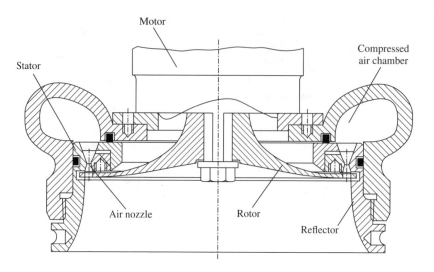

Figure 1.25   Design of a pneumatic sound generator [111]

the metal profile (a horn). This horn serves as an amplitude transformer to amplify the displacement of the driver and to match the transducer impedance with the impedance of the material to be processed. The mechanical vibrations from the horn are further coupled to the processed material either directly (the horn is inserted into a liquid or pasty material) or indirectly through a membrane.

1.3.3   Fundamental Theory for Ultrasound-Assisted Regeneration

Since the desiccant regeneration is essentially a heat and mass transfer process, the theory of ultrasonic-assisted regeneration may be illustrated by the following possible mechanisms of heat-and-mass-transfer alterations in an ultrasonic field:

1. Alternating compression and expansion due to high-frequency pressure pulsation create surface cavitation that breaks the boundary layer and allows liquid to evaporate under partial vacuum [114].
2. Intensive circulation flows (induced by sound pressure) on the drying surface promote surface evaporation [115].
3. Pressure pulses from the sound waves increase the turbulence which reduces the thickness of the laminar sublayer [116].
4. An increase in moisture diffusivity and a decrease in viscosity [117].
5. A pulsating partial vacuum transmitted into the material affects water vapor transport, possibly by decreasing or overcoming the attraction forces between the water and solid molecules; Shear reduction in boundary film surrounding the material [118].
6. Expansion of the vapor bubbles inside capillaries yields a migration of the water filament (sonic diffusion current) [119].
7. Increase in the bulk temperature due to the acoustic energy dissipation, also called "thermal effect" [120].

The data from the review of literature indicates that cavitation enhanced turbulence of the gas phase at the solid (or liquid droplet) surface and mechanical effects due to internal stresses appear to be the major contributors to enhanced moisture removal during drying. It is also reasonable to expect the liquid viscosity to be lowered in ultrasonic fields, which promotes diffusion of moisture toward the evaporation surface. The term cavitation in general refers to the formation and subsequent dynamic behavior of vapor bubbles in liquids. Acoustic cavitation occurs when high-intensity sound waves are coupled to the liquid surface, which results in the propagation of alternating regions of compression and expansion, and thus in the formation of micron-size vapor bubbles. If the bubbles are of a critical size (determined by acoustic frequency), they may implode violently, releasing energy in the form of impulses ($t < 0.1$ second) with local point temperature and pressure in the order of 5000 K and 1000 atm, respectively [121]. Since the effective temperature zone is confined to about 0.2 m from the surface of the collapsing bubble, the bulk of the liquid remains practically at the same temperature. However, as the bubbles at or near the surface implode, micron-size liquid droplets can be released into the surrounding air and evaporated instantaneously. Because of the cavitation threshold, larger acoustic pressure amplitudes at a given frequency are required for highly viscous liquids.

It is reasonable to assume that pressure fluctuation caused by sound waves disrupts the boundary layer at the solid or liquid surface, so that they should affect the inter-phase transfer rates and thereby accelerating drying rates. In fact, a substantial decrease (up to 15%) in the boundary layer thickness has been found when conventional spray drying of blood plasma was complemented by high-intensity sound at $f = 15$ kHz and $I = 155$ dB [116].

Aside from cavitation, the enhanced mass-transfer rates in acoustic fields can be attributed to the plug flow of the capillary liquid as well as to the enhanced dispersion of the liquid and vapor moisture due to alternating compression and expansion cycles, which result in reduced viscosity of the liquid–vapor mixture. In fact, a substantial increase in the amount of liquid diffusing through porous solids has been noted in the presence of ultrasound [122]. The enhanced diffusion appears to be of the directional type as mass transfer was hindered when ultrasound irradiation was opposed to the direction of diffusive flow [123].

According to Kardashev [124], the mechanism of acoustic drying of capillary-porous materials depends on the level of moisture content. When the material is very wet (200–500%), the effect of an ultrasound field is truly mechanical, and moisture is removed due to better dispersion of liquid water, especially at the antinodes of a standing wave. When the material moisture content is much lower (10–70%), but drying takes place during the constant-rate period, the sound waves reduce the thickness of the boundary layer, which alters moisture evaporation. In the falling rate period, the sound waves enhance only the moisture diffusivity due to temperature rise because the sonic energy is dissipated as heat. In the case of dispersed materials, the positive effect of sound waves on a drying rate appears for the SPL above a certain threshold value.

Under real drying conditions, heat generated within the material volume is transported away by conduction and convection; hence, the final equilibrium temperature tends to be determined by the heat balance. The equilibrium temperature for the bulk of the material is in the order of a few degrees [125], which justifies moisture evaporation due to thermal effects of sound irradiation to be neglected. This obviously favors sonic drying as a method for processing heat-sensitive materials. Considering heat generation on the micro scale, thermal effects may become important as the localized temperature increase is likely to affect fluid properties and solid–fluid interactions (e.g., lower surface tension or viscosity). According to the study by

Moy and DiMarco [126], for example, 7% of the sound-enhanced freeze-drying (F-D) rate of liquid food could be attributed to the thermally induced mechanical effects on the gas phase, resulting in friction, and adiabatic compression. Detailed discussions of these and other phenomena activated by the sound energy can be found elsewhere [127, 128].

1.4 Summary

This chapter mainly deals with the background of the topic relevant to this book and the corresponding technologies including desiccant materials, desiccant dryer systems, and regeneration methods based on the literature review. Afterwards, the basic knowledge about ultrasound and the sound generating methods as well as the fundamental theory for ultrasonic-assisted regeneration have been introduced for a better understanding of the novel regeneration method to be illustrated in the following chapters.

References

[1] Soleyn, K. (2003) Humidity control: preventing moisture contamination. *Chemical Engineering*, **110** (11), 50–51.

[2] Chen, X.Y., Jiang, Y., Li, Z. and Qu, K.Y. (2005) Field study on independent dehumidification air-conditioning system-i: performance of liquid desiccant dehumidification system. *ASHRAE Transactions*, **111** (2), 271–276.

[3] Chen, X.Y., Jiang, Y., Li, Z. and Qu, K.Y. (2005) Field study on independent dehumidification air-conditioning system ii: performance of the whole system. *ASHRAE Transactions*, **111** (2), 277–284.

[4] Zhang, L.Z. (2006) Energy performance of independent air dehumidification systems with energy recovery measures. *Energy*, **31** (8-9), 1228–1242.

[5] Cowie, M., Liao, X.H. and Radermacher, R. (2003) Performance comparison of waste heat-driven desiccant systems. *ASHRAE Transactions*, **109** (2), 572–579.

[6] Techajunta, S., Chirarattananon, S. and Exell, R.H.B. (1999) Experiments in a solar simulator on solid desiccant regeneration and air dehumidification for air conditioning in a tropical humid climate. *Renewable Energy*, **17** (4), 549–568.

[7] Demakos, P.G. (2009) Improving IAQ with liquid-desiccant dehumidification. *Heating/Piping/Air Conditioning HPAC Engineering*, **81** (9), 36–42.

[8] Chang, K.S., Wang, H.C. and Chung, T.W. (2004) Effect of regeneration conditions on the adsorption dehumidification process in packed silica gel beds. *Applied Thermal Engineering*, **24** (5–6), 735–742.

[9] Czanderna, A.W., Tillman, N.T. and Herdt, G.C. (1995) Polymers as advanced materials for desiccant applications part 3: alkali salts of PSSA and polyAMPSA and copolymers of polyAMPSASS. *ASHRAE Transactions*, **101** (1), 697–712.

[10] Azuara, E., Garcia, H.S. and Beristain, C.I. (1996) Effect of the centrifugal force on osmotic dehydration of potatoes and apples. *Food Research International*, **29** (2), 195–200.

[11] Chakraborty, M. and Kiran, K.T.S. (2002) Pressure swing adsorption: principles, processes and applications. *Chemical Engineering World*, **37** (9), 98–100.

[12] Yamamoto, T. and Yang, C.L. (1998) Plasma desorption and decomposition. Proceedings of the IEEE-IAS Annual Meeting, Saint Louis, MO, October 12–16,1998, pp. 1877–1883.

[13] Yamamoto, T., Tanioka, G. and Kuroki, T. (2007) Water vapor desorption and adsorbent regeneration for air conditioning unit using pulsed corona plasma. *Journal of Electrostatics*, **65** (4), 221–227.

[14] Qi, R., Tian, C. and Shao, S. (2010) Experimental investigation on possibility of electro-osmotic regeneration for solid desiccant. *Applied Energy*, **87** (7), 2266–2272.

[15] Qi, R., Tian, C., Shao, S. *et al.* (2011) Experimental investigation on performance improvement of electro-osmotic regeneration for solid desiccant. *Applied Energy*, **88** (8), 2816–2823.

[16] Yao, Y. and Liu, S.Q. (2008) Ultrasonic–a new regeneration technology for dehumidizer. Fourth International Conference on Cryogenics and Refrigeration, Shanghai, China, April 5–9, 2008, pp. 984–990.

[17] Breitbach, M., Bathen, D. and Schmidt-Traub, H. (2002) Desorption of a fixed-bed adsorber by ultrasound. *Ultrasonics*, **40** (5), 679–682.

[18] Hamdaoui, O., Naffrechoux, E. and Tifouti, L. (2005) Ultrasonic desorption of p-chlorophenol from granular activated carbon. *Chemical Engineering Journal*, **106** (2), 153–161.

[19] Lim, J. and Okada, M. (2005) Regeneration of granular activated carbon using ultrasound. *Ultrasonics Sonochemistry*, **12** (4), 277–282.

[20] Oualid, H., Rabiaa, D. and Emmanuel, N. (2005) Desorption of metal ions from activated carbon in the presence of ultrasound. *Industrial and Engineering Chemistry Research*, **44** (13), 4737–4744.

[21] Juang, R.S., Lin, S.H. and Cheng, C.H. (2006) Liquid-phase adsorption and desorption of phenol onto activated carbons with ultrasound. *Ultrasonics Sonochemistry*, **13** (3), 251–260.

[22] Yao, Y., Zhang, W. and Liu, S.Q. (2009) Feasibility study on power ultrasound for regeneration of silica gel-A potential desiccant used in air-conditioning system. *Applied Energy*, **86** (11), 2394–2400.

[23] Yao, Y., Zhang, W. and Liu, S.Q. (2009) Parametric study of high-intensity ultrasonic for silica gel regeneration. *Energy and Fuels*, **23** (6), 3150–3158.

[24] Yao, Y., Liu, S.Q. and Zhang, W. (2009) Regeneration of silica gel using ultrasonic under low temperatures. *Energy & Fuels*, **23** (1), 457–463.

[25] Yao, Y., Zhang, X., and Guo, Y. (2010) Experimental study on heat transfer enhancement of water-water shell-and-tube heat exchanger assisted by power ultrasonic. The 13th International Refrigeration and Air Conditioning Conference at Purdue, West Lafayette, Indiana, July 12–15, 2010, No. 2400 (eds Groll, E.A. and Braun, J.E.).

[26] Yao, Y, Zhang, W, Peng, Y et al. (2010) Modeling of silica gel dehydration assisted by power ultrasonic. The 13th International Refrigeration and Air Conditioning Conference at Purdue, West Lafayette, Indiana, July 12–15, 2010 (eds Groll, E.A. and Braun, J.E.), p. 2423.

[27] Yao, Y. (2010) Using power ultrasound for the regeneration of dehumidizers in desiccant air-conditioning systems: a review of prospective studies and unexplored issues. *Renewable and Sustainable Energy Reviews*, **14** (7), 1860–1873.

[28] Yao, Y., Zhang, W. and He, B. (2011) Investigation on the kinetic models for the regeneration of silica gel by hot air combined with power ultrasonic. *Energy Conversion and Management*, **52** (10), 3319–3326.

[29] Yao, Y., Zhang, W., Yang, K. and Liu, S.Q. (2012) Theoretical model on the heat and mass transfer in silica gel packed beds during the regeneration assisted by high-intensity ultrasound. *International Journal of Heat and Mass Transfer*, **55** (23-24), 7133–7143.

[30] Yao, Y., Yang, K., and Guo, H. (2013) Regeneration of liquid desiccant assisted by ultrasonic atomizing. Cryogenics and Refrigeration Proceedings of ICCR2013, Hangzhou, China, April 6–8, 2013, Paper ID: D2-40.

[31] Yao, Y., Lian, Z.W., and Liu, S.Q. (2007) Desiccant air-conditioning system with ultrasonic-assisted regeneration. China Patent 200510110441.8, Apr. 10 2007.

[32] Yao, Y., Liu, S.Q., and Chen, J. (2009) Shell-and-tube heat exchanger enhanced by ultrasound. China Patent 200710173262.8, May 15 2009.

[33] Yao, Y., Liu, S.Q., and Chen, J. (2009) Rotary desiccant dehumidifier assisted by high-intensity ultrasound. China Patent 200710173264.7, July 20 2009.

[34] Yao, Y., Liu, S.Q., and Chen, J. (2009) Evaporative cooling air-conditioning system based on ultrasonic technology. China Patent 200710173263.2, July 25 2009.

[35] Yao, Y., Liu, S.Q., and Chen J. (2012) Desiccant wheel air-conditioning system using power ultrasound for desiccant regeneration and heat pipe for heat recovery. China Patent 201110209559.1, Oct. 20 2012.

[36] Yao, Y., Liu, S.Q., and Chen, J (2012) A high-power ultrasonic transducer combined with heat pipe cooling system. China Patent 201110226180.1, Sept. 20 2012.

[37] Yao, Y., Liu, S.Q., and Chen J. (2013) Ultrasonic-assisted regenerator for the liquid desiccant. China Patent 201110224732.5, Nov. 05 2013.

[38] Zhang, W., Yao, Y. and Wang, R. (2010) Influence of ultrasonic frequency on the regeneration of silica gel by applying high-intensity ultrasound. *Applied Thermal Engineering*, **30** (14), 2081–2087.

[39] Zhang, W., Yao, Y., He, B. and Wang, R. (2011) Specific energy consumption of silica gel regeneration with high-intensity ultrasound. *Applied Energy*, **88** (6), 2146–2156.

[40] Yang, K. and Yao, Y. (2012) Effect of applying ultrasonic on the regeneration of silica gel under different air conditions. *International Journal of Thermal Sciences*, **61** (11), 67–78.

[41] Yang, K. and Yao, Y. (2013) Investigation on applying ultrasonic to the regeneration of a new honeycomb desiccant. *International Journal of Thermal Sciences*, **72** (10), 159–171.

[42] Yang, K., Yao, Y., and He B. (2013) Quantitative study on contributions of highintensity ultrasound to the enhancement of regeneration of silica gel. Cryogenics and Refrigeration Proceedings of ICCR2013, Hangzhou, China, April 6–8, 2013, Paper ID: B3-40.

[43] Liu S.Q., Lin S. Effects of geometry dimensions of the tool on the ultrasonic torsional vibration system. *Applied Acoustics*, 2004, **23**(6): 16–19, (in Chinese).

[44] Liu S.Q., Lin S. Study on the local resonance in torsional ultrasonic vibration systems. *Applied Acoustics*, 2004, **23**(2): 11–14, (in Chinese).

[45] Liu S.Q., Lin S. Study on the equivalent circuit of uniform cross section acoustic waveguide. *Journal of Southwest China Normal University*, 2005, **45**(6): 648–651, (Natural Science, in Chinese).

[46] Liu S.Q., Lin S, Guo J. Study on the radial vibration and equivalent circuit of thin elastic annular plate with tapered thickness along radial direction. *Piezoelectric and Acoustooptics*, 2006, **28**(3): 347–349, (in Chinese).

[47] Liu S.Q., Qiu Y. Ultrasonic torsional vibration and equivalent circuit of thin annular resonator. *Journal of Zhejiang Normal University*, 2007, **30**(3): 277–281, (Natural Science, in Chinese).

[48] Liu S.Q., Yao Y. Radial vibration characteristics of composite pipe power piezoelectric ultrasonic transducer. *Journal of Mechanical Engineering*, 2008, **44**(10): 239–244, (in Chinese).

[49] Liu S.Q., Lin S. Radial vibration frequency equation of composite disc piezoelectric ultrasonic transducer. *Journal of Mechanical Engineering*, 2008, **44**(9): 65–69, (in Chinese).

[50] Liu, S.Q. and Lin, S. (2009) The analysis of the electro-mechanical model of the cylindrical radial composite piezoelectric ceramic transducer. *Sensors and Actuators A*, **155** (2), 175–180.

[51] Liu S.Q., Yao Y, Lin S. Radial-torional vibration mode of disc-type piezoelectric transducer. *Journal of Mechanical Engineering*, (in Chinese), 2009, **45**(6): 176–180.

[52] Liu S.Q., Yao Y, Lin S. Study on the conversion mechanism of radial torsional vibration mode of piezoelectric hybrid disk with multiple diagonal slots. *Journal of Shanghai Jiaotong University*, 2009, **43**(8): 1312–1316, (Natural Science, in Chinese).

[53] Liu S.Q., Weizhong P. Load characteristics of composite tubular piezoelectric ultrasonic transducer. *Journal of Zhejiang Normal University*, 2009, **32**(4): 421–425, (Natural Science, in Chinese).

[54] Liu S.Q., Su C, Yao Y. Radial vibration characteristics of an annular ultrasonic concentrator with n-th power thickness variations. *Journal of Shanghai Jiaotong University*), 2011, **45** (6): 940–944, (Natural Science, in Chinese.

[55] Liu, S.Q. and Yao, Y. (2010) A type of piezoelectric ultrasonic transducer. China Patent 200710071087.1.

[56] Liu, S.Q., Zhang, Z., and Fang, J. (2011) A type of composite tubular ultrasonic transducer. China Patent 200910102269.X.

[57] Liu, S.Q., Zhang, Z., and Fang J. (2011) A type of tubular piezoelectric ultrasonic transducer. China Patent 200910102269.X.

[58] Liu, S.Q., Zhang, Z., and Fang, J. (2011) A type of cylindrical ultrasonic transducer. China Patent 200920196021.X.

[59] Alam, K.C.A., Saha, B.B., Kang, Y.T. *et al.* (2000) Heat exchanger design effect on the system performance of silica gel adsorption refrigeration systems. *International Journal of Heat and Mass Transfer*, **43** (24), 4419–4431.

[60] Liu, Y.L., Wang, R.Z. and Xia, Z.Z. (2005) Experimental performance of a silica gel–water adsorption chiller. *Applied Thermal Engineering*, **25** (2–3), 359–375.

[61] Lu, Z.S., Wang, R.Z. and Xia, Z.Z. (2013) Experimental analysis of an adsorption air conditioning with micro-porous silica gel–water. *Applied Thermal Engineering*, **50** (1), 1015–1020.

[62] Wu, J.Y. and Li, S. (2009) Study on cyclic characteristics of silica gel–water adsorption cooling system driven by variable heat source. *Energy*, **34** (11), 1955–1962.

[63] Enteria, N., Yoshino, H., Mochida, A. *et al.* (2012) Performance of solar-desiccant cooling system with Silica-Gel (SiO2) and titanium dioxide (TiO2) desiccant wheel applied in East Asian climates. *Solar Energy*, **86** (5), 1261–1279.

[64] Hougen, O.A. and Marshall, W.R. (1947) Adsorption from a fluid stream flowing through a stationary granular bed. *Chemical Engineering Progress*, **43** (4), 197–208.

[65] Aristov, Y.I., Tokarev, M.M. and Gordeeva, L.G. (1996) Selective water sorbents for multiple applications, 1. CaCl2 confined in mesopores of silica gel: sorption properties. *Reaction Kinetics and Catalysis Letters*, **59** (2), 325–333.

[66] Aristov, Y.I., Tokarev, M.M. and Restuccia, G. (1996) Selective water sorbents for multiple applications, 2. CaCl2 confined in micropores of silica gel: sorption properties. *Reaction Kinetics and Catalysis Letters*, **59** (2), 335–342.

[67] Aristov, Y.I., Tokarev, M.M., Gordeeva, L.G. *et al.* (1999) New composite sorbents for sorlordriven technology of fresh water production from atmosphere. *Solar Energy*, **66** (2), 165–168.

[68] Aristov, Y.I., Restucia, G., Tokarev, M.M. *et al.* (2000) Selective water sorbents for multiple applications. 11. CaCl2 confined to expanded vermiculite. *Reaction Kinetics and Catalysis Letters*, **71** (2), 377–384.

[69] Tokarev, M.M. and Aristov, Y.I. (1997) Selective water sorbents for multiple applications, 4. CaCl2 confined in silica gel pores: sorption / desorption kinetics. *Reaction Kinetics and Catalysis Letters*, **62** (1), 143–150.

[70] Thoruwa, T.F.N., Johnstone, C.M., Grant, A.D. and Smith, J.E. (2000) Novel, low cost CaCl2 based desiccants for solar crop drying applications. *Renewable Energy*, **19** (5), 513–520.

[71] Liu, Y.F. and Wang, R.Z. (2003) Pore structure of new composite SiO2·xH2O·yCaCl2 with uptake of water air. *Science in China, Serials E*, **46** (5), 551–559.

[72] Jia, C.X., Dai, Y.J., Wu, J.Y. and Wang, R.Z. (2007) Use of compound desiccant to develop high performance desiccant cooling system. *International Journal of Refrigeration*, **30** (2), 345–353.

[73] Misha, S., Mat, S., Ruslan, M.H. and Sopian, K. (2012) Review of solid/liquid desiccant in the drying applications and its regeneration methods. *Renewable and Sustainable Energy Reviews*, **16** (7), 4686–4707.

[74] Jeong, J., Yamaguchi, S., Saito, K. and Kawai, S. (2011) Performance analysis of desiccant dehumidification systems driven by low-grade heat source. *International Journal of Refrigeration*, **34** (4), 928–945.

[75] Thoruwa, T.F.N., Smith, J.E., Grant, A.D., and Johnstone, C.M. (1996) Development in solar drying using forced ventilation and solar regenerated desiccant materials. WREC-IV World Renewable Energy Congress, No. 4, Denver, Colorado, ETATS-UNIS (15/06/1996), 1996, Vol. 9(1-4), pp. 686–689.

[76] Thoruwa, T.F.N., Grant, A.D., Smith, J.E. and Johnstone, C.M. (1998) A solar-regenerated desiccant dehumidifier for the aeration of stored grain in the humid tropics. *Journal of Agricultural Engineering Research*, **71** (4), 257–262.

[77] Shanmugam, V. and Natarajan, E. (2006) Experimental investigation of forced convection and desiccant integrated solar dryer. *Renewable Energy*, **31** (8), 1239–1251.

[78] Nagaya, K., Li, Y., Jin, Z. *et al.* (2006) Low-temperature desiccant-based food drying system with airflow and temperature control. *Journal of Food Engineering*, **75** (1), 71–77.

[79] Madhiyanon, T., Adirekrut, S., Sathiruangsak, P. and Soponron-Narit, S. (2007) Integration of a rotary desiccant wheel into a hot-air drying system: drying performance and product quality studies. *Chemical Engineering and Processing*, **46** (4), 282–290.

[80] Gandhidasan, P.A. (2004) Simplified model for air dehumidification with liquid desiccant. *Solar Energy*, **76** (4), 409–416.

[81] Rane, M.V., Reddy, S.V.K. and Easow, R.R. (2005) Energy efficient liquid desiccant-based dryer. *Applied Thermal Engineering*, **25** (5-6), 769–781.

[82] Barati, A., Kokabi, M. and Famili, N. (2003) Modeling of liquid desiccant drying method for gelcast ceramic parts. *Ceramic International*, **29** (2), 199–207.

[83] Barati, A., Kokabi, M. and Famili, N. (2003) Drying of gelcast ceramic parts via the liquid desiccant method. *Journal of the European Ceramic Society*, **23** (13), 2265–2272.

[84] Zheng, Z., Zhou, D. and Gong, S. (2008) Studies of drying and sintering characteristics of gelcast BaTiO3-based ceramic parts. *Ceramics International*, **34** (3), 551–555.

[85] Trunec, M. (2011) Osmotic drying of gelcast bodies in liquid desiccant. *Journal of the European Ceramic Society*, **31** (14), 2519–2524.

[86] Mandegari, M. and Pahlavanzadeh, H. (2009) Introduction of a new definition for effectiveness of desiccant wheels. *Energy*, **34** (6), 797–803.

[87] Bassuoni, M.M. (2011) An experimental study of structured packing dehumidifier / regenerator operating with liquid desiccant. *Energy*, **36** (5), 2628–2638.

[88] Zaltash, A., Petrov, A.Y., Rizy, D.T. *et al.* (2006) Laboratory R&D on integrated energy systems (IES). *Applied Thermal Engineering*, **26** (1), 28–35.

[89] Lu, S.M., Shyu, R.J., Yan, W.J. and Chung, T.W. (1995) Development and experimental validation of two novel solar desiccant–dehumidification–regeneration systems. *Energy*, **20** (8), 751–757.

[90] Xiong, Z.Q., Dai, Y.J. and Wang, R.Z. (2009) Investigation on a two-stage solar liquid-desiccant (LiBr) dehumidification system assisted by CaCl2 solution. *Applied Thermal Engineering*, **29** (5-6), 1209–1215.

[91] Alosaimy, A.S. and Hamed, A.M. (2011) Theoretical and experimental investigation on the application of solar water heater coupled with air humidifier for regeneration of liquid desiccant. *Energy*, **36** (7), 3992–4001.

[92] Wang, W.C., Calay, R.K. and Chen, Y.K. (2011) Experimental study of an energy efficient hybrid system for surface drying. *Applied Thermal Engineering*, **31** (4), 425–431.

[93] Ania, C.O., Menendez, J.A., Parra, J.B. and Pis, J.J. (2004) Microwave-induced regeneration of activated carbons polluted with phenol: a comparison with conventional thermal regeneration. *Carbon*, **42** (7), 1377–1381.

[94] Liu, X., Gang, L. and Han, W. (2007) Granular activated carbon adsorption and microwave regeneration for the treatment of 2,4,5-trichlorobiphenyl in simulated soil-washing solution. *Journal of Hazardous Materials*, **147** (3), 746–751.

[95] Kim, K.J. and Ahn, H.G. (2008) A study on adsorption characteristics of benzene over activated carbons coated with insulating materials and desorption by microwave irradiation. *Jour nal of Korean Industrial and Engineering Chemistry*, **19** (4), 445–451.

[96] Polaerta, I., Estel, L., Huygheb, R. and Thomasb, M. (2010) Adsorbents regeneration under microwave irradiation for dehydration and volatile organic compounds gas treatment. *Chemical Engineering Journal*, **162** (3), 941–948.

[97] Mitsuhiro, K., Takuya, H., Satoshi, Y. *et al.* (2011) Water desorption behavior of desiccant rotor under microwave irradiation. *Applied Thermal Engineering*, **31** (8–9), 1482–1486.

[98] Mitsuhiro, K., Takuya, H., Satoshi, Y. and Hitoki, M. (2013) Regeneration characteristics of desiccant rotor with microwave and hot-air heating. *Applied Thermal Engineering*, **50** (2), 1576–1581.

[99] Rautenbach, R. and Albrecht, R. (1989) *Membrane Processes*, John Wiley & Sons, Inc., New York.

[100] Al-Farayedhi, A.A., Gandhidasan, P. and Ahmed, S.Y. (1999) Regeneration of liquid desiccants using membrane technology. *Energy Conversion & Management*, **40** (13), 1405–1411.

[101] Al-Sulaiman, F.A., Gandhidasan, P. and Zubair, S.M. (2007) Liquid desiccant based two-stage evaporative cooling system using reverse osmosis (RO) process for regeneration. *Applied Thermal Engineering*, **27** (14–15), 2449–2454.

[102] Alexander, S.R., Ananda, K.N., Srinivas, G. and Thomas, F.F. (2011) Modeling of a flat plate membrane-distillation system for liquid desiccant regeneration in air-conditioning applications. *International Journal of Heat and Mass Transfer*, **54** (15–16), 3650–3660.

[103] Yao, S.H., Hertzog, D.E. and Zeng, S.L. (2003) Porous glass electro osmotic pumps: design and experiments. *Journal of Colloid Interface Science*, **268** (1), 143–153.

[104] Kirby, B.J. and Hasselbrink, E.F. (2004) Zeta potential of micro fluidic substrates: 1. Theory, experimental techniques, and effects on separations. *Electrophoresis*, **25** (1), 187–202.

[105] Santiago, J.G. (2001) Electroosmotic flows in microchannels with finite inertial and pressure forces. *Analytical Chemistry*, **73** (10), 2353–2365.

[106] Li, B., Lin, Q.Y. and Yan, Y.Y. (2012) Development of solid desiccant dehumidification using electro-osmosis regeneration method for HVAC application. *Building and Environment*, **48** (1), 128–134.

[107] Ogata, A., Mizuno, K., Ito, D. *et al.* (2001) Removal of dilute benzene using zeolite-hybrid plasma reactor. *IEEE Transactions on Industry Applications*, **37** (4), 959–964.

[108] Malik, M.A., Kolb, J.F., Sun, Y. and Schoenbach, K.H. (2011) Comparative study of NO removal in surface-plasma and volume-plasma reactors based on pulsed corona discharges. *Journal of Hazardous Materials*, **197** (12), 220–228.

[109] Song, Y.H., Kim, S.J., Choi, K.I. and Yamamoto, T. (2002) Effects of adsorption and temperature on a non-thermal plasma process for removing VOCs. *Journal of Electrostatics*, **55** (4), 189–201.

[110] Wikipedia http://en.wikipedia.org/wiki/Ultrasound (accessed 9 April 2014).

[111] Kudra T, Mujumdar AS in *Advanced Drying Technologies*, Chapter 13 S. Drying (Ed.) 2nd edn Bosa Roca, FL: CRC Press, Taylor & Francis, 2009.

[112] Yiquan, Y. (1992) *Ultrasonic Transducer*, Nanjing University Press, Najing.

[113] Fridman, V.M. (1967) *Ultrasonic Chemical Equipment*, Mashinostroyennie, Moscow.

[114] Boucher, R.M.G. (1959) Drying by airborne ultrasonics. *Ultrasonic News*, **2**, 8–16.

[115] Borisov, Y.Y. and Gynkina, N.M. (1962) On acoustic drying in a standing sound wave. *Soviet Physics Acoustics*, **8** (1), 129–131.

[116] Zayas, Y. and Pento, V. (1975) Drying of thermally labile solutions in acoustic field. *Myasnaya Industriya*, **6** (1), 31–33.

[117] Bartolome, L.G., Hoff, J.E. and Purdy, K.R. (1969) Effect of resonant acoustic vibrations on drying rates of potato cylinders. *Food Technology*, **23**, 47–50.

[118] Muralidhara, H.S. and Ensminger, D. (1986) Acoustic drying of green rice. *Drying Technology*, **4** (1), 137–143.

[119] Greguss, P. (1963) The mechanism and possible applications of drying by ultrasonic irradiation. *Ultrasonics*, **1**, 83–86.

[120] Muralidhara, H.S., Chauhan, S.P., Senapati, N. *et al.* (1988) Electro-acoustic dewatering (EAD): a novel approach for food processing, and recovery. *Separation Science and Technology*, **23** (12, 13), 2143–2158.

[121] Suslick, K.S. (1988) *Ultrasound: Its Chemical, Physical and Biological Effects*, VCH Publishers, New York.

[122] Fairbanks, H.V. and Chen, W.I. (1969) Influence of ultrasonic upon liquid flow through porous media. *Ultrasonics*, **7** (3), 195–196.

[123] Kuznetsov, V.V. and Subbotina, N.I. (1965) An effect of ultrasound on the diffusion of electrolytic hydrogen through the ion membrane. *Elektrokhimia*, **1** (9), 1096–1098.

[124] Kardashev, G.A. (1990) *Physical Methods of Process Intensification in Chemical Technologies*, Khimiya, Moscow.

[125] Strumillo, C. and Kudra, T. (1986) *Drying: Principles, Applications and Design*, Gordon and Breach Science Publishers, New York.

[126] Moy, J.H. and Dimarco, G.R. (1970) Exploring airborne sound in a non-vacuum freeze drying process. *Journal of Food Science*, **35** (10), 811–817.

[127] Ensminger, D.E. (1988) *Ultrasonics: Fundamentals, Technology, Applications*, Marcel Dekker, Inc., New York and Basel.

[128] Gallego-Juarez, J.A., Rodriguez-Corral, G., Galvez-Moraleda, J.C. and Yang, T.S. (1999) A new high-intensity ultrasonic technology for food dehydration. *Drying Technology*, **17** (3), 597–608.

2

Ultrasound-Assisted Regeneration of Silica Gel

Silica gel is a granular, vitreous, porous form of silicon dioxide made synthetically from sodium silicate. As a desiccant, silica gel has been utilized for dehumidification processes in industrial and residential applications due to its large pore surface area (around $800\,m^2/g$) and good moisture adsorption capacity. Generally, process air flows through the silica gel bed, and the moisture in the air is absorbed by the silica gel. After the silica gel is saturated with moisture, the bed is heated, and purged off its moisture for regeneration. Currently, thermal heating is a common method used in the regeneration of silica gel. The regeneration temperature of silica gel is usually as high as over $100\,°C$ in order to satisfy the demand of real applications. However, such high regeneration temperature may have a fatal disadvantage in that they cause silica gel to utilize the lower-grade thermal energy (i.e., preferably the temperature is less than $80\,°C$), which is easily available in the world. On the other hand, a higher regeneration temperature will lead to more energy losses because the process of cooling the regenerated silica gel before it is reused for moisture adsorption needs more energy. Thus it can be seen that the regeneration temperature is one of the crucial factors that affect energy utilization efficiency during the regeneration process. To decrease the regeneration temperature of silica gel, the nonheating regeneration method with ultrasound is hence proposed.

2.1 Theoretical Analysis

As mentioned in Chapter 1, the high-intensity ultrasound will produce a variety of effects, such as micro-oscillating, cavitation, streaming, and radiation heating. These effects may influence mass transfer processes by producing changes in concentration gradients, diffusion coefficients, or boundary layer. The effects may vary for different physical states of a system. For the solid desiccant regeneration, the heat and mass transfer takes place in a solid-to-gas system. To probe into the mechanism of enhancement of heat and mass transfer by the use of power ultrasound in solid-gas system, the moisture transport in silica gel is illustrated here.

Silica gel has very high moisture adsorption capacity because of its micro-porous structure of internal interlocking cavities that are of a high internal surface area to a maximum $800\,m^2/g$.

Ultrasonic Technology for Desiccant Regeneration, First Edition. Ye Yao and Shiqing Liu.

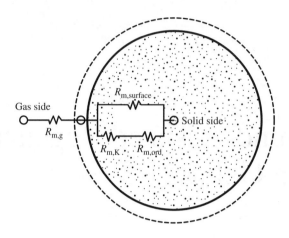

Figure 2.1   Detailed model of resistance of mass transfer for silica gel particles

During regeneration, the water vapor pressure in the drying air is normally lower than that at or near the external surface of the silica gel particles. This kind of pressure difference drives the water molecules in silica gel to diffuse to the surface and then enter the air. So, the mass transport should have two basic processes: one is the moisture diffusion in the solid side; another is the convective mass transfer between silica gel and air. The schematic diagram for the resistance model of mass transfer for silica gel is given in Figure 2.1. "$R_{m,g}$" in the figure stands for the gas-side mass transfer resistance; "$R_{m,surface}$," "$R_{m,K}$," and "$R_{m,ord}$" denote three kinds of mass transfer resistance on the solid side.

The three main transport mechanisms on the solid side are: the ordinary, Knudsen, and surface diffusion. The ordinary diffusion occurs when the molecules of the gas collide with each other more often than with the pore walls of a porous medium, and the Knudsen diffusion occurs when the gas molecules collide more frequently with pore walls than with each other [1]. For water vapor–air mixtures, the ordinary diffusion coefficient (D_{ord}: m²/s) and the Knudsen diffusion coefficient (D_K: m²/s) can be respectively given by Equations (2.1) and (2.2) [2].

$$D_{ord} = 1.735 \times 10^{-9} \frac{(t + 273.15)^{1.685}}{p_w}. \qquad (2.1)$$

$$D_K = 22.86 \times (t + 273.15)^{0.5} \cdot r_{pore}, \qquad (2.2)$$

where t is the gas temperature in degree Celsius (°C); p_w is the water vapor pressure in bar; and r_{pore} is the average pore radius in meters.

The pore void diffusion coefficient (D_{void}: m²/s) then may be approximately represented by an addictive resistance based on the ordinary and the Knudsen coefficient [2]:

$$D_{void} = \left[1/D_{ord} + 1/D_K\right]^{-1}. \qquad (2.3)$$

Equations (2.1)–(2.3) are valid only for long, uniform radius capillaries. For the real porous medium, the pore void diffusion coefficient should be modified as [3]:

$$D_{void,eff} = \frac{\varepsilon_{particle}}{\varsigma_g} D_{void}, \qquad (2.4)$$

where $D_{void,eff}$ is called the effective pore void diffusion coefficient given in m^2/s; $\varepsilon_{particle}$ is particle porosity which accounts for the reduction of free area for diffusion due to presence of solid phase; ς_g stands for gas tortuosity factor that accounts for the increase in diffusion length due to tortuous paths of real pores.

Surface diffusion is the transport of adsorbed molecules on the pore. The main mechanisms to explain surface flow include the hopping model, which takes into account that the gas molecules move on the surface by jumping from site to site with a specific velocity [4]. Based on the mechanistic hopping model, the expression Equation (2.5) for the effective surface diffusion coefficient ($D_{surface,eff}$: m^2/s) was obtained by Sladek *et al.* [5] as:

$$D_{surface,eff} = \frac{1.6 \times 10^{-6}}{\varsigma_{surface}} \exp \left(\frac{-0.974 H_{ads}}{t + 273.15} \right), \tag{2.5}$$

where $\varsigma_{surface}$ is surface tortuosity factor that accounts for the increase in diffusion length due to tortuous paths of real pores and H_{ads} is the heat of adsorption (kJ/kg), which may be considered as a function of moisture ratio (q: kg water/(kg dry sample)) in solid desiccant. For regular density silica gel with certain moisture ratio, H_{ads} may be experimentally given by Pesaran and Mills [6]:

$$H_{ads} = -12400q + 3500; \quad q \le 0.05. \tag{2.6a}$$

$$H_{ads} = -1400q + 2950; \quad q > 0.05. \tag{2.6b}$$

Since the pore void diffusion and the surface diffusion are a parallel process, the total diffusivity in solid side (D_{solid}: m^2/s) can be written as:

$$D_{solid} = D_{surface,eff} + \frac{g'}{\rho_p} D_{void,eff}, \tag{2.7}$$

where ρ_p is particle density in kg/m^3 and g' is derivative of the equilibrium isotherm, varying from 0 to 0.4 for regular density silica gel [3].

The lumped parameter model for the solid-side mass transfer coefficient ($K_{m,solid} = 1/R_{m,solid}$, kg/m^2 s) may be expressed by:

$$K_{m,solid} = \frac{\rho_p D_{solid}}{r_p}, \tag{2.8}$$

where r_p is the particle radius in meters.

Known from Equations (2.1)–(2.7), the factors that influence the solid-side diffusivity in porous medium mainly include the pore size, the pressure, the temperature, and the moisture ratio. Using the data published in Ref. [6], these influential factors are plotted in Figures 2.2 and 2.3, respectively. The basic calculation parameters include: $r_p = 2 \times 10^{-3}$ m; $\rho_p = 721.1$ kg/m^3; $\varsigma_{surface} = \varsigma_g = 2.8$; $\varepsilon_p = 0.716$; and $g' = 0.2$.

It can be seen from Figure 2.2 that the water vapor pressure in the porous medium has little influence on the solid-side diffusivity when the pore size is smaller than 200 Å (1 Å = 10^{-10} m). So, for those micro-porous solid desiccants like silica gel whose pore size is not bigger than 100 Å, the contribution of the "micro-oscillation effect" of power ultrasound to the solid-side diffusivity may be negligible. However, for those macro-porous media like green rice whose pore size may be larger than 1000 Å, the micro-oscillation effect that induces a series of pulsation partial vacuum in the medium (which equals to lowering the water vapor pressure) may

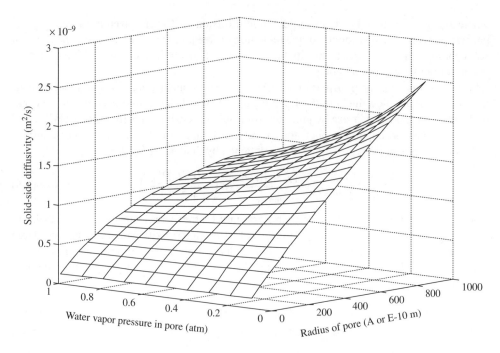

Figure 2.2 Effect of pore size and water vapor pressure in pore on the solid-side diffusivity ($t_{\text{solid}} = 40\,^\circ\text{C}$)

affect the transport of water vapor [7]. As shown in Figure 2.2, the influence of water vapor pressure on diffusivity becomes significant as the pore radius attains 1000 Å.

Figure 2.3 shows that both the temperature and moisture ratio of the solid medium have a great impact on the solid-side diffusivity which will increase with a corresponding increase of the two factors. The influence of the temperature on diffusivity becomes more significant under a higher moisture ratio in silica gel. As the sonic energy is dissipated as heat and makes the temperature of the material increase, the thermal effect of power ultrasonic is sure to improve the moisture transport in a porous medium.

The gas-side mass transfer coefficient ($K_{\text{m,g}} = 1/R_{\text{m,g}}$, kg/m^2 s) may be defined as:

$$K_{\text{m,g}} = \frac{\rho_g D_g}{L}, \tag{2.9}$$

where ρ_g is the gas density in kg/m^3; D_g is the gas-side diffusivity in m^2/s; and L is the thickness of particle surface layer where there is a gradient of moisture concentration.

Since D_g is mainly affected by the difference in moisture concentration between the particle surface and the gaseous phase, the increase in the gas-side mass transfer coefficient ($K_{\text{m,g}}$) brought by the increase in the air mass flow rate may be largely due to the reduction of the thickness of the laminar sub-layer surrounding the particle (L) because of the increased gas velocity. The effect of micro-oscillation induced by power ultrasound may have a similar result of increasing the air mass flow rate, which intensifies the turbulence of gas stream flowing over

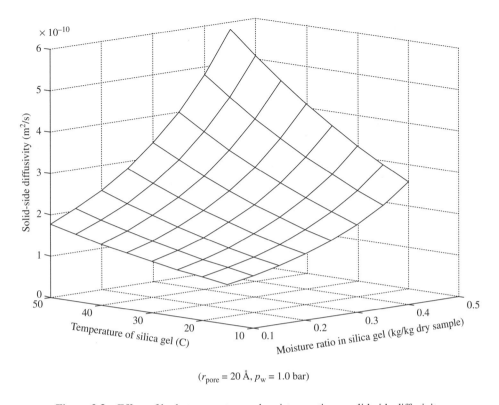

$(r_{pore} = 20\,\text{Å},\ p_w = 1.0\ \text{bar})$

Figure 2.3 Effect of body temperature and moisture ratio on solid-side diffusivity

the material and reduces the thickness of the boundary layer. It is reported that a substantial decrease (up to 15%) in the thickness of the boundary layer may be achieved by a sound of high intensity between 15 kHz and 155 dB during the drying process [8]. The gas-side mass transfer enhancement by power ultrasound may as well be explained by the theory of acoustic streaming [9]. The acoustic streaming is a steady circular airflow occurring in a high-intensity sound field, which is attributed to the friction between a fluid medium and a vibrating object. So, when the solid is vibrated by the high-intensity ultrasound, there will be an oscillatory tangential relative velocity between the solid surface and the fluid passing by, which is especially effective in promoting certain kinds of rate process occurring on the solid and fluid interface including the heat and mass transfer [10].

The above analysis indicates the following possible mechanisms of mass transfer enhancement in a solid-gas system by the power ultrasound:

1. For the micro-porous materials (like silica gel and molecular sieves) whose pore sizes are normally smaller than 200 Å, the sound waves enhance the moisture diffusivity inside a medium mainly due to temperature rise caused by its thermal effect. But for the macro-porous materials with pore sizes larger than 200 Å, the effect of an ultrasound field on the solid-side moisture diffusivity may be both thermal and mechanical.

2. The gas-side mass-transfer alteration is mainly due to the mechanical effect of power ultrasound that causes high oscillation velocities and micro-streaming at solid/gas interfaces which may affect the diffusion boundary layer.

2.2 Experimental Study

The following issues are to be addressed through experiments to technically illustrate the feasibility of the novel regeneration method by using power ultrasound:

1. How does power ultrasound impact the regeneration of silica gel?
2. How does power ultrasound influence the diffusivity of moisture in silica gel and the activation energy of moisture desorption from silica gel?
3. How about the energy consumption of regeneration assisted by ultrasound?

2.2.1 Experimental Setup

The experimental setup (as shown in Figure 2.4) mainly consists of a silica gel packed bed, an ultrasonic transducer, an ultrasonic generator, a fan, an electric heater with a power controller and air ducts. The silica gel packed bed is a cylindrical container with a height of about 95 mm. It is made up of two steel cylindrical shells with numerous orifices (about 2.5 mm in size) in the surface and two round steel plates. The two cylindrical shells, about 20 and 50 mm, respectively, in diameter, are concentrically placed and fixed by the two plates at both ends. The sample (i.e., silica gel) is then filled in the space enveloped by the two cylindrical shells and the two plates. The ultrasonic transducer is fixed tightly onto one plate through which the ultrasound propagates into the silica gel in the bed. The regenerating air from the duct

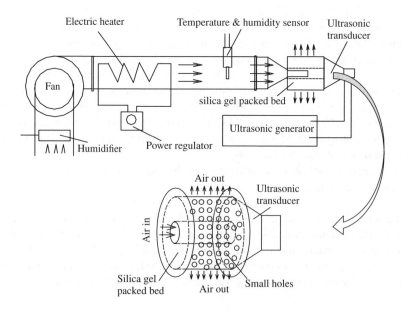

Figure 2.4   The schematic diagram of the setup for the experiment

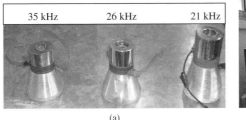

(a)   (b)

Figure 2.5   Photographs of ultrasonic transducers and ultrasonic generator for this study. (a) Ultrasonic transducers with different frequencies and (b) ultrasonic generator

first enters into the inner cylindrical shell, then passes through the silica gel in the bed and finally exhausts outside from the orifices of the outer cylindrical shell. The positive/negative electrode of the ultrasonic transducer is of active connection with the positive/negative output of the ultrasonic generator that can produce high-energy ultrasound with the power range of 0–300 W and different frequencies ranging from 16 to 100 kHz. The photographs of ultrasonic transducers and ultrasonic generator are shown in Figure 2.5. The electric heater, which is used for producing different experimental temperatures of regenerating air, is installed in the upward stream of the air duct. A temperature and humidity sensor (type: HMT100; measurement precision: ±2% in humidity and ±0.2 °C in temperature) is placed at the outlet of the air duct to monitor the conditions of regeneration air during the experiments. A copper-constantan thermocouple (measurement precision: ±0.2 °C) is placed in the central position of the bed where the surface temperature of silica gel is observed during the regeneration. A humidifier is used to wet the silica gel in the bed to an initial moisture ratio for the regeneration experiment. The other instruments include an electronic balance (measurement precision: ±0.1 g) for measuring the moisture change in silica gel, a dry-wet bulb thermometer (measurement precision: ±0.5 °C) for observing the ambient air conditions and a digital anemometer (measurement precision: ±1% of reading data) for measuring the airflow rate in the air duct.

The sample of silica gel used in this experimental study has the particle size distribution of 3.5 ± 0.5 mm in diameter. The physical properties which are provided by the manufacturer mainly include the following aspects: Specific surface area $\geq 600 \, \text{m}^2/\text{g}$; Pore diameter $= 20$–30 Å; pore volume $= 0.35$–0.45 ml/g; bulk density $= 750$ g/l.

2.2.2 Procedure for Experiments

Experiments have been performed under a series of regeneration temperatures to investigate the effect of ultrasound on the regeneration. Different acoustic power levels combined with different acoustic frequencies were employed for this experimental study. During the experiments, the environmental conditions were basically kept stable, that is, the air temperature and relative humidity were at about 28 ± 1 °C and $80 \pm 5\%$, respectively. And the airflow rate in the air duct was kept at about 0.3 ± 0.05 m/s.

The mass change of moisture ratio in the silica gel during the regeneration was measured by the weighing method. The basic procedure for the experiment was summarized as below:

- To begin with, a certain amount of fresh silica gel was fully filled in the bed. The total weight of the bed together with silica gel and the ultrasonic transducer was measured and recorded.

- Then the bed was connected to the air duct through a funneled connection. The humidifier and the fan were started up to wet the silica gel in the bed until the experimental moisture ratio was achieved.
- And then, the bed was temporarily moved away from the air duct. The humidifier was turned off, while the heater was turned on. The experimental regeneration temperature (e.g., 45 °C) was produced by adjusting the power controller that controls input power of the electric heater.
- Afterwards, the bed was reconnected with the air duct and the experiment with and without ultrasound was performed at the experimental temperature. During the regeneration experiment, the bed was weighed with an electronic balance after every 8 minutes in order to observe the moisture change in the silica gel. The total experimental time for each condition lasted until no measurable weight loss was observed in the sample.
- Finally, the silica gel in the bed was fully dried by an electronic oven at a baking temperature of 300 °C to get the mass of the dry sample which is a crucial parameter in order to analyze the drying kinetics of the sample under different regeneration conditions.

2.2.3 Methods

2.2.3.1 Analysis Parameters

To evaluate the possible benefits brought by the power ultrasound, three indicators including the moisture ratio in the silica gel, the regeneration degree (RD), the enhanced ratio (ER) of regeneration, and the energy-saving rate, are suggested here.

Moisture ratio in silica gel (q_s) is written as:

$$q_s = \frac{m_{s,wet} - m_{s,dry}}{m_{s,dry}}, \tag{2.10}$$

where $m_{s,wet}$ and $m_{s,dry}$ denote the mass of the wet and the dry silica gel, respectively, in kilograms.

Regeneration degree is defined as the ratio of the mass of moisture desorption ($m_{w,loss}$ in kg) to the initial mass of moisture ($m_{w,ini}$: kg) in the sample.

$$RD = \frac{m_{w,loss}}{m_{w,ini}}. \tag{2.11}$$

ER of regeneration assisted by ultrasound is evaluated by:

$$ER = \frac{(MRS)_U - (MRS)_{NU}}{(MRS)_{NU}}, \tag{2.12}$$

where MRS denotes the mean regeneration speed (kg/s), that is, the average moisture desorption rate in a period of regeneration time. The subscript "U" and "NU" denote the case in the presence and absence of ultrasonic radiation, respectively.

Energy-saving ratio (ESR) brought by power ultrasound is defined as:

$$ESR = \frac{E_{NU}(RD) - E_U(RD)}{E_{NU}(RD)}, \tag{2.13}$$

where E (RD) denotes the energy consumption (J) used for the regeneration when certain RD of silica gel is achieved.

2.2.3.2 Determination of Moisture Diffusivity and Desorption Activation Energy

Since the pore diameter of silica gel is about 20–$30\,\text{Å}$, the kinetics could possibly be pore-diffusion controlled. Meanwhile, the power ultrasound doesn't change the mechanism of moisture diffusion in silica gel. Therefore, the diffusion model developed by Crank [11], as shown in Equation (2.14), can be employed in this study.

$$MR = \frac{q_\tau - q_e}{q_{ini} - q_e} = \frac{6}{\pi^2} \sum_{n=1}^{\infty} \frac{1}{n^2} \exp\left(-\frac{n^2\pi^2 D_e \tau}{r^2}\right), \tag{2.14}$$

where MR is the dimensionless moisture ratio; q_τ is the instant moisture ratio during the regeneration, kg/(kg dry sample); and q_{ini} and q_e stand for initial and equilibrium moisture ratio, respectively. D_e denotes effective diffusivity, m^2/s, which is defined as "the amount of a particular substance that diffuses across a unit area in 1 second under the influence of a gradient of one unit"; τ is the time, seconds and r is the radius of particle, meters.

Equation (2.14) is derived from the solution of Fick's second law, in spherical geometry, nonsteady state, and with constant surface concentration. Through fitting the experimental data, the moisture diffusivity of silica gel can be obtained by Equation (2.14).

Equation (2.14) assumes that the effective diffusivity (D_e) is constant and no shrinkage occurs in the spherical sample. When the time, τ is long enough, Equation (2.14) could be simplified to a linear equation as [12]:

$$\ln(MR) = \ln\left(\frac{6}{\pi^2}\right) - \left(\frac{\pi^2 D_e}{r^2}\right)\tau. \tag{2.15}$$

The effective diffusivity (D_e) can be gotten from a straight line of $\ln(MR)$ versus the time (τ) that is plotted using the experimental data. Using the slope of the straight line (κ), the effective diffusivity (D_e) can be calculated by:

$$D_e = \frac{\kappa \cdot r^2}{\pi^2}. \tag{2.16}$$

In the following calculation of D_e, the particle radius, r is assumed to be 0.00175 m.

The temperature dependence on the effective diffusivity (D_e) is also subject to the Arrhenius equation [13]:

$$D_e = A_o \exp\left(-E_a/RT\right), \tag{2.17}$$

where, A_o is the pre-exponential factor of Arrhenius equation (m^2/s); E_a is the activation energy (kJ/mol); T is the drying temperature (K); and R is the gas constant (kJ/mol·K).

Equation (2.17) can be written as:

$$\ln D_e = \ln D_o - \left(\frac{1}{RT}\right)E_a. \tag{2.18}$$

Thus, the activation energy (E_a) can be determined with respect to the slope of the straight line in which the values of $\ln\left(D_e\right)$ are plotted against $1/T$.

2.2.4   Results and Discussions

2.2.4.1   RD and ER under Different Regeneration Conditions

Figure 2.6 shows the variations of RD against the time under different regeneration conditions. Obviously, the values of RD during the regeneration with ultrasound are higher than those without ultrasound. Since the indicator, RD, directly reflects how much proportion of moisture

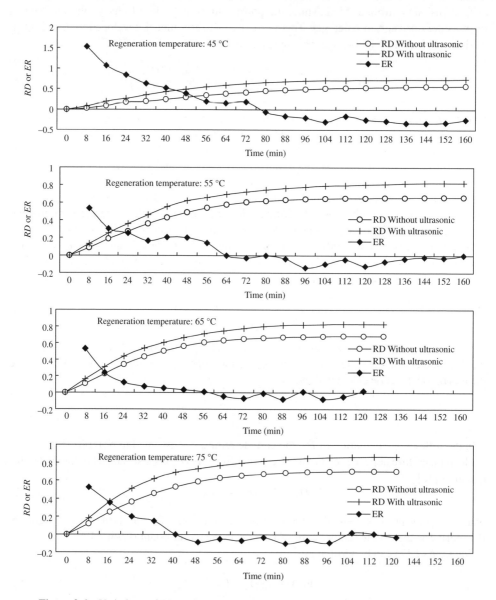

Figure 2.6   Variations of *RD* and *ER* during the regeneration with and without ultrasound

is removed from silica gel during the regeneration, the higher the RD achieved means the better effect of the regeneration. Hence, the RD curves in Figure 2.6 can well prove the effect of power ultrasound on silica gel regeneration. To investigate the role of ultrasound in the whole stage of regeneration, the values of ER assisted by ultrasound have been investigated every 8 minutes. As also shown in Figure 2.6, the ER tends to drop with regeneration time. Particularly, negative ER occurs in the latter stage of regeneration. It means that ultrasound will play a more important role in the starting stage of regeneration in which the silica gel is of relatively higher moisture ratio. This can be inferred as well from the results in Figure 2.7 which show that the gap between $(MRS)_U$ and $(MRS)_{NU}$ becomes smaller as the moisture ratio in silica gel decreases.

Figure 2.8 shows the influence of ultrasonic power on the ER of regeneration during the first 16-minute regeneration. It can be seen that a higher ER brought by ultrasound will be achieved at a lower regeneration temperature. Taking 21 kHz, for example, the 20-W ultrasound brings about 0.60, 0.45, 0.30, and 0.20 in ER, at the regeneration temperature of 45, 55, 65, and 75 °C, respectively. This may be explained by the following two reasons: Firstly, the higher regeneration temperature means the more thermal energy being put into the regeneration, which results in a small proportion of ultrasonic energy in the total energy applied to the regeneration. Therefore, the role of ultrasound in the enhancement of regeneration will be weakened. The second reason may be that the high regeneration temperature leads to the high working temperature of ultrasonic transducer that will decrease the working efficiency of the machine. Another interesting phenomenon to be found from Figure 2.8 is that the curves of ER versus the ultrasonic power bend upwards at regeneration temperatures of over 65 °C and downwards at regeneration temperatures of below 55 °C. This indicates that the increase of ultrasonic power will be more effective for the enhancement of regeneration under lower regeneration temperatures.

The ER of regeneration against the ultrasonic frequency under different regeneration temperatures are plotted in Figure 2.9. The curve trends indicate that there should be a good acoustic frequency for a specific regeneration temperature under which the highest ER brought by ultrasound can be achieved.

2.2.4.2   Moisture Diffusivity under Different Regeneration Conditions

The slopes (κ) of the straight lines of $\ln(MR)$ plotted against the time τ at each temperature used for the determination of the effective diffusivity (D_e), are given in Table 2.1. The high coefficients of determination (R^2) for κ (above 0.95 for all) indicate the validity of the ultimate results of D_e.

The variation trend of the effective diffusivity (D_e) against the ultrasonic power is plotted in Figure 2.10, which shows that D_e increases significantly with the increase of the ultrasound power applied. It convincingly proves that the power ultrasound can enhance the moisture transfer in the silica gel. Due to the heating effect of the power ultrasound, there will be a temperature rise in the media. Thus, the moisture diffusivity in the silica gel increases.

The acoustic frequency may also have an influence on the moisture diffusivity in the media. As shown in Figure 2.11, the moisture diffusivity tends to decrease with the increase of the acoustic frequency, especially under a higher ultrasound power.

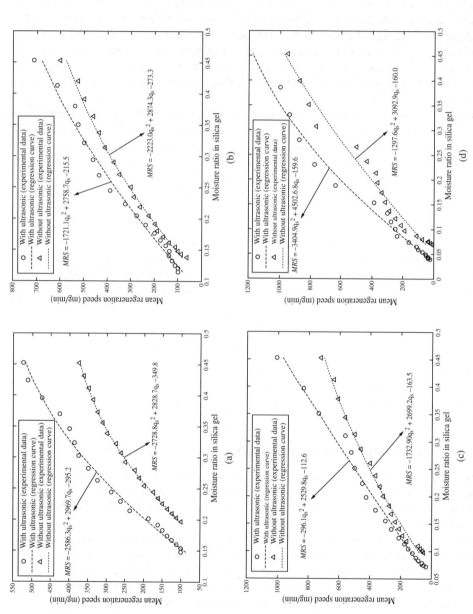

Figure 2.7 Comparisons between $(MRS)_U$ and $(MRS)_{NU}$ under different regeneration temperatures. (a) Regeneration temperature: 45 °C, (b) regeneration temperature: 55 °C, (c) regeneration temperature: 65 °C, and (d) regeneration temperature: 75 °C

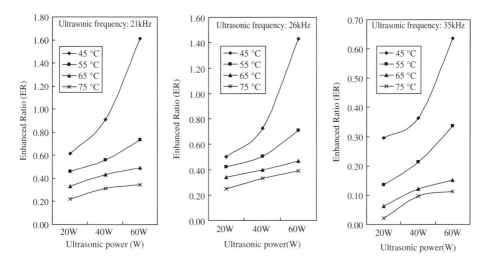

Figure 2.8 ER versus ultrasonic power for the first 16-minute regeneration

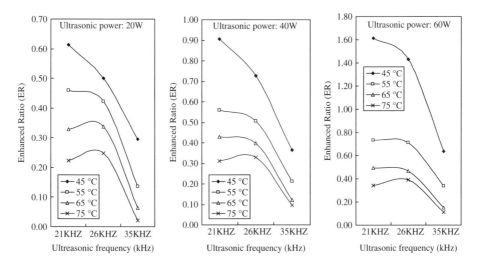

Figure 2.9 ER versus ultrasonic frequency for the first 16-minute regeneration

2.2.4.3   Activation Energy of Moisture Desorption

Taking 40 W and 26 kHz as ultrasonic parameters, for example, the possible impact of power ultrasound on the activation energy of moisture desorption from the silica gel is investigated. As mentioned above, the activation energy can be obtained from the slope of the fitting line about the values of $\ln(D_e)$ against $1/T$. According to Figure 2.12, the activation energy of moisture desorption from silica gel is found to be 29.93 and 33.13 kJ/mol for the presence and absence of ultrasonic radiation respectively. In the calculation, the gas constant of water vapor

Table 2.1 Slope (κ) of the straight lines of $\ln(MR)$ versus time τ for the determination of D_e

Regeneration temperature (°C)	$UF = 21\,\text{kHz},\ UP = 20\,\text{W}$		$UF = 21\,\text{kHz},\ UP = 40\,\text{W}$		$UF = 21\,\text{kHz},\ UP = 60\,\text{W}$	
	$\kappa\ (\times 10^{-4})$	R^2	$\kappa\ (\times 10^{-4})$	R^2	$\kappa\ (\times 10^{-4})$	R^2
45	−3.862	0.9973	−5.095	0.9956	−7.313	0.9829
55	−5.804	0.9834	−8.074	0.9735	−9.799	0.9628
65	−8.385	0.9723	−10.111	0.9836	−11.343	0.9728
75	−10.867	0.9913	−12.592	0.9726	−13.332	0.9682
	$UF = 26\,\text{kHz},\ UP = 20\,\text{W}$		$UF = 26\,\text{kHz},\ UP = 40\,\text{W}$		$UF = 26\,\text{kHz},\ UP = 60\,\text{W}$	
45	−3.788	0.9686	−4.734	0.9983	−6.857	0.9725
55	−6.050	0.9829	−8.103	0.9782	−9.553	0.9836
65	−8.632	0.9519	−10.230	0.9725	−11.137	0.9894
75	−11.193	0.9682	−12.839	0.9793	−13.524	0.9593
	$UF = 35\,\text{kHz},\ UP = 20\,\text{W}$		$UF = 35\,\text{kHz},\ UP = 40\,\text{W}$		$UF = 35\,\text{kHz},\ UP = 60\,\text{W}$	
45	−3.665	0.9685	−4.355	0.9785	−6.115	0.9838
55	−5.609	0.9737	−6.543	0.9729	−7.980	0.9724
65	−7.892	0.9835	−8.829	0.9902	−9.170	0.9538
75	−10.231	0.9599	−11.113	0.9743	−11.606	0.9677

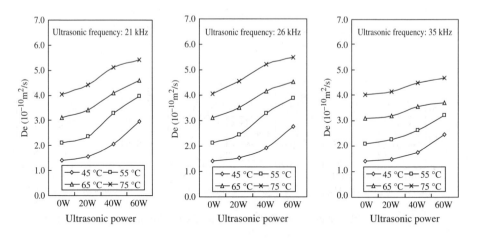

Figure 2.10 Influence of ultrasound power on D_e in silica gel during the regeneration

is given as 0.008 298 kJ/(mol·K), and the particle radius, r, is assumed to be 0.00175 m. The result confirms that by applying ultrasound, the activation energy of moisture desorption from silica gel decreases to some degree. Less activation energy means lower thermal temperature required for the regeneration. From this point of view, the regeneration temperature of silica gel can be reduced in the presence of power ultrasound. Although power ultrasound causes pressure variation in media, it can not affect the adsorbate species on the molecular level, that

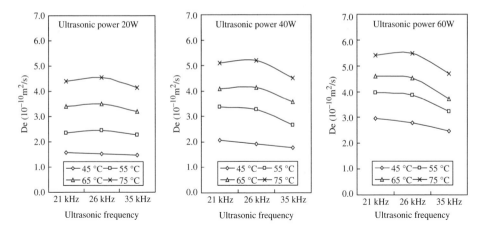

Figure 2.11   Influence of ultrasound frequency on D_e in silica gel during the regeneration

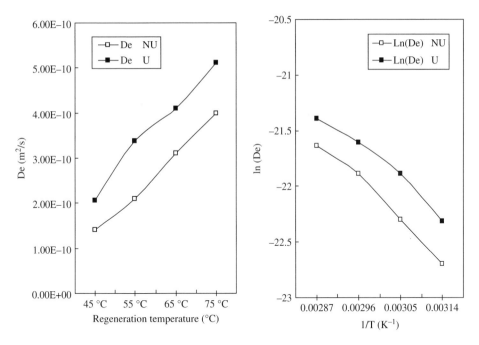

Figure 2.12   Moisture diffusivity against regeneration temperature ("U" denotes the regeneration assisted by ultrasound with 40 W in power and 26 kHz in frequency)

is, the mechanical force induced by ultrasound can not overcome the bonding force among water molecules in silica gel. Only when the frequency of the external mechanical vibration exceeds the critical value (about 1333 GHz) can it break the molecular bond force [14]. Hence, the decrease of activation energy due to the ultrasonic radiation can only be explained by the fact that part of activation energy is substituted by acoustic energy. In Rege's study [15], the researchers owed the lowering of activation energy of phenol on macroreticular resin to the acoustic cavitation energy that exists in liquids. In a solid medium like silica gel, however, the special "heating effect" of the power ultrasound may be responsible for the lowering of the activation energy of moisture desorption.

2.2.4.4 Regeneration Time

The conditioned regeneration time (CRT), which refers to the time required for obtaining a certain RD, is suggested here to investigate the reduction of regeneration time brought by ultrasound. As shown in Figure 2.13, ultrasound can significantly reduce the regeneration time of silica gel. Taking 45 °C (the regeneration temperature), for example, the CRT ($RD = 0.5$) in the presence of ultrasound (21 kHz in frequency and 20 W in power) is about 40 minutes less than that without ultrasonic radiation. The CRT will be further reduced as the ultrasonic power applied to the regeneration increases. The effect of ultrasonic power on CRT may also be affected by the regeneration temperature and the ultrasonic frequency. Comparing the curves of the CRT versus the ultrasonic power in Figure 2.13, it can be found that equal increases of ultrasound power will bring about a shorter CRT under a lower regeneration temperature or the lower ultrasonic frequency.

2.2.4.5 Energy Consumption

The energy consumptions are compared between the "NU" (No ultrasound) case and the "U" (ultrasound existing) case under different conditions. The energy consumptions required for different RD achieved ($RD = 0.3$, $RD = 0.4$, and $RD = 0.5$) are calculated based on the CRT and the total power used by the heater and the ultrasonic generator, respectively. The results of energy consumption and the ESR brought by the ultrasound for these cases are presented in Figures 2.14 and 2.15, respectively, which confirm that the regeneration energy can be

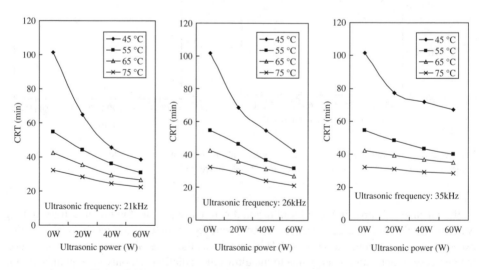

Figure 2.13   Influence of ultrasonic power on CRT at $RD = 0.50$

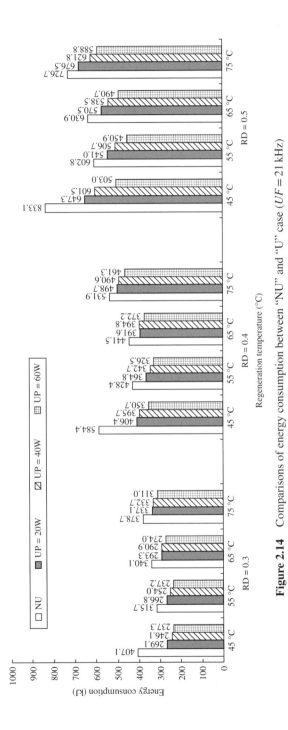

Figure 2.14 Comparisons of energy consumption between "NU" and "U" case ($UF = 21$ kHz)

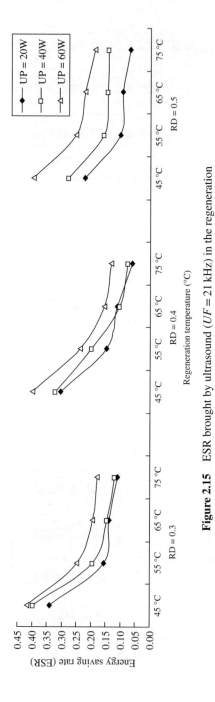

Figure 2.15 ESR brought by ultrasound ($UF = 21$ kHz) in the regeneration

significantly reduced after the ultrasonic power is applied to the regeneration. There are several reasons for the energy savings. Firstly, the heat and mass transfer can be enhanced by ultrasound, which will improve the utilization efficiency of heating energy during regeneration. Secondly, the CRT can be reduced due to the presence of ultrasound, and hence, less heat will be dissipated into the environment during regeneration. Lastly, ultrasonic energy is capable of penetrating through the whole body of the material immediately once ultrasound begins to work for the regeneration. Therefore, a higher efficiency of energy utilization will be achieved by ultrasonic regeneration compared with the heating method in which the heat energy only works on the medium surface.

The higher the ultrasonic power tends to gain, the higher is the ESR for the same RD. As can be seen from Figure 2.15, the ultrasound with 60 W in power achieves a higher ESR than that with lower power levels (20 and 40 W). Besides, the ESR increases with the regeneration temperature decreasing. Taking RD = 0.3 (i.e., the final RD of the silica gel arrives at 0.3), for example, the ESR brought by ultrasound with 60 W exceeds 0.40 at the regeneration temperature of 45 °C, while it drops to 0.25, 0.19, and 0.18, as the regeneration temperature rises to 55, 65, and 75 °C respectively. This indicates that the performance of energy saving brought by the ultrasound-assisted regeneration is better under the lower regeneration temperatures.

2.3 Empirical Models for Ultrasound-Assisted Regeneration

2.3.1 Model Overviews

Quantitative understanding of the dehydration process is crucial for the design and control of the regeneration process. It needs an appropriate model to describe the drying kinetics of silica gel. For practical applications, people often prefer empirical or semi-empirical models. They are generally derived by simplifying general series solutions (Equation (2.14)) of Fick's second law with the assumptions of diffusion-controlled moisture migration, negligible shrinkage, and constant diffusion. These empirical models mainly include the Lewis model (Equation (2.19)), Henderson and Pabis model (Equation (2.20)), Logarithmic model (Equation (2.21)), Weibull model (Equation (2.22)), and Gaussian model (Equation (2.23)). They have been widely used for dehydration modeling of different agricultural and industrial products [16–22]. In this section, these models are employed to model the drying kinetics of silica gel during ultrasound-assisted regeneration.

$$MR = \frac{q_\tau - q_e}{q_{ini} - q_e} = \exp(-K\tau) \tag{2.19}$$

$$MR = \frac{q_\tau - q_e}{q_{ini} - q_e} = A\exp(-K\tau) \tag{2.20}$$

$$MR = \frac{q_\tau - q_e}{q_{ini} - q_e} = A\exp(-K\tau) + C \tag{2.21}$$

$$MR = \frac{q_\tau - q_e}{q_{ini} - q_e} = \exp\left(-\left(\frac{\tau}{\beta}\right)^n\right) \tag{2.22}$$

$$MR = \frac{q_\tau - q_e}{q_{ini} - q_e} = a\exp\left(-\left(\frac{\tau - b}{c}\right)^2\right), \tag{2.23}$$

where A, K, C, a, b, c, n, β are model coefficients determined by experimental data; MR is the dimensionless moisture ratio; and q_τ, q_0, and q_e stand for the instant, the initial and the equilibrium moisture ratio, respectively. τ is the time, seconds.

Equilibrium moisture content (q_e) is one of the key parameters in the drying models. It is closely related to regeneration conditions. In some studies [20, 21], the equilibrium moisture content was assumed to be zero during the modeling of some food drying, based on the fact that the equilibrium moisture content is much less than the initial one. This assumption is valid only at the beginning of drying when the moisture content is relatively high. Figure 2.16 shows different curves of moisture ratio (MR) caused by different values of equilibrium moisture content (q_e) according to the experimental data under the drying condition of 45 °C (drying air temperature) combined with the 26-kHz-and-40-W ultrasound. As indicated from Figure 2.16, the influence of equilibrium moisture content on the calculated moisture ratio becomes increasingly significant as the drying time goes on (i.e., the moisture content approaches the equilibrium state). To decrease such error, the regeneration time should be long enough for each experimental condition so that the silica gel arrives at its equilibrium state as far as possible.

Figure 2.17 gives the quasi-equilibrium moisture contents under different experimental drying conditions. It indicates that besides the drying temperature, both the acoustic power and the frequency affect the equilibrium moisture content.

2.3.2 Model Analysis

By using the Matlab's curve fitting tool (Nonlinear Least Squares Regression), the empirical constants in these models have been gotten with respect to the experimental data under different

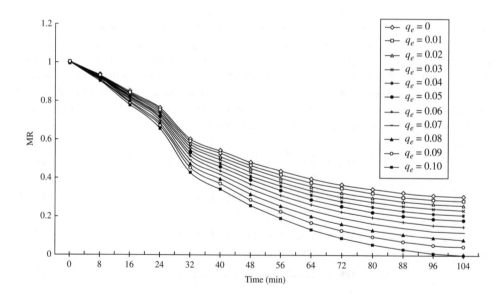

Figure 2.16 Influence of equilibrium moisture content (q_e) on the dimensionless moisture ratio (MR)

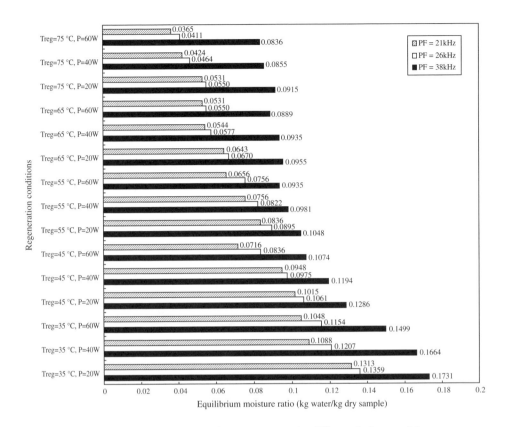

Figure 2.17 Equilibrium moisture content under different drying conditions

drying conditions. The results are listed in Table 2.2. In the Lewis, the Henderson and Pabis, and the Logarithmic models, the constant K (also called the drying rate constant) is directly related to the drying rate, and a higher dehydration rate will result in a bigger value of K in these models. Therefore, it is reasonable to see that the parameter, K, increases with the increase of the drying air temperature or the ultrasonic power level. The change trends of K versus the acoustic frequency of power ultrasonic are presented in Figure 2.18, which indicate that a lower frequency will be favorable for improving the drying rate since a larger value of K is obtained at a lower frequency. This is because a higher acoustic frequency will cause more energy dissipation of ultrasound during the sound propagation in the porous medium, and hence, reduce the effect of power ultrasound on the drying rate.

The constant, β in the Weibull model, however, is found to be inversely linked to the drying rate. As shown in Figure 2.19, β decreases with the increase of the drying air temperature which positively boosts the drying rate. It is reasonable to see that a lower acoustic frequency results in a smaller value of β in the Weibull model. The parameter "c" in the Gaussian model seems to have little relation with the drying rate. As presented in Figure 2.20, the parameter "c" changes in an irregular pattern with the drying air temperature and the two acoustic parameters (i.e., frequency and power level).

Table 2.2 Estimated coefficients in the empirical models

Acoustic parameters		Drying air temperature (°C)	Lewis model	Henderson and Pabis model		Logarithmic model			Weibull model		Gaussian model		
UF (kHz)	UP (W)		K	A	K	A	K	C	n	β	a	b	c
21	20	35	0.0126	1.0450	0.0138	1.7232	0.0064	−0.7232	1.2847	66.414	1.1921	−40.811	99.551
		45	0.0201	1.0571	0.0217	1.9371	0.0096	−0.9159	1.3343	47.940	1.2621	−34.781	73.810
		55	0.0284	1.0331	0.0295	1.2210	0.0211	−0.2110	1.3041	35.143	2.0181	−63.671	76.071
		65	0.0304	1.0270	0.0310	1.1590	0.0242	−0.1518	1.2316	33.345	2.6131	−78.041	79.601
		75	0.0387	1.0410	0.0402	1.1360	0.0315	−0.1171	1.1962	27.223	1.8631	−43.451	55.261
	40	35	0.0150	1.043	0.0162	1.6311	0.0084	−0.6081	1.2909	61.641	1.7251	−85.351	117.610
		45	0.0218	1.0510	0.0232	1.6381	0.0111	0.6218	1.3256	41.792	1.3581	−40.990	74.777
		55	0.0306	1.0404	0.0320	1.2231	0.0223	−0.2089	1.2754	32.742	1.5651	−41.951	62.571
		65	0.0331	1.0311	0.0341	1.1721	0.0251	−0.1651	1.2125	30.340	2.2951	−64.120	70.351
		75	0.0403	1.0341	0.0415	1.1081	0.0341	−0.0913	1.2136	25.221	2.3011	−53.131	58.310
	60	35	0.0177	1.0290	0.0185	1.1681	0.0149	−0.1483	1.1445	55.024	5.5871	−206.71	158.310
		45	0.0263	1.0381	0.0275	1.3560	0.0164	−0.3487	1.2256	37.142	1.6461	−52.012	73.681
		55	0.0336	1.0620	0.0357	1.3021	0.0225	−0.2738	1.3521	29.742	1.2701	−23.690	48.991
		65	0.0346	1.0361	0.0358	1.1431	0.0279	−0.1269	1.1925	28.521	2.3191	−61.412	67.181
		75	0.0427	1.0350	0.0441	1.1141	0.0355	−0.0976	1.2066	24.221	1.9311	−41.721	51.480
26	20	35	0.0112	1.0421	0.0122	1.804	0.0053	−0.8041	1.3481	72.940	1.2910	−58.777	119.100
		45	0.0180	1.0501	0.0195	1.5081	0.0089	−0.4812	1.3034	51.243	1.4371	−47.841	80.161
		55	0.0276	1.0371	0.0287	1.2311	0.0199	−0.2173	1.2641	36.742	2.0231	−65.121	77.880
		65	0.0271	1.0211	0.0277	1.0841	0.0236	−0.0747	1.1225	35.521	9.8301	−184.41	122.11
		75	0.0363	1.0490	0.0395	1.2120	0.0274	−0.1923	1.1621	28.271	2.4871	−62.251	65.571
	40	35	0.0146	1.0350	0.0155	1.3721	0.0078	−0.3512	1.2255	64.940	2.4031	−133.21	143.81
		45	0.0211	1.0490	0.0225	1.5801	0.0097	−0.5610	1.2872	45.241	1.4371	−47.840	80.160
		55	0.0296	1.0371	0.0308	1.2031	0.0218	−0.1891	1.2124	35.041	1.8851	−56.521	71.081
		65	0.0319	1.0310	0.0332	1.1471	0.0246	−0.1361	1.1725	30.521	4.3290	106.21	87.941
		75	0.0390	1.0370	0.0404	1.0851	0.0333	−0.0595	1.2302	26.120	3.3661	−73.781	67.331

	60	35	0.0171	1.0321	0.0179	1.1801	0.0142	−0.1604	1.2993	57.942	5.0341	−201.41	159.21
		45	0.0261	1.0490	0.0028	1.4461	0.0151	−0.4313	1.3182	38.243	1.4211	−39.411	66.921
		55	0.0329	1.0451	0.0348	1.2471	0.0211	−0.2329	1.2824	31.041	1.5110	−36.512	56.871
		65	0.0338	1.0281	0.0342	1.0921	0.0262	−0.0772	1.1825	29.523	5.3481	−114.81	88.871
		75	0.0412	1.0461	0.0431	1.1231	0.0347	−0.0944	1.2602	26.019	1.7671	−37.901	50.512
38	20	35	0.0102	1.0440	0.0113	2.5215	0.0034	−1.5215	1.2843	80.940	1.1821	−44.580	113.41
		45	0.0166	1.0571	0.0181	2.3951	0.0058	−1.3721	1.2532	56.241	1.2551	−39.040	84.610
		55	0.0252	1.0481	0.0252	1.3641	0.0164	−0.3449	1.2324	38.841	1.4991	−44.221	70.041
		65	0.0262	1.0291	0.0271	1.1141	0.0225	−0.0966	1.1725	36.721	5.4081	−143.11	110.511
		75	0.0362	1.0401	0.0390	1.0891	0.0262	−0.0587	1.2303	30.821	3.1851	−73.121	68.391
	40 W	35	0.0133	1.0391	0.0144	1.7810	0.0068	−0.7603	1.2543	68.941	1.6671	−89.612	127.510
		45C	0.0174	1.0471	0.0186	1.9890	0.0073	−0.9723	1.4232	54.240	1.3490	−47.841	88.721
		55	0.0268	1.0460	0.0282	1.3251	0.0177	−0.3084	1.2924	37.741	1.5371	−44.921	68.910
		65	0.0301	1.0340	0.0313	1.1341	0.0237	−0.1026	1.1825	33.220	3.6221	−98.971	87.661
		75	0.0375	1.0410	0.0390	1.0891	0.0322	−0.0607	1.1893	28.121	3.1851	−73.111	68.390
	60	35C	0.0145	1.0381	0.0155	1.8571	0.0069	−0.8421	1.2746	61.941	1.5412	−73.841	113.900
		45C	0.0189	1.0340	0.0198	1.4361	0.0116	−0.4244	1.3233	47.640	1.8860	−82.621	104.441
		55	0.0285	1.0481	0.0301	1.3790	0.0174	−0.3655	1.2659	35.042	1.3761	−34.531	61.051
		65	0.0319	1.0341	0.0331	1.1961	0.0245	−0.0737	1.1562	32.220	4.7421	−110.51	88.990
		75	0.0396	1.0441	0.0413	1.1221	0.0337	−0.0957	1.2402	27.121	1.8321	−41.051	53.142

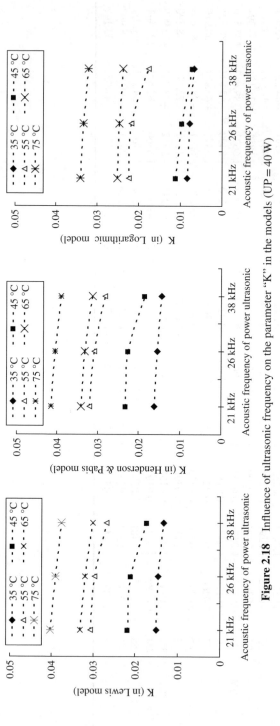

Figure 2.18   Influence of ultrasonic frequency on the parameter "K" in the models (UP = 40 W)

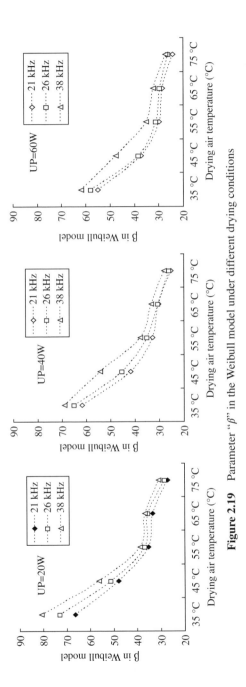

Figure 2.19 Parameter "β" in the Weibull model under different drying conditions

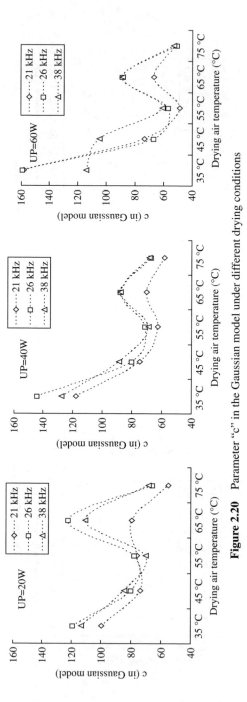

Figure 2.20   Parameter "c" in the Gaussian model under different drying conditions

To evaluate the validity of the models, the prediction error (*PE*, %) of the model results compared with the experimental data is defined as below:

$$PE\,(\%) = \frac{100}{n} \sum_{i=1}^{n} \left[\frac{\left| MR_{\text{experiment},i} - MR_{\text{model},i} \right|}{MR_{\text{experiment},i}} \right], \qquad (2.24)$$

where n is the number of experiments for the model validation.

Figure 2.21 represents the results of *PE* obtained by these models under different drying conditions. It can be easily found that among these models, the Gaussian model and the Weibull model have relatively favorable results (the lowest *PE*) for predicting the change of moisture ratio in silica gel during the regeneration assisted by power ultrasound, and the Henderson and Pabis model and the Logarithmic model have the poorest prediction results. The Gaussian function is a probabilistic model that can provide information about predicting uncertainties which are difficult to be described by a nonlinear parametric model, and it can, like neural networks, be used for modeling the static nonlinearities and the dynamic characteristics of the nonlinear dynamic systems [23]. The Gaussian model is also highlighted in constructing the estimation function of RBF (Radial Basis Function) neural network that has been proved to be fairly accurate in predicting the moisture content evolution in microwave-assisted drying process [24].

As shown in Figure 2.21, the prediction errors (PEs) caused by the Gaussian model are less than 20% in most cases, while those by the Henderson and Pabis model and the Logarithmic model mostly exceed 50%, and some are even more than 80%. Although the Weibull model is inferior to the Gaussian model, its accuracy is still acceptable in describing this special drying process. As mentioned above, the parameter β, in the Weibull model changes with the drying conditions (i.e., the drying temperature, the ultrasonic power level, and frequency) in a regular way, in which it is easier to find a specific relationship between β and the drying conditions through experiments. Compared with the Gaussian model, the Weibull model appears simpler since it has fewer parameters (only two), which will to some extent reduce workload of modeling. Based on these reasons, we recommend the Weibull model as the process model for the ultrasound-assisted regeneration of silica gel.

2.4 Theoretic Model for Ultrasound-Assisted Regeneration

Although the empirical models are simple and practical, they are only suitable for some specific situations and difficult to be generalized. In contrast, a theoretic model may have many advantages. For example, it can assist us to understand the mechanism of the ultrasound-assisted regeneration, probe into the change patterns of some key parameters that are difficult to be observed during regeneration, and optimize the ultrasonic parameters according to different actual situations.

As shown in Figure 2.22, the mechanism of ultrasound-assisted regeneration may be summarized as below:

1. The special effect of oscillating and acoustic micro-streaming induced by high-intensity ultrasound helps to reduce the boundary layer thickness of moisture film on the surface of the solid media and to increase the air turbulent velocity around the media. Thus, the external heat and mass transfer is enhanced.

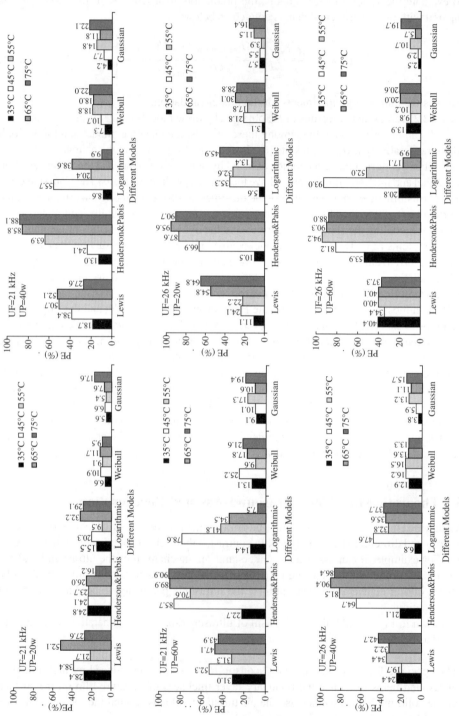

Figure 2.21 Error analysis of models under different regeneration conditions

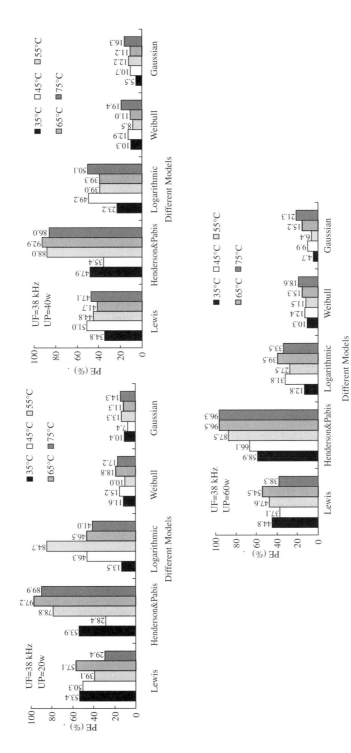

Figure 2.21 (*continued*)

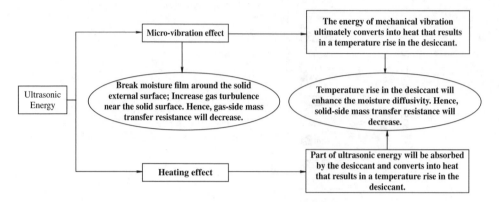

Figure 2.22   Mechanism of ultrasonic effects on the improvement of solid desiccant regeneration

2. The heating effect of ultrasound causes a temperature rise in the solid desiccants, which results in an increase in internal moisture diffusion. As a result, the moisture ratio or moisture pressure on the external surface of the material increases, and the driving force for mass transfer between the desiccant and the air is enhanced.

According to the above mechanism, a theoretic model for the regeneration process of silica gel assisted by power ultrasound is to be established in the following section.

2.4.1   Physical Model

According to the actual applications of ultrasound-assisted regeneration, the desiccant-packed bed can be designed in the form of radial and transverse flow. The physical models of the two types of bed configurations are presented in Figure 2.23. For the radial-flow bed, the regeneration air flows radically from the center of the bed. And for the transverse-flow bed, the regeneration air flows transversely from right to left of the bed (x direction). The ultrasonic waves propagate in the vertical direction (y) of the bed.

2.4.2   Mathematical Model for Ultrasonic Wave Propagation

2.4.2.1   Assumptions

Several assumptions are necessarily made for the model development:

1. The silica gel bed is considered as a porous-medium, which is packed with numerous rigid solid particles.
2. The one-dimension model is developed here. So, the additional velocity of air induced by the ultrasonic vibration in the bed $u_s(y, \tau)$ is a mean velocity that can be defined as the volumetric flow rate per unit cross-section area in the bed along the bed axis. By introducing void fraction (ε: the fraction of volume that is not occupied by the solid), the mean velocity $u_s(y, \tau)$ will be continuous.

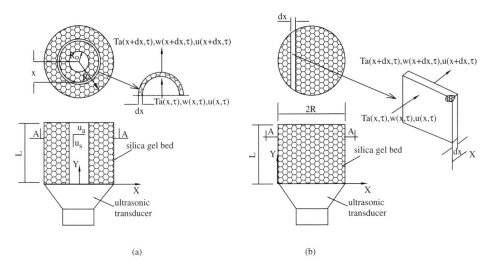

Figure 2.23 Physical models for two types of silica gel bed. (a) Radial-flow bed and (b) transverse-flow bed

3. The bed is a vertical cylindrical bed with an opening at one end for the incidence of ultrasound and a rigid boundary at the other end. The sound waves propagate in an adiabatic one-dimensional way. Meanwhile, the effect of air flow and temperature gradient on the patterns of the standing acoustic waves in the bed is negligible.

2.4.2.2 Basic Equations

The linearized-acoustic continuity, state, and momentum equations for a porous media can be written by Equations (2.25)–(2.27), respectively [25]:

Continuity equation:

$$\rho_a \frac{\partial u_s}{\partial y} = -\varepsilon \frac{\partial \rho_{a,u}}{\partial \tau}. \tag{2.25}$$

State equation:

$$p_u = c_0^2 \left(\rho_{a,u} - \rho_a \right). \tag{2.26}$$

Momentum equation:

$$-\frac{\partial p_u}{\partial y} = \frac{\varsigma}{\varepsilon} \rho_a \frac{\partial u_s}{\partial \tau} + R \cdot u_s. \tag{2.27}$$

In Equations (2.25)–(2.27), u_s is the induced velocity of air due to ultrasonic oscillation, m/s; ρ_a and $\rho_{a,u}$ are air density (kg/m^3) in the absence and presence of an ultrasonic field, respectively; p_u is the macroscopic excess pressure (Pa) in the bed due to the ultrasonic radiation, that is, sound pressure; ε is void fraction, that is, the fraction of volume not occupied by the solid medium in the bed; c_0 is the adiabatic sound velocity in the air, m/s; ς is the tortuosity factor of the packed bed; R is dynamic flow resistance, kg/(m^2·s); and τ is time, s.

The continuity equation (Equation (2.25)) reflects the relationship between the variation of the air density and the gradient of inductive air velocity along the bed axis caused by acoustic vibration. The state equation (Equation (2.26)) reflects the relationship between the acoustic pressure and the air density. And the momentum equation (Equation (2.27)) reflects the relationship between the rate of change of inductive air velocity caused by ultrasound and the gradient of acoustic pressure along the bed axis. The third term, $R \cdot u_s$ in the momentum equation denotes the air resistance along the bed axis.

2.4.2.3 Solution for Acoustic Pressure and Air Velocity in the Bed

Taking partial derivative of Equation (2.26) with respect to the time (τ), the following equation can be obtained:

$$\frac{\partial \rho_{a,u}}{\partial \tau} = \frac{1}{c_0^2} \frac{\partial p_u}{\partial \tau}. \tag{2.28}$$

Manipulating Equations (2.25) and (2.28), Equation (2.29) can be yielded:

$$\rho_a \frac{\partial u_s}{\partial y} = -\frac{\varepsilon}{c_0^2} \frac{\partial p_u}{\partial \tau}. \tag{2.29}$$

Taking partial derivative of Equation (2.29) with respect to the time (τ), we have Equation (2.30):

$$\rho_a \partial \left(\frac{\partial u_s}{\partial y} \right) \Big/ \partial \tau = -\frac{\varepsilon}{c_0^2} \frac{\partial^2 p_u}{\partial \tau^2}. \tag{2.30}$$

Taking partial derivative of Equation (2.27) with respect to the space (y), we have Equation (2.31):

$$-\frac{\partial^2 p_u}{\partial y^2} = \frac{\varsigma}{\varepsilon} \rho_a \partial \left(\frac{\partial u_s}{\partial \tau} \right) \Big/ \partial y + R \frac{\partial u_s}{\partial y}. \tag{2.31}$$

Combining Equations (2.30) and (2.31), Equation (2.32) can be derived:

$$\frac{\partial^2 p_u}{\partial y^2} = \frac{\varsigma}{c_0^2} \frac{\partial^2 p_u}{\partial \tau^2} - R \frac{\partial u_s}{\partial y}. \tag{2.32}$$

Substituting Equation (2.29) into Equation (2.32), we get Equation (2.33):

$$\frac{\partial^2 p_u}{\partial y^2} = \frac{\varsigma}{c_0^2} \frac{\partial^2 p_u}{\partial \tau^2} + \frac{R\varepsilon}{\rho_a c_0^2} \frac{\partial p_u}{\partial \tau}. \tag{2.33}$$

Assuming the following equation is valid:

$$p_u (y, \tau) = p_u (y) \, e^{i\omega\tau}, \tag{2.34}$$

where ω is the angular frequency ($\omega = 2\pi f$, f is sound frequency) and i is the unit of the imaginary number.

Equation (2.35) can be deduced:

$$\frac{\partial^2 p_u (y)}{\partial y^2} + \left(\frac{\varsigma \omega^2}{c_0^2} + \frac{R\varepsilon\omega}{i\rho_a c_0^2} \right) p_u (y) = 0. \tag{2.35}$$

Let $k' = \dfrac{\omega}{c_0}\left(\varsigma - \dfrac{iR\varepsilon}{\rho_a \omega}\right)^{1/2}$; Equation (2.35) can be written as:

$$\frac{\partial^2 p_u(y)}{\partial y^2} + k'^2 p_u(y) = 0,\qquad(2.36)$$

where k' is a modified complex wave number.

The solution for Equation (2.36) is given by:

$$p_u(y) = C_1 \sin k'y + C_2 \cos k'y,\qquad(2.37)$$

where C_1, C_2 are constants depending on the boundary conditions. Thus, the acoustic pressure can be expressed as:

$$p_u(y, \tau) = C_3 e^{-i(\omega\tau - ky)},\qquad(2.38)$$

where C_3 is a constant depending on the boundary conditions; k is sound wave number, which is the real part of k'.

The problem treated in this study involves sound transmission in the two single-phase media in a vertical cylindrical bed. The incident ultrasonic waves travel through the silica gel bed, and then a fraction of the incident energy will be reflected when the incident waves reach the rigid boundary (Refer to Figure 2.24). Thus, the "standing wave" will appear in the bed because of the superposition of the incident waves and the reflected waves. In such a case, the total acoustic pressure in the bed can be expressed as [26]:

$$p_u(y, \tau) = p_0\left(e^{-\alpha y}e^{i(\omega\tau - ky)} + e^{\alpha x}e^{i(\omega\tau + ky)}\right),\qquad(2.39)$$

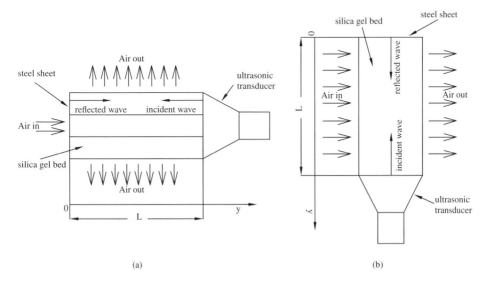

(a)   (b)

Figure 2.24 Schematic diagram of the sound propagation in silica gel bed. (a) Radial-flow bed and (b) transverse-flow bed

The real values of acoustic pressure equals the amplitude of Equation (2.39) $|p_u(y, \tau)|$, which can be written as:

$$|p_u(y, \tau)| = 2p_0\left(\cosh^2 \alpha y \cos^2 ky + \sinh^2 \alpha y \sin^2 ky\right)^{1/2}. \tag{2.40}$$

In Equations (2.39) and (2.40), p_o is acoustic pressure on the radiation surface of the ultrasonic transducer; k is the sound wave number; α is the acoustic attenuation factor.

For the high-frequency ultrasound transmitting in the porous media, the attenuation factor, a, and the wave number, k, can be calculated by Equations (2.41) and (2.42) respectively [27]:

$$\alpha = \left(\varsigma v\omega \Big/ \left(2\rho_a c_0^2\left(\psi\frac{d_s}{2}\right)^2\right)\right)^{1/2} \tag{2.41}$$

$$k = -2\varsigma^{\frac{3}{2}} v\omega^2 \rho_a c_0, \tag{2.42}$$

where v is the coefficient of dynamic viscosity, Pa·s; ψ is the structure factor of the packed bed; and d_s is the diameter of solid particle, meters.

The schematic diagram of acoustic pressure and air velocity in the bed is presented in Figure 2.25. The acoustic pressure at different locations of the bed, $p(y_1), p(y_2), \ldots, p(y_n)$, can be computed by Equation (2.40). The constant parameter p_0 can be obtained according to the boundary condition. When $y = L$, the acoustic pressure $|p_u(L)|$ can be written as:

$$|p_u(L)| = 2p_0\left[\cosh^2(\alpha L)\cos^2(kL) + \sinh^2(\alpha L)\sin^2(kL)\right]^{1/2}. \tag{2.43}$$

At the same time, the acoustic pressure at position L can be estimated by [28]:

$$|p_u(L)| = \sqrt{I_0 Z_s}, \tag{2.44}$$

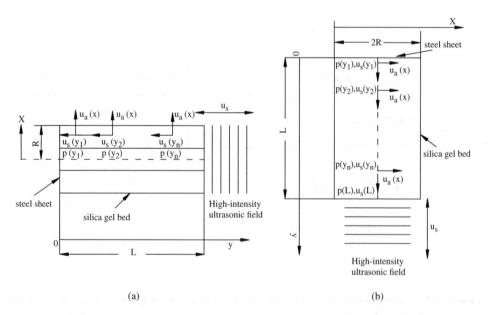

Figure 2.25 Schematic diagram of sound pressure and air velocity in silica gel bed. (a) Radial-flow bed and (b) transverse-flow bed

where I_0 is the sound intensity on the radiation surface of ultrasonic transducer, W/m^2 and Z_s is the acoustic impedance in medium, Pa·s/m^3.

Thus, constant parameter p_0 can be expressed as:

$$p_0 = \frac{\sqrt{2I_0 Z_s}}{2\left[\cosh^2(\alpha L)\cos^2(kL) + \sinh^2(\alpha L)\sin^2(kL)\right]^{1/2}}, \quad (2.45)$$

where $\cosh(\alpha L) = \dfrac{e^{\alpha L} + e^{-\alpha L}}{2}$; $\sinh(\alpha L) = \dfrac{e^{\alpha L} - e^{-\alpha L}}{2}$.

The inductive air velocity in porous media caused by sound pressure can be calculated by [29]:

$$u_s(y, \tau) = \frac{|p_u(y, \tau)|}{Z_s}. \quad (2.46)$$

The resultant velocity, $U_{as}(x, y)$, in the location (x, y) of bed can be obtained by:

$$U_{as}(x, y) = \sqrt{\left[u_s(y)\right]^2 + \left[u_a(x)\right]^2}. \quad (2.47)$$

2.4.3 Mathematical Model for Heat and Mass Transfer in Silica Gel Bed

2.4.3.1 Assumptions

1. The process of heat and mass transfer between the air and the silica gel in the bed is treated as one dimensional (please refer to Figure 2.23), and mainly controlled by a convection method.
2. Air inside the bed is assumed to be an ideal gas mixture, and the properties of the particles are homogeneous and isotropic.
3. The moisture ratio adjacent to the external surface of silica gel particles is only related to the temperature of silica gel and the moisture content in the material.
4. In the gas-phase equations, the rate of energy and moisture accumulation in the microelement of air is neglected, that is, the air in a microelement is assumed to be steady in the short time during the heat and mass transfer process.

2.4.3.2 Basic Equations

According to the assumptions above, the following equations can be developed based on the fundamental law of energy and mass conservation.

Energy conservation on the solid phase:

$$(1 - \varepsilon)\rho_s c_s \frac{\partial t_s}{\partial \tau} = H_m S_V (t_a - t_s) - H_{des} K_m S_V (w_s^* - w_a) + \frac{a_u \eta_T P_u}{(1 - \varepsilon) V}. \quad (2.48)$$

Mass conservation on solid-side:

$$-(1 - \varepsilon)\rho_s \frac{\partial q_s}{\partial \tau} = K_m S_V (w_s^* - w_a). \quad (2.49)$$

Energy conservation on the gas phase:

$$\rho_a c_{p,a} u_a(x) \frac{\partial t_a}{\partial x} = H_m S_V (t_s - t_a) + K_m S_V c_{p,v}(w_s^* - w_a)(t_s - t_a). \quad (2.50)$$

Mass conservation on the gas phase:

$$\rho_a u_a(x)\left(-\frac{\partial w_a}{\partial x}\right) = (1-\varepsilon)\rho_s\frac{\partial q_s}{\partial \tau}. \tag{2.51}$$

Boundary and initial conditions:

$$t_a(0,\tau) = t_{a0} \tag{2.52}$$

$$w_a(0,\tau) = w_0 \tag{2.53}$$

$$t_s(r,0) = t_{s0} \tag{2.54}$$

$$q_s(r,0) = q_{s0}. \tag{2.55}$$

In Equations (2.48)–(2.55), ρ_s is density of silica gel, kg/m^3; c_s is specific heat of silica gel, J/(kg·°C); t_s and t_a are temperature of silica gel and air respectively, °C; H_m is coefficient of heat transfer, W/(m^2 · °C); K_m is coefficient of mass transfer, kg/(m^2 · s); S_V is volumetric surface area in the packed bed, m^2/m^3; H_{ads} is adsorption heat of moisture on silica gel, kJ/(kg water); w_a is air humidity, kg/(kg dry air); $w_s{}^*$ is humidity of air on the surface of solid (e.g., air humidity equilibrium with the main stream), kg/(kg dry air); a_u is ultrasonic absorptivity by media; η_T is the electromechanical conversion efficiency of ultrasonic transducer; P_u is the power of ultrasound in W; V is volume of the packed bed, m^3; $c_{p,a}$ and $c_{p,v}$ are specific heat of air and water vapor at constant pressure, respectively, J/(kg·°C); and q_s is moisture ratio in silica gel, kg/(kg dry material).

2.4.3.3   Discrete Numerical Model

For the convenience of computer solution, Equations (2.48)–(2.51) are converted into discrete forms by using the finite volume method.

Energy conservation on the solid phase:

$$(1-\varepsilon)\rho_s c_s\left(t_{s,N}^1 - t_{s,N}^0\right) = H_m S_V\left(t_{a,N}^1 - t_{s,N}^1\right)\Delta\tau - H_{des}K_m S_V\left(w^*{}_{s,N}^1 - w_{a,N}^1\right)\Delta\tau$$

$$+ \frac{a_u\eta_T P_u}{(1-\varepsilon)V}\Delta\tau. \tag{2.56}$$

Mass conservation on solid-side:

$$q_{s,N}^1 = q_{s,N}^0 - \frac{K_m S_V\left(w^*{}_{s,N}^1 - w_{a,N}^1\right)\Delta\tau}{(1-\varepsilon)\rho_s}. \tag{2.57}$$

Energy conservation on the gas phase:

$$t_{a,N}^1 = \frac{u_a(x)\rho_a c_{p,a} t_{a,N-1}^1 + \left[H_m S_V t_{s,N}^1 + K_m S_V c_{p,v}\left(w^*{}_{s,N}^1 - w_{a,N}^1\right)t_{s,N}^1\right]\Delta\tau}{u_a(x)\rho_a c_{p,a} + \left[H_m S_V + K_m S_V c_{p,v}\left(w^*{}_{s,N}^1 - w_{a,N}^1\right)\right]\Delta\tau}. \tag{2.58}$$

Mass conservation on the gas phase:

$$w_{a,N}^1 = \frac{u_a(x)\rho_a w_{a,N-1}^1 + K_m S_V w^*{}_{s,N}^1\Delta\tau}{u_a(x)\rho_a + K_m S_V\Delta\tau}. \tag{2.59}$$

In Equations (2.56)–(2.59), $N = [1, n]$, n is the number of space discrete along the x-direction. The superscript "0" and "1" stand for the current time and the next time of calculation respectively.

2.4.3.4 Determination of Key Model Parameters

1. Propagation velocity of sound in the air (c_o: m/s) [28]

$$c_o = 331.6\sqrt{1 + \frac{t_a}{273.15}}, \tag{2.60}$$

where t_a is air temperature, °C.

2. Air density in the packed bed under the ultrasonic oscillation ($\rho_{a,u}$: kg/m^3) [29].

$$\rho_{a,u} = \frac{\varsigma}{\varepsilon}\rho_a\left[1 + \sqrt{16v/\left(i\omega\rho_a d_e^2\right)}\right], \tag{2.61}$$

where

$$\rho_a = 0.003484B/t_a - 0.001315p_w/t_a, \tag{2.62}$$

$$v_a = 1.83 \times 10^{-5} - 4.93 \times 10^{-7}\left(23 - t_a\right), \tag{2.63}$$

$$d_e = \frac{2\varepsilon d_s}{3\left(1 - \varepsilon\right)}. \tag{2.64}$$

In Equations (2.61)–(2.64), ρ_a is air density, kg/m^3; ς is the tortuosity factor of the packed bed, which is suggested to describe the actual length of air passage. If the air passages in the packed bed are considered as tubes that are oriented uniformly in a given direction, the tortuosity factor, ς, equals 3.0 [27]. v_a is the dynamic viscosity of air given as Pa·s; i is the unit of the imaginary number; ω is the acoustic angular frequency, $= 2\pi f$; B is standard atmosphere pressure, Pa; p_w is the water vapor pressure in the ambient air, Pa; and d_e is the equivalent diameter (m) of pore in the packed bed; the void fraction in the bed ε can be obtained by Equation (2.65)

$$\varepsilon = \left(V - V_s\right)/V, \tag{2.65}$$

$$V_s = m_s/\rho_s, \tag{2.66}$$

where V is space volume of silica gel packed bed, m^3; V_s is space volume occupied by silica gel, m^3; m_s is mass of silica gel, kg; and ρ_s is density of silica gel, kg/m^3.

3. Acoustic impedance in media (z_s: Pa·s/m^3) [29].

$$z_s = \sqrt{C_4\rho_{a,u}}, \tag{2.67}$$

where

$$C_4 = c_{Y,a}p_a\left[1 - 2\left(\sqrt{c_{Y,a}} - \frac{1}{\sqrt{c_{Y,a}}}\right)\sqrt{\frac{\lambda_a}{i\omega\left(\frac{d_e}{2}\right)^2\rho_{a,u}c_{p,a}}}\right], \tag{2.68}$$

where ρ is density of medium through which sound propagates, kg/m^3; c_Y is specific heat ratio of air, $c_Y = c_{a,p}/c_{a,v}$; λ_a is thermal conductivity of air, W/m·°C; and $c_{p,a}$ and $c_{v,a}$ is specific heat of air at constant pressure and constant volume, J/(kg·°C).

4. Flow resistance in the packed bed (R: kg/(m^2·s)) [27].

$$R = \frac{\varsigma\left(2v_a\omega\rho_a\right)^{1/2}}{\varepsilon\psi d_s},$$

(2.69)

where d_s is diameter of silica gel, meters; ψ is structure factor of silica gel packed bed, which is used to describe the gap between the structure of air passage in the packed bed and that of air tubular type. The factor $\psi = \sqrt{8}$ has been suggested in Ref. [30].

5. Saturated water vapor pressure ($p_{w,qb}$: p$_a$) and air humidity (w_a: kg/(kg dry air)) [31].

$$p_{w,qb} = 6.1121e^{\frac{\left(18.678 - \frac{t_a}{234.5}\right)t_a}{257.14 + t_a}},$$

(2.70)

$$w_a = \frac{0.622\varphi_a p_{w,qb}}{B - \varphi_a p_{w,qb}},$$

(2.71)

where φ_a is air relative humidity, %.

6. Specific heat of silica gel (c_s: kJ/(kg·°C)).

$$c_s = 4.178 \times q_s + 0.921,$$

(2.72)

where q_s is moisture ratio in silica gel, kg/(kg dry material).

7. Volumetric surface of silica gel in the bed (S_V: m^2/m^3) [32].

$$S_V = \frac{6\left(1 - \varepsilon\right)}{d_e}.$$

(2.73)

8. Desorption heat of silica gel (H_{des}: kJ/(kg water desorption)).The moisture desorption heat of silica gel (H_{des}) can be expressed by the equation dependent on the moisture ratio in the material (q_s). Normally, the desorption heat is equal to the adsorption heat. Normally Equations (2.6a) and (2.6b) can be employed as well to calculate the desorption heat of silica gel.

9. Electromechanical conversion efficiency of ultrasonic transducer (η_T).The electromechanical conversion efficiency of ultrasonic transducer, η_T, can be determined by using the method proposed by Lin and Zhang [33]. The electromechanical conversion efficiency of ultrasonic transducers used for this study has been measured with respect to different acoustic frequencies and power levels, respectively. The results are listed in Table 2.3. The results show that the working efficiency of ultrasonic transducer decreases with the increase of the acoustic frequency and power level.

10. Ultrasonic absorptivity (a_u).

The absorptivity of ultrasound, a_u is defined as how much acoustic energy is absorbed directly by silica gel. It can be determined through experiments by using Equation (2.74)

$$m_s c_s \frac{\mathbf{d}t_s}{\mathbf{d}\tau} = a_u P_u \eta_T,$$

(2.74)

Table 2.3 Electromechanical conversion efficiency of ultrasonic transducers

Ultrasonic frequency (kHz)	Ultrasonic power (W)	η_T
21	20	0.92
	40	0.87
	60	0.84
26	20	0.89
	40	0.86
	60	0.81
38	20	0.85
	40	0.81
	60	0.78

where m_s is the mass of silica gel, kg; c_s is the specific heat of silica gel, J/(kg·°C); t_s is the temperature,°C; P_u is power of ultrasound applied, W. Experimental data shows that a_u ranges from 0.15 to 0.20.

11. Moisture ratio on the external surface of silica gel (w^*: kg/(kg dry air)).The relative humidity of air on the surface of silica gel can be expressed as a function dependent on the temperature (t_s) and moisture content of silica gel (q_s) [34].

$$w^* = s_1 t_s q_s^2 + s_2 t_s q_s + s_3 q_s^4 + s_4 q_s^3 + s_5 q_s^2 + s_6 q_s + s_7 \mathrm{Ln}\left(q_s\right),\qquad(2.75)$$

where the parameters, $s_1, s_2, s_3, s_4, s_5, s_6, s_7$, can be gotten by fitting experimental data. By using Equation (2.75), the air humidity adjacent to the external surface of silica gel particles (w^*) can be calculated.

To get the empirical parameters ($s_1, s_2, s_3, s_4, s_5, s_6, s_7$), a series of experiments at different inlet air conditions (temperature and humidity) have been performed. In the experiments, certain mass of silica gel (whose initial moisture content is known) in the bed was humidified by the moist air at steady temperature and humidity. When the silica gel gets saturated, it is assumed that the surface of silica gel attains equilibrium state with the moist air. So, the relative humidity of the air on the surface of silica gel and the temperature of silica gel are equal, respectively, that of the moist air. Meanwhile, the moisture content in the saturated silica gel can be gotten by the weighing method. The value is 82.9, −4.64, 252 148.6, −125 044.6, 16 001.7, −173.4, and 0.863, for $s_1, s_2, s_3, s_4, s_5, s_6$, and s_7 respectively. As shown in Figure 2.26, the fitting data agrees well with the experimental data.

12. Coefficients of heat and mass transfer (H_m, K_m) [3].

Basically, the heat and mass transfer coefficient model can be expressed by Equations (2.76) and (2.77) respectively.

$$H_m = h_a G_{a,m} c_{a,m} \mathrm{Re}_m^{-h_b},\qquad(2.76)$$

$$K_m = k_a G_{a,m} \mathrm{Re}_m^{-k_b},\qquad(2.77)$$

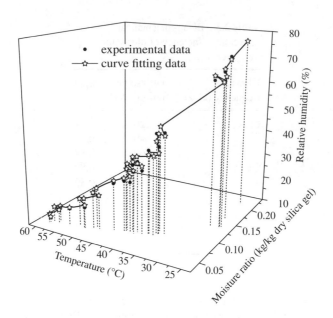

Figure 2.26 Comparisons of the fitting data with the experimental one for Equation (2.75)

where G_a is air flow rate passing though the desiccant bed, kg/s and Re is the Reynolds number, $Re = \rho_a u_a d_e / \mu_a$; h_a, h_b, k_a, and k_b are undetermined parameters obtained through experimental data. The subscript, m, denotes the average value of corresponding variables.

To obtain the undetermined parameters from the experimental data, Equations (2.76) and (2.77) are written, respectively, as below:

$$\lg \left(\frac{H_m}{G_{a,m} c_{a,m}} \right) = \lg \left(h_a \right) - h_b \lg \left(Re_m \right), \tag{2.78}$$

$$\lg \left(\frac{K_m}{G_{a,m}} \right) = \lg \left(k_a \right) - k_b \lg \left(Re_m \right). \tag{2.79}$$

The mass transfer coefficient, H_m, and the heat transfer coefficient, K_m, can be determined by fitting the simulation moisture concentration and temperature breakthrough profiles with the experimental ones. The dynamic model equations for the gas-phase concentration and temperature, the solid-phase moisture ratio and temperature as well as the gas velocity can be solved to obtain the simulation breakthrough profiles. And the experimental breakthrough profiles can be gotten based on the experimental method illustrated in Ref. [35]. The adsorption and desorption breakthrough profiles have been simultaneously matched to improve the reliability of the heat and mass transfer coefficients. So, at least two experimental breakthrough profiles are required for the determination of the two coefficients under each regeneration condition.

Thus, the parameters, h_a and h_b in Equation (2.76), can be determined with respect to the intercept and slope of the straight line in which the values of $\lg \left(H_m / G_{a,m} c_{a,m} \right)$ are plotted against $\lg \left(Re_m \right)$. And so is the method for the determination of the parameters (k_a, k_b) in Equation (2.77).

A series of experiments have been performed to determine the parameters (h_a, h_b, k_a, and k_b) under different conditions. The results are shown in Figures 2.27 and 2.28, respectively.

To develop a more adaptive model for the heat and mass transfer coefficients, the empirical equations for the parameters, h_a, h_b, k_a, and k_b, are recommended as below:

$$h_a = a_1 + a_2 \cdot t + a_3 \cdot UP + a_4 \cdot UF + a_5 \cdot t \cdot UP + a_6 \cdot t \cdot UF + a_7 \cdot UP \cdot UF$$
$$+ a_8 \cdot t^2 + a_9 \cdot UP^2 + a_{10} \cdot UF^2 + a_{11} \cdot t \cdot UP \cdot UF, \tag{2.80}$$

$$h_b = b_1 + b_2 \cdot t + b_3 \cdot UP + b_4 \cdot UF + b_5 \cdot t \cdot UP + b_6 \cdot t \cdot UF + b_7 \cdot UP \cdot UF$$
$$+ b_8 \cdot t^2 + b_9 \cdot UP^2 + b_{10} \cdot UF^2 + b_{11} \cdot t \cdot UP \cdot UF, \tag{2.81}$$

$$k_a = c_1 + c_2 \cdot t + c_3 \cdot UP + c_4 \cdot UF + c_5 \cdot t \cdot UP + c_6 \cdot t \cdot UF + c_7 \cdot UP \cdot UF$$
$$+ c_8 \cdot t^2 + c_9 \cdot UP^2 + c_{10} \cdot UF^2 + c_{11} \cdot t \cdot UP \cdot UF, \tag{2.82}$$

$$k_b = d_1 + d_2 \cdot t + d_3 \cdot UP + d_4 \cdot UF + d_5 \cdot t \cdot UP + d_6 \cdot t \cdot UF + d_7 \cdot UP \cdot UF$$
$$+ d_8 \cdot t^2 + d_9 \cdot UP^2 + d_{10} \cdot UF^2 + d_{11} \cdot t \cdot UP \cdot UF. \tag{2.83}$$

By fitting the experimental data into the equations, the empirical constants in Equations (2.80)–(2.83) are obtained, as listed in Table 2.4. Figure 2.29 shows the goodness of fitting for h_a, h_b, k_a, and k_b, and the corresponding degree of fitting is estimated as 0.87, 0.94, 0.89, and 0.96, respectively.

2.4.3.5 Model Solution

According to the discrete equations Equations (2.56)–(2.59) and the relevant parameters, the coupled, nonlinear, partial differential ones (Equations (2.48)–(2.51)) can be solved by computer. To begin with, the necessary parameters including the bed dimensions, the silica gel properties (density, mean particle diameter) and the ultrasonic parameters (acoustic intensity and frequency) are given. Meanwhile, the initial conditions including the moisture content in the silica gel and its surface temperature, the inlet air temperature and humidity as well as the air flow rate are also provided. Initially ($\tau = 0$), temperature and moisture ratio in the gas

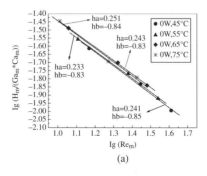

Figure 2.27 h_a and h_b in the heat transfer equation under different regeneration conditions. (a) For nonultrasonic regeneration and (b) for ultrasound-assisted regeneration

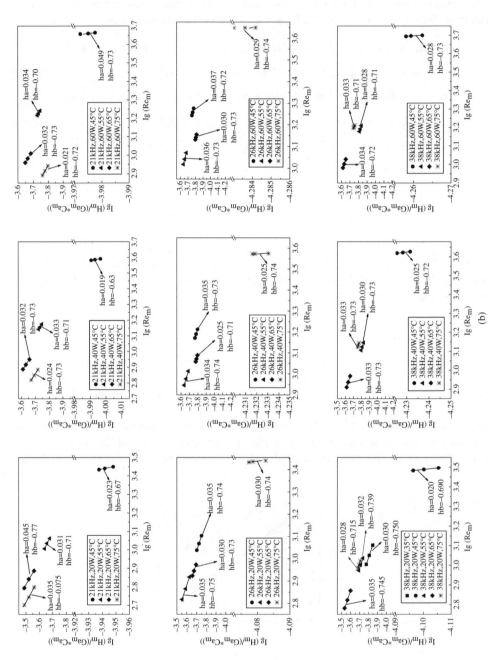

Figure 2.27 (*continued*)

phase and the solid phase is uniform. Then, the models of heat and mass transfer are solved in a sequential manner using an overall iterative loop with respect to the presence or absence of an ultrasonic field. Within this iterative loop, internal iterations are performed for each separate equation. The main dependent variables (t_a, w_a, t_s, q_s) are updated to make the program converge quickly. When calculations at all the grid points are done, and the maximum relative difference between the current values and those of the previous iteration for (t_a, w_a, t_s, q_s) are less than a very small value ε (E-05), the program proceeds to the next time step until the termination time is reached. The flow chart for the calculation is presented in Figure 2.30.

2.4.4 Model Validation

2.4.4.1 Radial-Flow Bed

The experimental system for the radial-flow bed has been illustrated in Section 2.2.1 (see Figure 2.4). The model is validated in terms of the mass of moisture desorption for a specific time period of regeneration.

2.4.4.2 Transverse-Flow Bed

The experimental system for the transverse-flow bed is shown in Figure 2.31. The model is validated in terms of the exit air temperature and humidity of the bed. A temperature-and-humidity sensor (type: HHC2-S; measurement precision: ±0.8% in humidity and ±0.1°C in temperature) is placed at the inlet and outlet of the packed bed respectively to observe the corresponding air conditions. A digital anemometer (measurement precision: ±1% of reading data) is used to measure the airflow rate passing through the packed bed. The parameters used for the model calculation mainly include the inlet air temperature and humidity of the bed, the air flow rate passing through the bed, the ultrasonic power and frequency, the initial moisture ratio and mass of the trial sample.

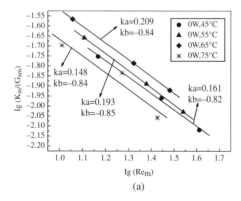

Figure 2.28 k_a and k_b in the mass transfer equation under different regeneration conditions. (a) For nonultrasonic regeneration and (b) for ultrasound-assisted regeneration

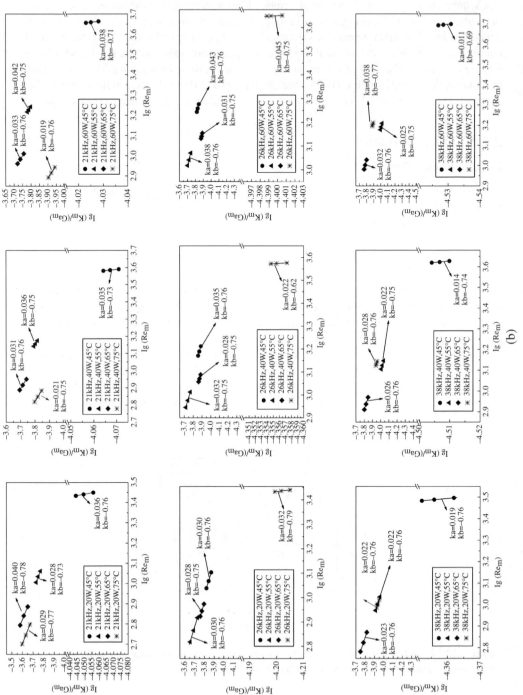

(b)

Figure 2.28 *(continued)*

Table 2.4 Constants for Equations (2.80)–(2.83)

a_1	a_2	a_3	a_4	a_5	a_6	a_7	a_8	a_9	a_{10}	a_{11}
0.173 09	1.804E–03	3.58E–04	–0.012 35	–3.39E–05	–2.90E–06	–3.02E–05	–1.10E–05	1.46E–05	1.94E–04	7.99E–07
b_1	b_2	b_3	b_4	b_5	b_6	b_7	b_8	b_9	b_{10}	b_{11}
–0.697 9	–0.004 60	0.006 60	0.007 19	–5.83E–05	–5.73E–05	–1.96E–04	3.75E–05	–2.83E–05	–4.98E–05	2.90E–06
c_1	c_2	c_3	c_4	c_5	c_6	c_7	c_8	c_9	c_{10}	c_{11}
0.079 60	0.003 523	0.000 717	–0.008 18	–3.60E–05	1.59E–06	–6.44E–05	–3.07E–05	1.32E–05	1.13E–04	1.36E–06
d_1	d_2	d_3	d_4	d_5	d_6	d_7	d_8	d_9	d_{10}	d_{11}
–0.608 32	–0.007 24	4.61E–04	–1.33E–04	4.18E–05	4.12E–05	1.38E–04	5.50E–05	–3.55E–05	–4.65E–05	–2.18E–06

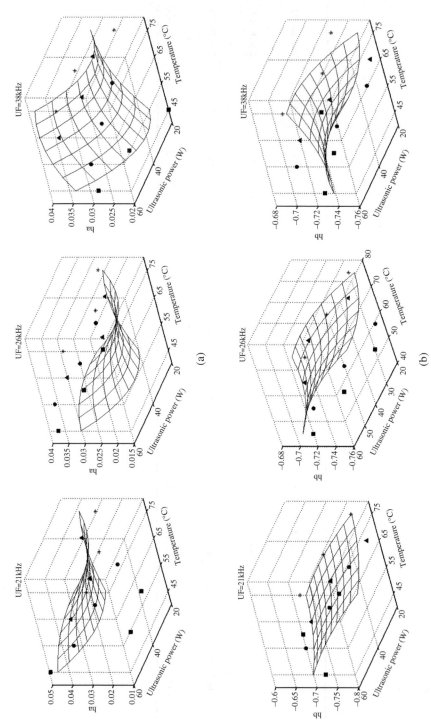

Figure 2.29 Comparisons of the fitting value (surface) to the experimental data (scatter). (a) Fitting value of h_a by Equation (2.80) (surface) versus the experimental data (scatter), degree of fitting: 0.87; (b) fitting value of h_a by Equation (2.81) (surface) versus the experimental data (scatter), degree of fitting: 0.94; (c) fitting value of h_a by Equation (2.82) (surface) versus the experimental data (scatter), degree of fitting: 0.89; and (d) fitting value of h_a by Equation (2.83) (surface) versus the experimental data (scatter), degree of fitting: 0.96

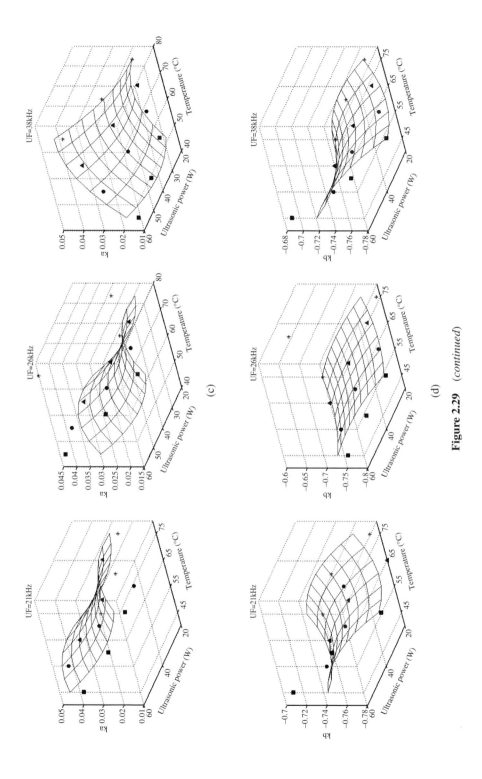

Figure 2.29 *(continued)*

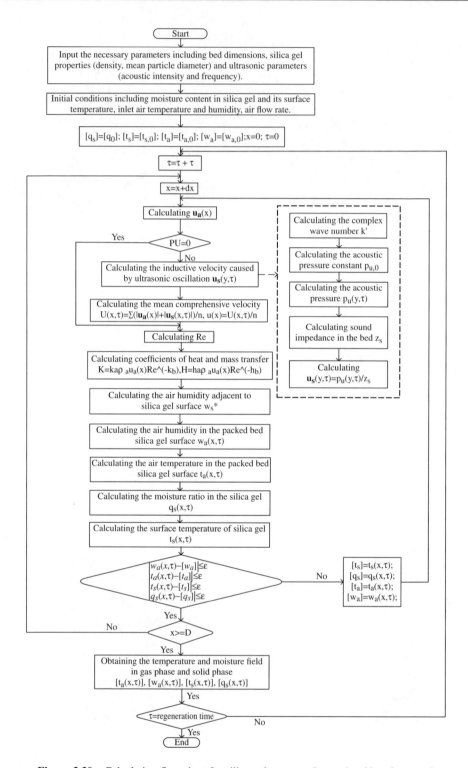

Figure 2.30   Calculation flow chart for silica gel regeneration assisted by ultrasound

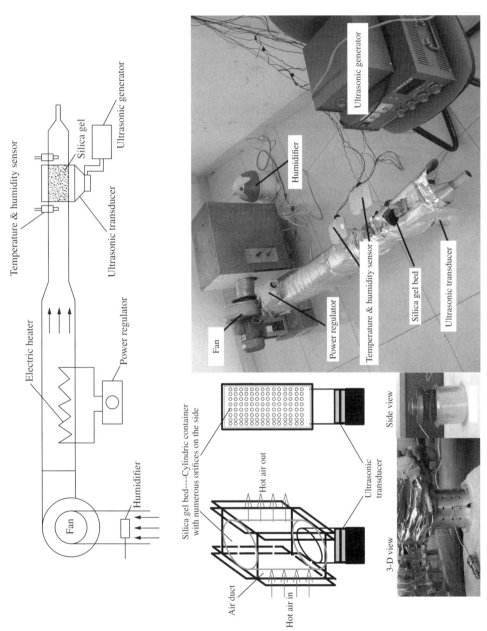

Figure 2.31 Experimental system for the transverse-flow bed model validation

2.4.4.3 Results and Discussion

The output of the theoretical model is, mainly, the transient variation of the packed bed thermo-physical properties including the solid-phase and gas-phase moisture content and temperature, the exit air conditions (humidity ratio and temperature), and the mass of moisture desorption. The validity of a model can be evaluated by Mean Relative Error (MRE, %) defined as below:

$$MRE\,(\%) = \frac{1}{n}\sum_{i=1}^{n}\left[\left|\frac{x_{\text{exp},i} - x_{\text{mod},i}}{x_{\text{exp},i}}\right|\right] \times 100, \qquad (2.84)$$

where $x_{\text{exp},i}$ and $x_{\text{mod},i}$ denote, respectively, the experimental and theoretical value at the ith time point during the regeneration process; n is the number of data sample for the model validation.

For the Radial-Flow Bed

Figure 2.32 gives comparisons between the calculated and experimental mass of moisture desorption for every 8 minutes during the silica gel regeneration for the radial-flow bed. The experimental conditions for model validation mainly include four regeneration temperatures (i.e., 45, 55, 65, and 75 °C), four ultrasonic power levels (i.e., 0, 20, 40, and 60 W) combined with three ultrasonic frequencies (i.e., 21, 26, and 38 kHz).

Although there appears some points in Figure 2.32 where the calculated value deviates a little far from the experimental one (the maximum error attains 24.4%), the calculation model is still valid in terms of statistical analysis. The results show that the *MRE* (Mean Relative Error) of calculated value compared with the experimental data for these experimental cases are all less than 10%.

For the Transverse-Flow Bed

Figure 2.33 demonstrates the transient variations of exit air temperature and humidity during the ultrasound-assisted regeneration process in the presence or absence of ultrasonic radiation. The exit air humidity (kg/kg dry air) is shown to increase in the first few minutes, and then followed by a gradual decrease. As we know, the apparent mass transfer happens where there is a gap in mass concentration or partial pressure. In the beginning, the temperature of silica gel surface is relatively low, and the moisture pressure on the surface is relatively low as well. Thus, the gap of moisture pressure between the surface of silica gel and the regeneration air is small, which results in a small increase in the exit air humidity (kg/kg dry air). With the regeneration going on, however, the silica gel temperature increases due to the heat gain from the hot air. In addition, more moisture inside the material moves to the surface, thus leading to an increase in the rate of mass transfer from the silica gel to air. As a result, the exit air humidity goes up. But the moisture ratio in the silica gel decreases with the regeneration time, which will reduce the surface moisture pressure and slow down the rate of mass transfer. Therefore, the change trend of the exit air humidity presented in Figure 2.33 depends on the two factors. When the factor in favor of the mass transfer dominates, the exit air humidity increases. Otherwise, the exit air humidity decreases.

As for the exit air temperature, it begins from a temperature lower than the inlet one and ultimately rises to the equilibrium value. It can be found as well from the curves of the temperature that the slope change in the first few minutes is obviously steeper than that in the following time. At the very beginning, the temperature of silica gel is much lower than the

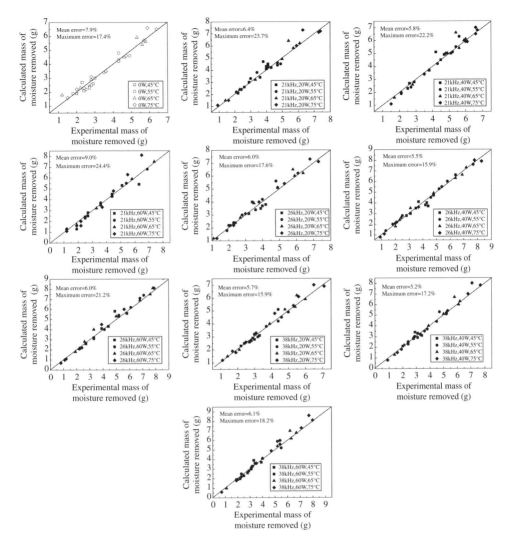

Figure 2.32　Calculated moisture desorption mass versus experimental one for radial-flow bed

inlet air temperature, and hence, more thermal energy transfers from the air to the silica gel, which results in a lower temperature in the exit air. Afterwards, the surface temperature of silica gel increases, and the temperature difference between the surface and the air decreases, which indicates a reduction of heat transfer from the air to the silica gel. Thus, the change rate of the exit air temperature becomes smaller.

It can be seen as well from Figure 2.33 that the calculated results of the exit air temperature and humidity have a favorite agreement with the experimental data including the trend. The comparisons of calculated and experimental exit air temperature and humidity during the regeneration are made as shown in Figure 2.34. The MRE of the calculated exit air temperature and humidity are all less than 2.5% for these experimental cases. The validity of the calculated

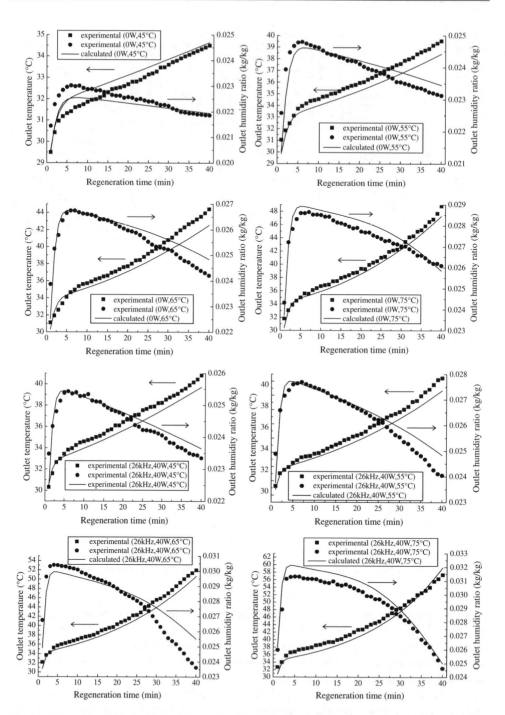

Figure 2.33 Variations of exit air temperature and humidity with the regeneration time under different conditions (calculated versus experimental)

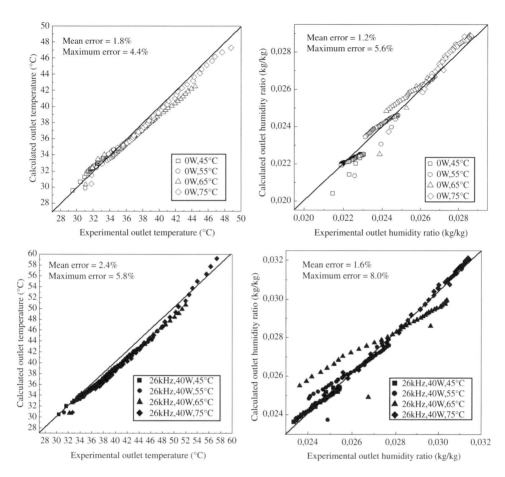

Figure 2.34 Comparisons of the calculated and the experimental exit air temperature and humidity during the regeneration process with or without ultrasound

moisture desorption for every 8 minutes during the regeneration are also investigated. The results in Figure 2.35 show that the MRE of the calculated moisture desorption for these cases are less than 6.0%.

2.4.5 Error Analysis for Experimental Data

According to the theory of error analysis, the measurement error of indirect parameter can be deduced based on that of direct one. Assuming the indirect measurement variable (y) is a specific function dependent of the direct measurement ones ($x_1, x_2, \cdots, x_n$), that is, $y = f(x_1, x_2, \cdots, x_n)$ and no covariance among these variables, the experimental relative error (ERE) of the indirect measurement variable (y) can be written as:

$$\frac{\Delta y}{y} = \left[\left(\frac{\partial f}{\partial x_1} \right)^2 \left(\frac{\Delta x_1}{y} \right)^2 + \left(\frac{\partial f}{\partial x_2} \right)^2 \left(\frac{\Delta x_2}{y} \right)^2 + \cdots + \left(\frac{\partial f}{\partial x_n} \right)^2 \left(\frac{\Delta x_n}{y} \right)^2 \right]^{1/2}. \quad (2.85)$$

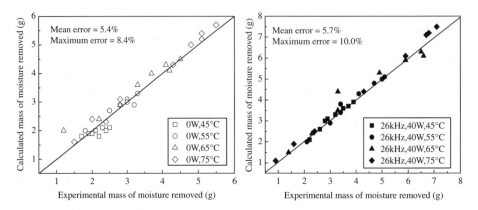

Figure 2.35   Calculated moisture desorption mass versus experimental one for transverse-flow bed

2.4.5.1   Radial-Flow Bed

The experimental system for the radial-flow bed observes the moisture desorption rate through directly measuring the mass change (Δm) of silica gel bed in a period of regeneration time.

$$\Delta m = m_1 - m_2, \tag{2.86}$$

where m_1 and m_2 stand for mass of silica gel bed at the time point "1" and "2," respectively. According to Equation (2.85), the relative error of Δm can be expressed as below:

$$\frac{\Delta(\Delta m)}{\Delta m} = \left[\left(\frac{\partial(\Delta m)}{\partial m_1} \right)^2 \cdot \left(\frac{\Delta M_1}{\Delta m} \right)^2 + \left(\frac{\partial(\Delta m)}{\partial M_2} \right)^2 \cdot \left(\frac{\Delta M_2}{\Delta m} \right)^2 \right]^{1/2}. \tag{2.87}$$

In this experimental system, the measurement error of the electronic balance is identified as 0.0001 kg. Thus, the *ERE* of Δm can be simplified as:

$$\frac{\Delta(\Delta m)}{\Delta m} \approx \frac{0.00014}{\Delta m}. \tag{2.88}$$

Figure 2.36 gives the EREs of moisture desorption for the radial-flow bed during the regeneration. The results show that the ERE of moisture desorption tends to rise with the regeneration time. This is because the moisture change (Δm) for a specific time period of regeneration drops with the regeneration time. Thus, the ERE increases with time as a result of Equation (2.88). In this experimental study, the EREs of moisture desorption are mostly less than 5% within the first 40 minutes of the regeneration.

2.4.5.2   Transverse-Flow Bed

In the experimental system for the transverse-flow bed, the direct measurement parameters include the air temperature (t_a, °C) and the air moisture ratio (ϕ_a, %), while the indirect measurement one is the air humidity (w_a, kg/(kg dry air)). Equation (2.71) gives the

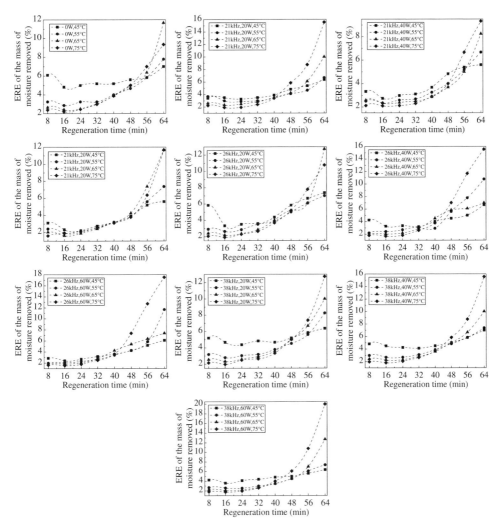

Figure 2.36 Experimental relative error (ERE) of moisture desorption for every 8 minutes during the regeneration for the radial-flow bed

relationship between w_a and (t_a, ϕ_a). Since the measurement precision of t_a and ϕ_a is given, the relative error of w_a can be calculated by Equation (2.89).

$$\frac{\Delta w_a}{w_a} = \left[\left\{ w_a + \frac{w_a^2}{0.622} \right\}^2 \left(\frac{\Delta \phi_a}{\phi_a \cdot w_a} \right)^2 + \left\{ \frac{w_a \left(0.622 + \phi_a \right)}{0.622} \cdot \frac{4802.86 - 2.2 t_a - \dfrac{t_a^2}{234.5}}{\left(257.14 + t_a \right)^2} \right\}^2 \left(\frac{\Delta t_a}{w_a} \right)^2 \right]^{1/2}. \qquad (2.89)$$

In this experimental study, the measure of precision of the air temperature and relative humidity sensor is ±0.1 °C and ±0.8%, respectively. The uncertainties of the experimental exit air temperature and humidity during the regeneration are presented in Figures 2.37 and 2.38, respectively. It can be seen that the ERE of exit air temperature decreases with the regeneration time, while the situation is reverse for the ERE of exit air humidity (kg/(kg dry air)). This is because the exit air temperature of silica gel bed increases, while the exit air humidity decreases during the regeneration. As known from Equation (2.85), the ERE has an opposite relationship with the magnitude of the measured value. The results show that the relative error of the experimental exit air temperature and humidity in this study can be controlled within 0.4 and 3.5%, respectively, during regeneration. In spite of this, the ERE of moisture desorption for the transverse-flow system is much larger than that for the radial-flow one. As shown in Figure 2.39, the EREs of the moisture desorption for the transverse-flow system are all over 10% during regeneration. For some cases (e.g., the case of "0 W, 45 °C"), the EREs are even more than 20%. Although the transverse-flow system is not as accurate as the radial-flow one in terms of the moisture desorption measurement during the regeneration, the former can provide

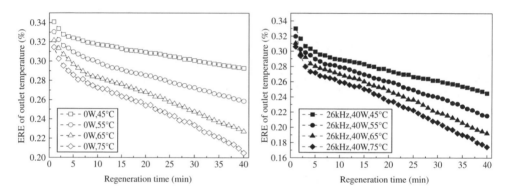

Figure 2.37   Experimental relative errors (EREs) of exit air temperature of transverse-flow bed

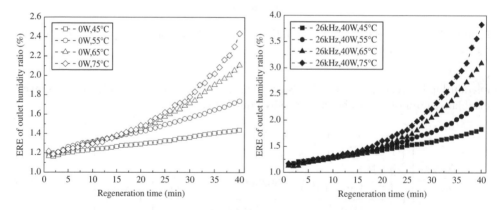

Figure 2.38   Experimental relative errors (EREs) of exit air humidity (kg/(kg dry air)) of the transverse-flow bed

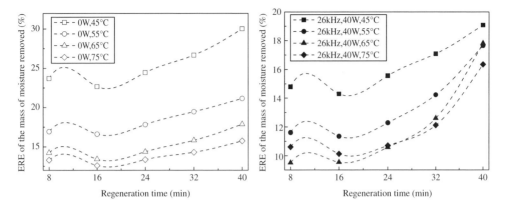

Figure 2.39 Experimental relative error (ERE) of moisture desorption for every 8 minutes during the regeneration for the transverse-flow bed

more information about the exit air conditions of bed and the moisture desorption during the regeneration which are very necessary for the transient model validation.

2.5   Parametric Study on Silica Gel Regeneration Assisted by Ultrasound

Taking the transverse-flow bed (refer to Figure 2.23b) as the study object, the key parameters are investigated with the theoretical model during the ultrasound-assisted regeneration under different conditions, including the acoustic pressure distribution and corresponding oscillation velocity induced in the packed bed, the temperature and moisture ratio distribution in the packed bed, the dynamic drying kinetics and the enhancement of moisture desorption brought by ultrasound.

The simulation conditions mainly include: inlet air temperature and humidity ranges from 35 to 80 °C and from 0.01 to 0.035 kg/(kg dry air) respectively; inlet air velocity from 0.1 to 1.5 m/s; acoustic frequency from 20 to 35 kHz combined with the power from 20 to 80 W; particle size of silica gel in the bed from 1.0 to 7.0 mm.

2.5.1   Acoustic Pressure and Oscillation Velocity in the Packed Bed

2.5.1.1   Different Ultrasonic Power Levels

Figure 2.40 shows the distributions of sound pressure and corresponding air oscillation velocity along the axial direction of silica gel bed under the exposure of 20-kHz ultrasonic radiation with different power levels (e.g., 20, 40, 60, and 80 W). The value "y/L" on the horizontal axis is a dimensionless number for the position along the axial direction of silica gel bed. "$y/L = 0.0$" is the farthest position from the ultrasonic transducer in the bed (refer to Figure 2.23b). It is obvious that a higher ultrasonic power will result in a higher sound pressure and air oscillation velocity in the bed. And they change along the axial direction of the bed in a periodical way.

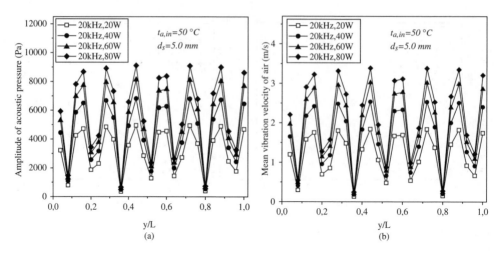

Figure 2.40   (a,b) Distributions of acoustic pressure and air vibration velocity along the axial direction of silica gel bed for different ultrasonic power levels

2.5.1.2   Different Ultrasonic Frequencies

Figure 2.41 shows the acoustic pressure and air vibration velocity distributions in the bed for different ultrasonic frequencies (e.g., 20, 25, 30, and 35 kHz). It can be seen that the acoustic pressure and the air vibration velocity under the 25-kHz ultrasonic radiation are much higher than that under the other ultrasonic frequencies. This indicates that for the bed with a specific depth, there should be a resonance frequency of ultrasound to achieve the greatest effect of ultrasonic oscillation. According to the acoustic standing-wave theory [36], the system reaches the highest sound pressure at resonance conditions for $kL = (2n - 1)\pi/2$, where k is the wave number and L is the bed height.

2.5.1.3   Different Temperatures of Regeneration Air

Figure 2.42 presents the acoustic pressure and air vibration velocity distributions in the bed for different regeneration temperatures (e.g., 35, 45, 55, 65, and 75 °C). It has been found that the regeneration air temperature will affect the ultrasonic oscillation to some degree, and the better effect of ultrasonic oscillation will be achieved under the higher regeneration temperature. The reason may be that the higher temperature is not conducive to the sound propagation. This has been proved by Ref. [37] in which the acoustic attenuation is found to increase with the media temperature rising.

2.5.1.4   Different Particle Sizes of Silica Gel

Figure 2.43 shows the influence of particle size on the acoustic pressure and air vibration velocity distributions in the bed. The results suggest that the effect of acoustic oscillation will be more significant under a larger particle size of the packing material. It is because the smaller particle size of packing material will cause the bigger sound attenuation in the packed bed. This point has been indicated as well in the study by Morse [26].

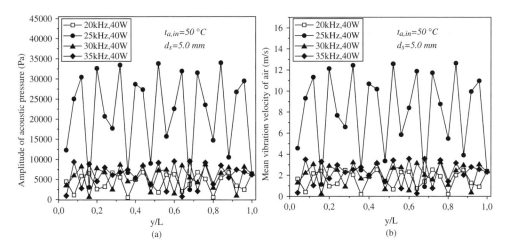

Figure 2.41 (a,b) Acoustic pressure and air vibration velocity distributions along the axial direction of silica gel bed for different ultrasonic frequencies

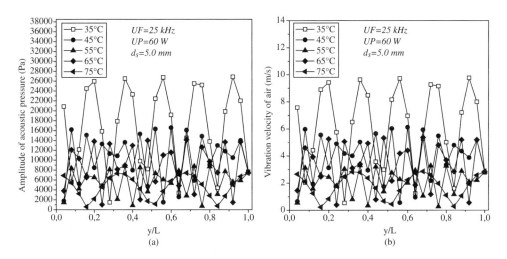

Figure 2.42 (a,b) Acoustic pressure and air vibration velocity distributions along the axial direction of silica gel bed for different regeneration temperatures

2.5.2 *Thermal Characteristics of the Bed during Ultrasound-Assisted Regeneration*

The thermal parameters of air and silica gel in the bed during regeneration are investigated based on the basic simulation conditions as below: 0.35 in the initial moisture ratio of silica gel ($q_{ini} = 0.35$); 0.3 m/s in the inlet air velocity of the bed ($u_{a,in} = 0.3$ m/s); 50 °C in the inlet air temperature ($t_{a,in} = 50$ °C); 0.02 kg/(kg dry air) in the inlet air humidity ($w_{a,in} = 0.02$ kg/(kg dry air)); 25 °C in the initial temperature of silica gel ($t_{s,ini} = 25$ °C); 5 mm in the diameter

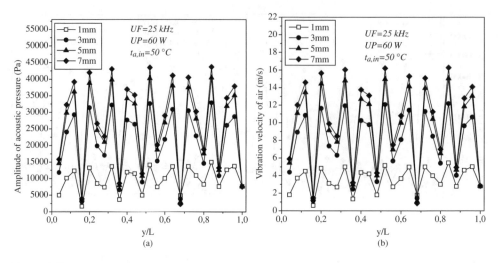

Figure 2.43   (a,b) Acoustic pressure and air vibration velocity distributions along the axial direction of silica gel bed for different particle sizes

of silica gel particle ($d_s = 5$ mm). The subscript "ini" stands for initial conditions, and "in" for the inlet air of bed.

2.5.2.1   Air and Silica Gel Temperatures in the Bed

Figure 2.44 gives the spatial distributions of air and silica gel temperature in the bed during the regeneration assisted by ultrasound with different power levels (i.e., 20, 40, 60, and 80 W). The value "x/D" on the horizontal axis is a dimensionless position along the x direction of silica gel bed (refer to Figure 2.23b). "$x/D = 0.0$" denotes the inlet of the bed, and "$x/D = 1.0$" denotes the outlet of the bed. It can be seen that the air and silica gel temperatures all decrease along the direction of air flow in the bed. The trend of change of the spatial air and silica gel temperature in the bed is easily understood because heat is transferred from the air to the silica gel during the regeneration, that is, the air is cooled by the silica gel or the silica gel is heated by the air. With the regeneration going on, the silica gel tends to have an equilibrium state with the air. So, the silica gel temperature rises gradually over the regeneration time.

The influence of the ultrasonic power on the air and the silica gel temperatures in the bed are presented in Figure 2.45. It can be seen that the air temperature in the bed decreases with the increase of ultrasonic power in the former stage of regeneration (before 32 minutes). However, as the regeneration lasts for 40 minutes, the increase of ultrasonic power causes the air temperature in the bed to rise. One of the main reasons for the above phenomena may be the reduction of the moisture content in the silica gel during regeneration. At the beginning of the regeneration, the moisture content in the silica gel is relatively high, which results in a bigger driving force for the heat and mass transfer. Therefore, the enhancement effect of heat and mass transfer brought by the ultrasonic vibration will be more significant and cause the air temperature to drop faster in the former stage of regeneration. But with regeneration continuing, the moisture content in the silica gel decreases, and the moisture desorption rate will

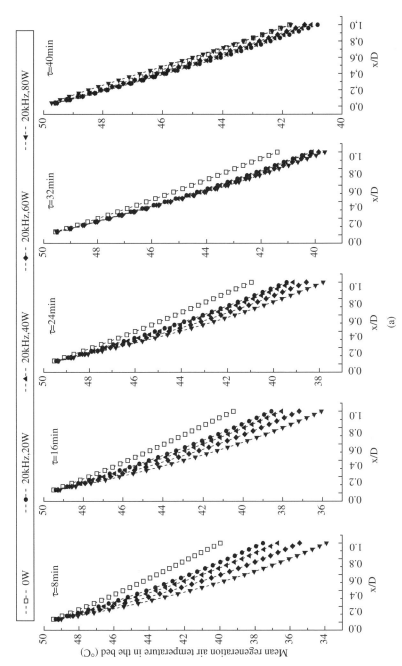

Figure 2.44 Spatial distributions of air and silica gel temperature in the bed during regeneration assisted by ultrasound with different power levels. (a) Air temperature distributions in the bed and (b) distributions of silica gel temperature in the bed

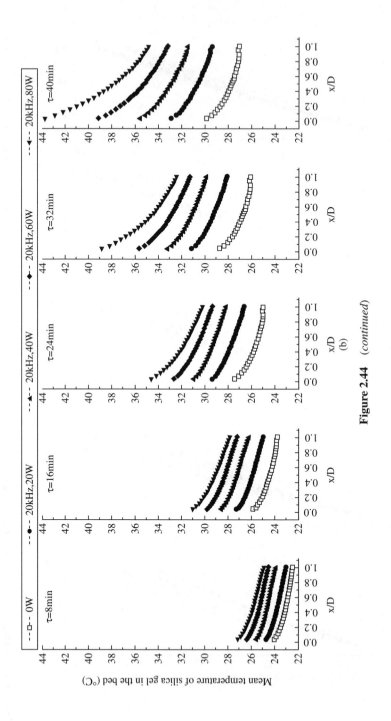

Figure 2.44 *(continued)*

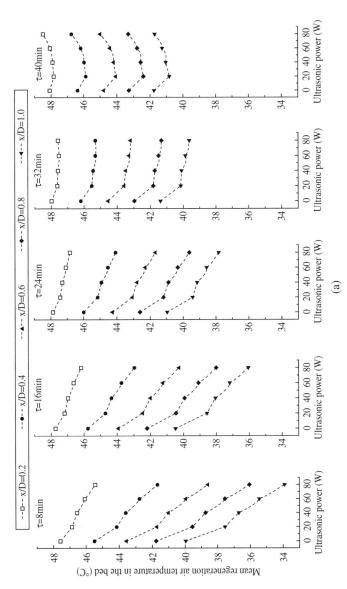

Figure 2.45 Influence of ultrasonic power on the air and the silica gel temperature in the bed during the regeneration. (a) Air temperatures under different ultrasonic power levels and (b) silica gel temperatures under different ultrasonic power levels

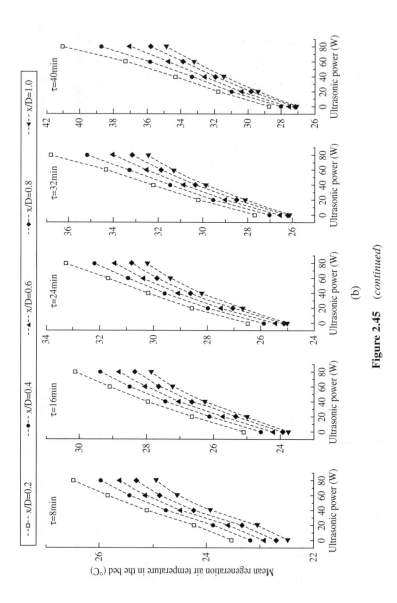

Figure 2.45 (*continued*)

decrease as well. Thus, more ultrasonic energy will be used for heating the silica gel instead of moisture desorption (see Figure 2.45b), which in turn causes the air temperature to rise. Since the heat conversion from ultrasonic energy is directly proportional to the power level, the air temperature increases with the increase of ultrasonic power.

Figure 2.46 gives the spatial distributions of air and silica gel temperature in the bed during the regeneration assisted by ultrasound with different ultrasonic frequencies (i.e., 20, 25, 30, and 35 kHz). As shown in Figure 2.46, the spatial variations of air and silica gel temperature all have a downward trend from the inlet (x/D = 0.0) to the outlet of the bed (x/D = 1.0), and the temperatures increase as the regeneration continues. Meanwhile, the air and the silica gel temperature both have the steepest descending slope along the direction of air flow for the case of 25 kHz. This is because among these frequencies, the 25-kHz ultrasound produces the greatest vibration according to the results shown in Figure 2.41, which brings about the best effect of heat and mass transfer enhancement during the regeneration.

In Figure 2.47, the effects of ultrasonic frequency on the air and the silica gel temperatures in the bed are clearly presented. It can be seen that the lowest air temperature in the bed occurs at the ultrasonic frequency of 25 kHz. This means that the 25-kHz ultrasonic radiation produces a greater effect on the air temperatures in the bed than the other frequencies. Meanwhile, the influence of ultrasonic frequency on the air temperature becomes less significant over the regeneration time. Taking the location of "x/D = 0.2," for example, the amplitude of variation of air temperature with the ultrasonic frequency for the 8-minute regeneration is much greater than that for the 40-minute regeneration, as shown in Figure 2.47a. However, the situation is opposite for the amplitude of variation of silica gel temperature with the ultrasonic frequency (see Figure 2.47b).

2.5.2.2   Air Humidity Ratio in the Bed

As shown in Figure 2.48, the air humidity ratio presents an upward trend from the inlet ($x/D = 0.0$) to the outlet of the bed ($x/D = 1.0$) for all the cases, and the higher ultrasonic power level results in the greater change in the air humidity ratio. This is reasonable because the higher ultrasonic power level brings about more moisture desorption from the silica gel during the regeneration. The results also show that the air humidity ratio in the bed decreases with the regeneration time. This is because the moisture content in the silica gel reduces as the regeneration continues further.

The effect of ultrasonic power on the air humidity in the bed during regeneration is presented in Figure 2.49. The results show that the air humidity ratio in the bed increases with increasing ultrasonic power, and the increase rate drops with the regeneration time. This indicates that with the moisture content in the silica gel decreasing, the effect of ultrasonic power on the enhancement of silica gel regeneration will become less significant.

The spatial distributions of air humidity ratio in the silica gel bed during regeneration assisted by ultrasound with different frequencies are presented in Figure 2.50. It can be seen that the air humidity ratio in the bed shares a similar spatial distribution with the air temperature in the bed. The effects of ultrasonic frequency on the air humidity in the silica gel bed during the regeneration are shown in Figure 2.51, and it can be found that the air humidity ratio is higher under the ultrasonic frequency of 25 kHz, and the effect of ultrasonic frequency on the air humidity ratio in the bed reduces with the regeneration time.

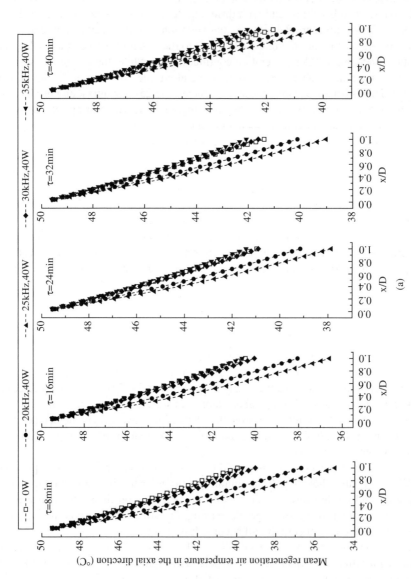

Figure 2.46 Spatial distributions of air and silica gel temperature in the bed during regeneration assisted by ultrasound with different power frequencies. (a) Air temperature distributions in the bed and (b) silica gel temperature distributions in the bed

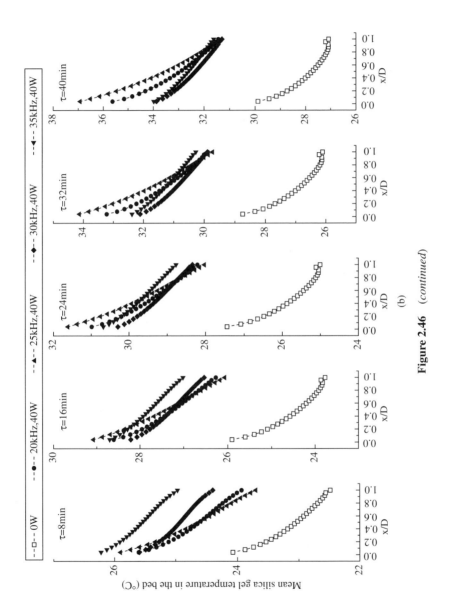

Figure 2.46 *(continued)*

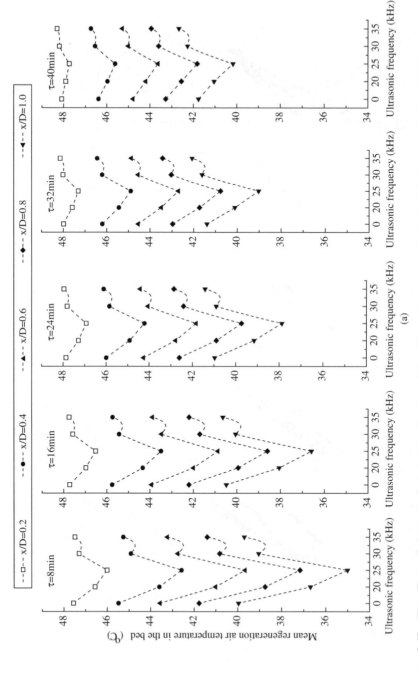

Figure 2.47 The effects of ultrasonic frequency on the air and the silica gel temperature in the bed during regeneration. (a) Air temperatures under different ultrasonic frequencies and (b) silica gel temperatures under different ultrasonic frequencies

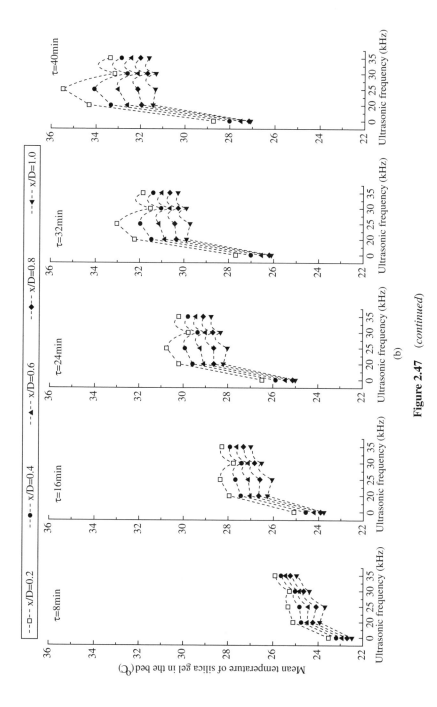

Figure 2.47 *(continued)*

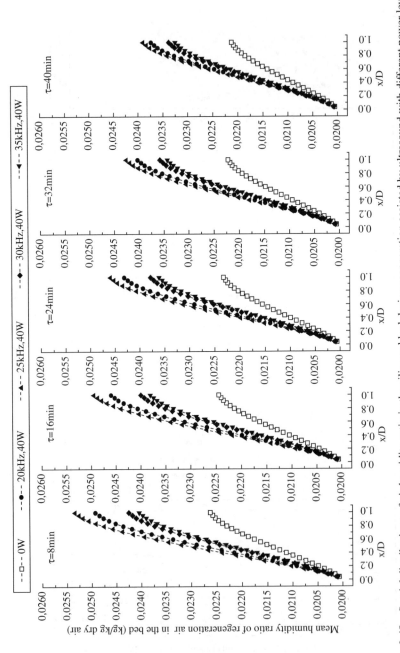

Figure 2.48 Spatial distributions of air humidity ratio in the silica gel bed during regeneration assisted by ultrasound with different power levels

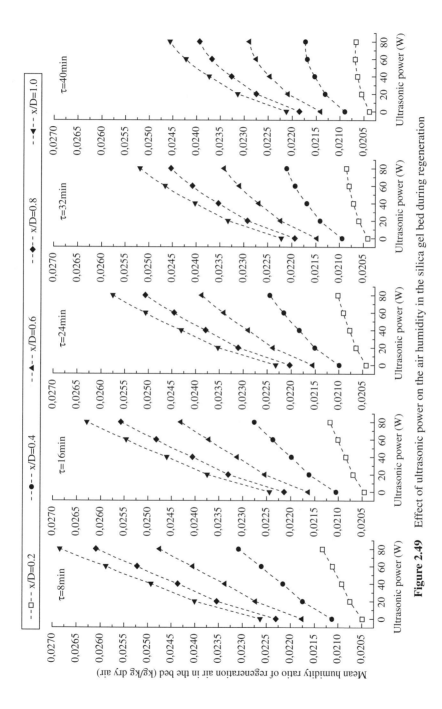

Figure 2.49 Effect of ultrasonic power on the air humidity in the silica gel bed during regeneration

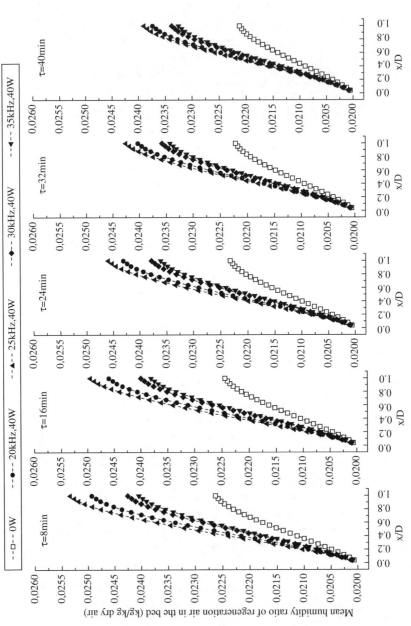

Figure 2.50 Spatial distributions of air humidity ratio in the silica gel bed during regeneration assisted by ultrasound with different frequencies

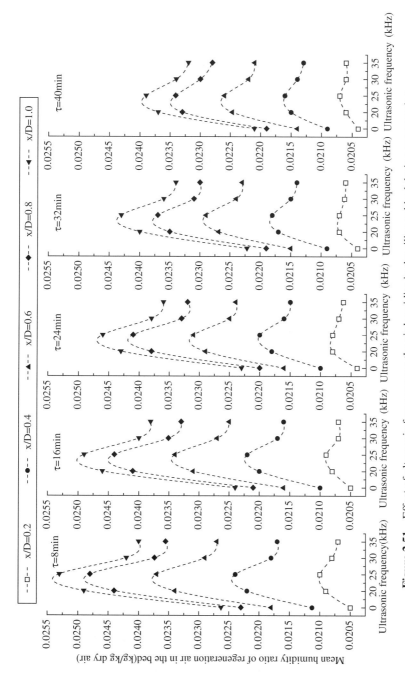

Figure 2.51 Effect of ultrasonic frequency on the air humidity in the silica gel bed during regeneration

2.5.3 Enhancement of Regeneration Assisted by Ultrasound

2.5.3.1 Different Ultrasonic Power Levels

The influence of acoustic power on the ER of silica gel regeneration brought about by ultrasound can be seen in Figure 2.52. It is clear to see that a higher ultrasonic power will bring about a higher ER of silica gel regeneration, and the influence of ultrasonic power on the ER of silica gel regeneration will be affected by the regeneration time. In addition, there should be an optimal regeneration time for which the highest ER brought about by the ultrasound can be acquired.

2.5.3.2 Different Ultrasonic Frequencies

Figure 2.53 shows the influence of ultrasonic frequency on the ER of silica gel regeneration brought about by the ultrasound. Among these cases, the highest ER occurs at the acoustic frequency of 25 kHz under which the best effect of ultrasonic vibration is acquired (refer to

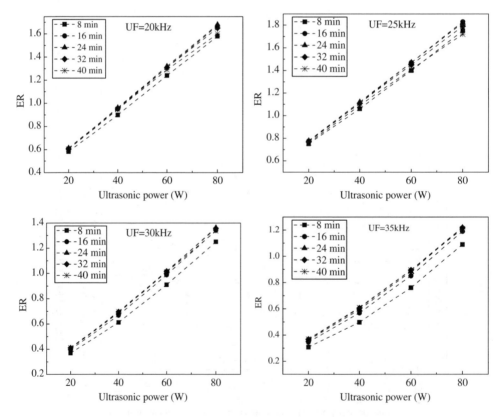

Figure 2.52 Influence of acoustic power on the ER of silica gel regeneration brought about by ultrasound. Other conditions: $q_{ini} = 0.35$; $u_{a,in} = 0.3$ m/s; $t_{a,in} = 50\,°C$; $w_{a,in} = 0.02$ kg/(kg dry air); $m_s = 0.2$ kg; $t_{s,ini} = 25\,°C$; and $d_s = 5$ mm

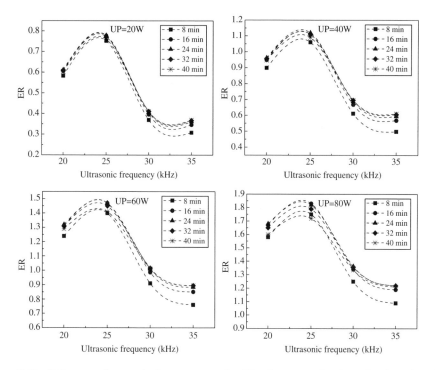

Figure 2.53 Influence of acoustic frequency on the ER of regeneration brought about by ultrasound. Other conditions: $q_{ini} = 0.35$; $u_{a,in} = 0.3$ m/s; $t_{a,in} = 50\,°C$; $w_{a,in} = 0.02$ kg/(kg dry air); $m_s = 0.2$ kg; $t_{s,ini} = 25\,°C$; and $d_s = 5$ mm

Figure 2.41). This indicates that ultrasonic frequency is one of the crucial parameters that greatly affect the ER of silica gel regeneration.

2.5.3.3 Different Particle Sizes of Silica Gel

In Figure 2.54, it can be seen that the particle size of silica gel produces a great influence on the ER of regeneration, and the ER increases with the particle size of silica gel increasing. Similar results can be found as well in the study by Sujith *et al.* [38]. According to Figure 2.43, the larger particle size results in the greater vibration velocity of air induced by ultrasound, which is conducive for the enhancement of heat and mass transfer during regeneration. As a result, the ER of silica gel regeneration brought by ultrasound will be higher for a larger particle size.

2.5.3.4 Different Initial Moisture Ratios of Silica Gel

Figure 2.55 shows the influence of initial moisture content in silica gel on the ER of regeneration brought about by ultrasound. It shows that the ER decreases with the increase of initial moisture content. Meanwhile, the declining slope of ER versus the initial moisture ratio of silica gel becomes smaller with the regeneration time. This indicates the influence of initial

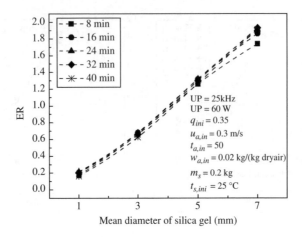

Figure 2.54 Influence of particle size of silica gel on the ER of regeneration brought about by ultrasound

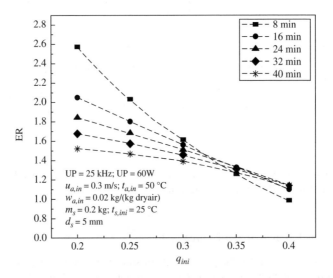

Figure 2.55 Influence of initial moisture content in silica gel on the ER of regeneration brought by ultrasound

moisture ratio on the ER brought about by ultrasound will become less significant with the regeneration carrying on.

2.5.3.5 Different Air Flow Rates

The influence of air velocity on the ER of silica gel regeneration brought about by ultrasound is presented in Figure 2.56. It can be seen that for a specific regeneration time, there ought to be an optimal air flow rate under which the highest ER can be achieved. Taking the regeneration

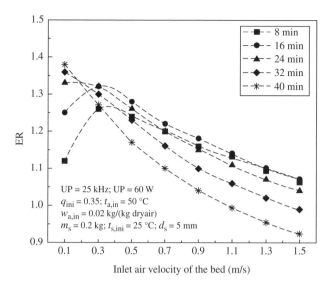

Figure 2.56 Influence of air velocity on the ER of silica gel regeneration brought about by ultrasound

time of 8 minutes, for example, the peak value of ER appears at around 0.3 m/s. In addition, the optimal air flow rate will decrease with the increase of regeneration time, for example, the optimal air flow rate in the case of 40-minute regeneration is obviously smaller than that in the case of 8-minute regeneration.

2.5.3.6 Different Regeneration Air Conditions

The regeneration air conditions will produce great influence on the ER as well. As shown in Figure 2.57a, there is apparently a decreased ER as the regeneration air temperature rises.

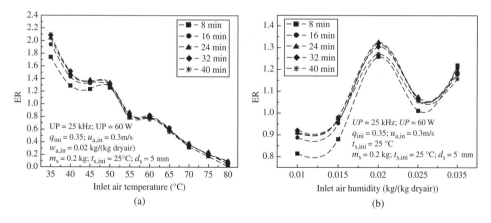

Figure 2.57 Influence of regeneration air conditions on the ER of silica gel regeneration brought about by ultrasound. (a) ER versus regeneration air temperature and (b) ER versus regeneration air humidity

This can be found as well in the study by García-Pérez *et al.* [39] It indicates that the higher air temperature is not preferred for the ultrasound-assisted regeneration in terms of *ER*. In this case, the ER almost approaches zero when the regeneration air temperature arrives at 80 °C. In addition, the results in Figure 2.57b demonstrate that a higher or a lower air humidity ratio will cause the *ER* to decline.

2.5.4 Comparisons between the Transverse- and Radial-Flow Beds

To investigate the influence of bed structure on the ultrasound-assisted regeneration, the *MRS* and the *ER* (brought about by ultrasound) for the transverse- and radial-flow bed are simulated with the theoretical model under the following conditions: $q_{ini} = 0.35$, $u_{a,in} = 0.3$ m/s, $t_{a,in} = 50$ °C, $w_{a,in} = 0.02$ kg/(kg dry air), $m_s = 0.2$ kg, $t_{s,ini} = 25$ °C, $d_s = 5$ mm, UP $= 40$ W, UF $= 25$ KHz. The results are presented in Figures 2.58 and 2.59, respectively.

As can be seen from Figure 2.58, the MRS for the transverse-flow bed is obviously larger than that for the radial-flow bed. Compared with the radial-flow bed, the transverse-flow one has a smaller cross-sectional area for the passage of regeneration air. Thus, the air flow velocity in the transverse-flow silica gel bed is higher, and the larger heat and mass transfer coefficients will be acquired in the transverse-flow bed. So, the transverse-flow bed should be a priority in the practice to gain a larger MRS during regeneration. The results of ER in Figure 2.59 show that the silica gel regeneration will benefit more from the ultrasonic radiation in the radial-flow bed. It also indicates that there will be a peak value of ER in the course of regeneration, and the occurrence time for the maximum ER will be affected by the regeneration conditions.

2.6 Quantitative Contribution of Ultrasonic Effects to Silica Gel Regeneration

2.6.1 Theoretical Analysis

As mentioned earlier, the special mechanical effect of power ultrasound will intensify the air turbulence near the surface of silica gel and enhance the process of heat and mass transfer during regeneration. Meanwhile, the heating effect of ultrasound causes a temperature rise in the media and makes moisture pressure on the surface of silica gel to increase. The mechanism of ultrasound-assisted regeneration is depicted in Figure 2.60. To have a better understanding of the mechanism, the quantitative contribution of ultrasonic effects to the enhancement of silica gel regeneration will be discussed in this section.

Basically, moisture desorption rate of desiccant during regeneration equals the product of contact area, mass transfer coefficient and difference of air humidity between the main stream and the surface of desiccant. Thus, the increased moisture desorption rate brought about by the ultrasound (ΔRR: kg/s) can be expressed by:

$$\Delta RR = S_V V \left(K_m + \Delta K_m \right) \left[\left(w_a{}^* + \Delta w_{a,teff}{}^* + \Delta w_{a,veff}{}^* \right) - w_a \right] - S_V V K_m \left(w_a{}^* - w_a \right).$$

$$(2.90)$$

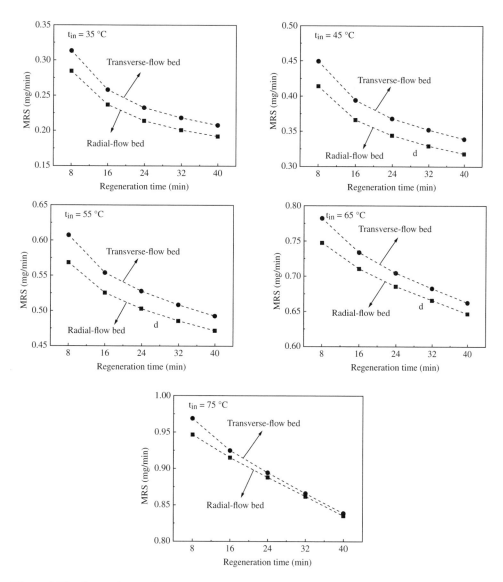

Figure 2.58 Comparisons of MRS between the transverse and the radial bed for ultrasound-assisted regeneration

To separate the heating and the mechanical effect of ultrasound on the moisture desorption rate, Equation (2.90) can be written as below:

$$\Delta RR = S_V V K_m \cdot \Delta w_{a,\text{teff}}{}^* + S_V V \left[\left(w_a{}^* + \Delta w_{a,\text{veff}}{}^* - w_a \right) \Delta K_m + K_m \Delta w_{a,\text{veff}}{}^* \right]$$
$$+ S_V V \Delta K_m \Delta w_{a,\text{teff}}{}^*, \tag{2.91}$$

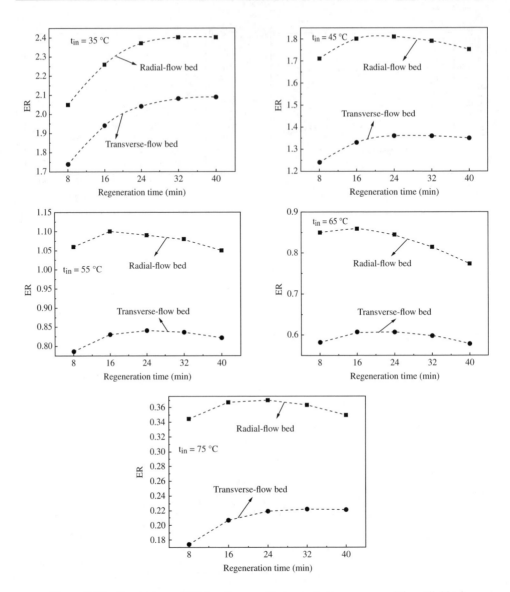

Figure 2.59 Comparison of ER by ultrasound between the transverse and the radial bed

where S_V is volumetric surface area in the packed bed, m^2/m^3; V is volume of the packed bed, m^3; K_m is mass transfer coefficient in $kg/(m^2 \cdot s)$; w_a and w_a^* denote the air humidity in the main stream and on the surface of desiccant, respectively, given in kg/(kg dry air); ΔK_m and $\Delta w_{a,veff}^*$ denote, the increased mass transfer coefficient and air humidity on the surface of desiccant due to the ultrasonic mechanical effect respectively; and $\Delta w_{a,teff}^*$ is the increased air humidity on the surface of desiccant due to the ultrasonic heating effect.

The first and second term on the right side of Equation (2.91) express the contribution of ultrasonic heating effect and mechanical effect to the moisture desorption rate during

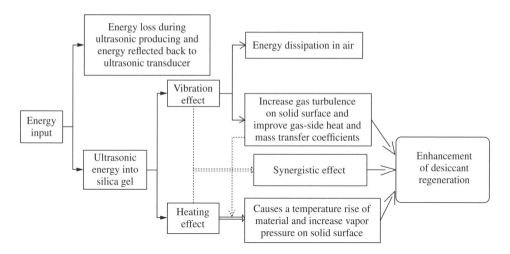

Figure 2.60   Mechanism of the ultrasound-assisted regeneration and its energy flow

regeneration, respectively. And the third term on the right side of Equation (2.91), which is related to both the heating and the mechanical effect of ultrasound, can be considered as "synergistic effect" arising between the two effects. It means that the two ultrasonic effects (i.e., the heating and the mechanical) may produce an effect greater than the summation of its individual effects on the enhancement of regeneration.

2.6.2   Method

The quantitative study on the contributions of heating and mechanical effects of ultrasound to enhancement is made in terms of the average moisture desorption rate (RR_{ave}) in a time period of regeneration (τ_f) as shown below:

$$RR_{ave} = \frac{1}{\tau_f} \int_0^{\tau_f} G_a \left[w_{a,o}(\tau) - w_{a,i}(\tau) \right] d\tau, \tag{2.92}$$

where τ_f is regeneration time required for the moisture ratio of silica gel arriving at a specific value in seconds (s); G_a is mass flow rate of regeneration air in kg/s; and $w_{a,i}(\tau)$ and $w_{a,o}(\tau)$ are inlet and outlet air humidity of bed (kg/(kg dry air)), respectively.

The regeneration enhancement (RE) and the enhancement ratio of average regeneration rate (ERARR) are used to evaluate the ultrasonic effect on regeneration:

$$RE = RR_{ave,U} - RR_{ave,NU}, \tag{2.93}$$

$$ERARR = \left(RR_{ave,U} - RR_{ave,NU} \right) / RR_{ave,NU} \times 100\%, \tag{2.94}$$

where the superscript "U" and "NU" stand for regeneration in the presence and absence of ultrasonic radiation, respectively.

To make a quantitative study on the contribution of heating and vibration effects to the regeneration enhancement, the ultrasound-assisted regeneration with the individual effect of ultrasound is simulated with the model. The regeneration enhancement brought by ultrasonic heating effect (RE_{teff}) is calculated by assuming that the vibration velocity induced by ultrasound be zero in the model, and that brought by ultrasonic mechanical effect (RE_{veff}) is calculated by assuming that the adsorption coefficient of ultrasound by media be zero in the model.

The contribution ratio (CR) of the heating, mechanical, and synergistic effect to the total enhancement of regeneration (RE) can be expressed, respectively, as below:

$$CR_{teff} = RE_{teff}/RE \times 100\%$$

$$= \frac{K_m \cdot \Delta w_{a,teff}{}^*}{K_m \cdot \Delta w_{a,teff}{}^* + \left[\left(w_a{}^* + \Delta w_{a,veff}{}^* - w_a\right)\Delta K_m + K_m \Delta w_{a,veff}{}^*\right] + \Delta K_m \Delta w_{a,teff}{}^*}$$
$$\times 100\%, \tag{2.95}$$

$$CR_{veff} = RE_{veff}/RE \times 100\%$$

$$= \frac{\left(w_a{}^* + \Delta w_{a,veff}{}^* - w_a\right)\Delta K_m + K_m \Delta w_{a,veff}{}^*}{K_m \cdot \Delta w_{a,teff}{}^* + \left[\left(w_a{}^* + \Delta w_{a,veff}{}^* - w_a\right)\Delta K_m + K_m \Delta w_{a,veff}{}^*\right] + \Delta K_m \Delta w_{a,teff}{}^*}$$
$$\times 100\%, \tag{2.96}$$

$$CR_{syn} = RE_{syn}/RE \times 100\%$$

$$= \frac{\Delta K_m \Delta w_{a,teff}{}^*}{K_m \cdot \Delta w_{a,teff}{}^* + \left[\left(w_a{}^* + \Delta w_{a,veff}{}^* - w_a\right)\Delta K_m + K_m \Delta w_{a,veff}{}^*\right] + \Delta K_m \Delta w_{a,teff}{}^*}$$
$$\times 100\%. \tag{2.97}$$

2.6.3   Results and Discussions

A series of conditions are simulated to discuss the influence of key parameters on the contributions of ultrasonic effects to the enhancement of regeneration. These parameters include regeneration air temperature and humidity, air flow rate, and ultrasonic power.

2.6.3.1   Different Regeneration Air Temperatures

Figure 2.61 gives the contribution ratio of ultrasonic effects on the enhancement of regeneration. The results show that the mechanical effect of ultrasound takes a dominant role in the enhancement of regeneration, and it becomes increasingly significant as the regeneration temperature rises. In contrast, the contribution ratio of ultrasonic heating effect on the regeneration drops with the increase of regeneration temperature. As seen from Figure 2.61a, the CR_{veff} increases from 68 to 86.8% while the CR_{teff} decreases from 30 to 12.7% when the regeneration temperature rises from 40 to 100 °C. This is because the driving force of heat and mass transfer becomes larger as the regeneration temperature increases and promotes the mechanical effect of ultrasound on the regeneration. The results in Figure 2.61b indicate that the "synergistic effect" (CR_{syn}) decreases as the regeneration temperature rises.

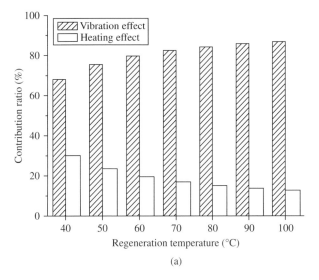

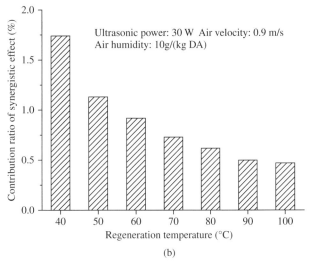

Figure 2.61 Contribution ratios of ultrasonic effects on the enhancement of regeneration under different regeneration air temperatures. (a) CR of ultrasonic heating and mechanical effect and (b) CR of ultrasonic synergistic effect

2.6.3.2 Different Regeneration Air Humidity Ratios

Figure 2.62 shows the contribution ratio of ultrasonic effects on the enhancement of regeneration under different air humidity ratios. The results manifest that the contribution ratios of ultrasonic mechanical and heating effect on the enhancement of regeneration (CR_{veff} and CR_{teff}) are hardly affected by the regeneration air humidity, and the contribution ratio of synergistic effect (CR_{syn}) has a slight increase as the air humidity ratio rises.

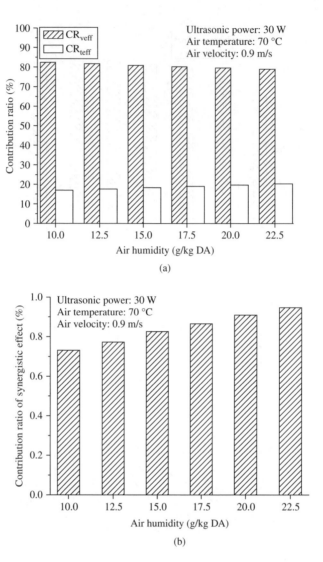

Figure 2.62 Contribution ratios of ultrasonic effects on the enhancement of regeneration under different regeneration air humidity ratios. (a) CR of ultrasonic heating and mechanical effect and (b) CR of ultrasonic synergistic effect

2.6.3.3 Different Regeneration Airflow Rates

Figure 2.63 demonstrates the contribution ratio of ultrasonic effects on the enhancement of regeneration under different air flow rates. It can be found that the contribution of the ultrasonic mechanical effect dominates when the regeneration air velocity is greater than 0.3 m/s. As shown in Figure 2.63a, the CR_{veff} increases rapidly from 66.9 to 85.2% when the regeneration air velocity increases from 0.3 to 2.5 m/s, and declines slowly afterwards, while the CR_{teff}

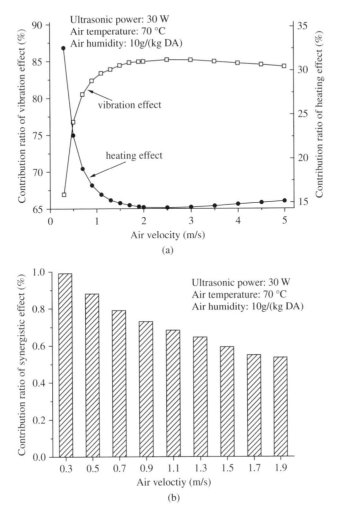

Figure 2.63   Contribution ratios of ultrasonic effects on the enhancement of regeneration under different regeneration airflow rates. (a) CR of ultrasonic heating and mechanical effect and (b) CR of ultrasonic synergistic effect

first drops from 32.1 to 14.4% and then increases slightly with the regeneration air velocity increasing. For the contribution ratio of the "synergistic effect" on the regeneration (CR_{syn}), it decreases with the air velocity increasing as shown in Figure 2.63b.

2.6.3.4   Different Ultrasonic Power Levels

In Figure 2.64, we can see that the contribution ratio of ultrasonic mechanical effect on the enhancement of regeneration decreases slightly with the increase of ultrasonic power, and the situation is opposite for the ultrasonic heating effect and synergistic effect.

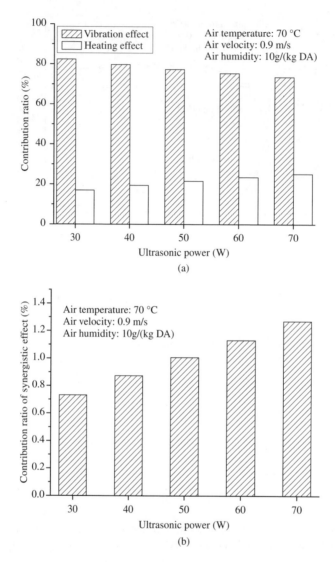

Figure 2.64 Contribution ratios of ultrasonic effects on the enhancement of regeneration under different ultrasonic power levels. (a) CR of ultrasonic heating and mechanical effect and (b) CR of ultrasonic synergistic effect

2.6.3.5 Brief Summary

Several conclusions can be drawn from the results above:

1. The mechanical (or vibration) effect of ultrasound takes a dominant role in the ultrasound-assisted regeneration. The contribution ratio of ultrasonic mechanical effect on the enhancement of regeneration is mainly affected by the regeneration air temperature and air flow rate.
2. The heating effect of ultrasound is the second most important factor that enhances the rate of regeneration. The contribution ratio of ultrasonic heating effect on the enhancement

of regeneration is also mainly influenced by the regeneration air temperature and air flow rate.

3. The synergy of the heating and mechanical effect produces extra effect on the regeneration enhancement (also called "synergistic effect"). Although the contribution of "synergistic effect" to the enhancement of regeneration is much smaller than the ultrasonic mechanical and heating effects, it does exist in the ultrasound-assisted regeneration. The results manifest the contribution ratio of the "synergistic effect" increases with the increase of ultrasonic power and regeneration air humidity, and decreases with the increase of the regeneration air temperature and air flow rate.

2.7 Energy-Saving Features of Silica Gel Regeneration Assisted by Ultrasound

In Section 2.2, the ultrasound-assisted regeneration has been proved to be energy-saving compared with the purely heating regeneration method. A simple analysis has been made on the energy saving of ultrasound-assisted regeneration in terms of the total energy consumption. However, the total energy consumption can not have a comprehensive description about the energy consumption characteristics of the regeneration process. More details about the energy consumption are needed for an overall analysis on the new regeneration method with ultrasound. The specific energy consumption (SEC) is considered a good index in comparing the energy efficiency of different drying processes and has been successfully employed to investigate the energy efficiency of new drying methods like microwave drying [40] and spray drying [41]. The index of SEC helps us probe into the efficiency of energy use in relation to the overall drying processes and understand the reason why one process is more energy saving than the others. This section deals with the energy-saving characteristic of silica gel regeneration assisted by ultrasound with the index SEC, and discusses about the influence of regeneration air temperature, ultrasonic power level and final moisture ratio in material on the SEC of the ultrasound-assisted regeneration.

2.7.1 Specific Energy Consumption

Three kinds of SEC indices are used for this study. These are the total specific energy consumption (TSEC), the adiabatic specific energy consumption (ASEC), and the excess specific energy consumption (ESEC). The TSEC reflects how much energy is totally used for per unit mass (kg) of moisture removal in a dehydration process. The ASEC refers to the SEC of a dehydration process under the adiabatic conditions. It depends on the enthalpy of regeneration air and ambient air. And the ESEC is the measurement of energy loss.

The total, the adiabatic and the ESEC for the desiccant regeneration are defined, respectively, by Equations (2.98)–(2.100).

$$TSEC = \frac{E_{\text{total}}}{m_{\text{w,desorption}}}, \tag{2.98}$$

$$ASEC = \frac{G_{\text{a,reg}}\left(h_{\text{a,reg}} - h_{\text{a,env}}\right)\tau}{m_{\text{w,desorption}}}, \tag{2.99}$$

$$ESEC = TSEC - ASEC, \tag{2.100}$$

where E_{total} is total energy consumption for a specific time period of regeneration in J; $m_{w,desorption}$ is the mass of moisture desorption during the regeneration in kg; $G_{a,reg}$ is the mass flow rate of regeneration air in kg/s; $h_{a,reg}$ is enthalpy of regeneration air in J/kg; $h_{a,env}$ is the enthalpy of environmental air in J/kg; and τ is the regeneration time in s.

The ESR of the regeneration assisted by ultrasound, *ESR*, is defined by Equation (2.101) in terms of the TSEC.

$$ESR = \frac{TSEC_{NU} - TSEC_{U}}{TSEC_{NU}}. \tag{2.101}$$

2.7.2 Results and Discussions

The influential factors that affect the SEC and the ESR are to be discussed in the following parts. These factors include the regeneration air temperature, the acoustic power level and the final moisture ratio in the sample after a specific time period of regeneration. The 21-kHz ultrasound with four power levels (i.e., 0, 20, 40, and 60 W) combined with different drying air temperatures (i.e., 35, 45, 55, and 65 °C) are investigated with the conditions as below: 0.05 kg/s in the air flow rate ($G_{a,reg} = 0.05$ kg/s), 28 °C and 80% in the ambient air temperature and air relative humidity ($t_{a,env} = 28$ °C, $\varphi = 80\%$), 0.35 in the initial moisture ratio of silica gel ($q_{ini} = 0.35$), 0.90 and 0.80 in the energy efficiency of heater and ultrasonic transducer, respectively.

2.7.2.1 Different Regeneration Air Temperatures

Figure 2.65 presents the change trend of the TSEC versus the regeneration air temperature. Basically, the TSEC decreases greatly as the regeneration air temperature increases from 35 to 45 °C, then gradually increases as the regeneration air temperature continues rising. The results indicate that there should be a best regeneration air temperature under which the lowest TSEC can be achieved for the silica gel regeneration. Taking the 40-W ultrasound, for example, the optimum temperature used for the minimum TSEC falls between 45 and 55 °C. The results in Figure 2.65 also manifest that it does not always bring about energy saving by applying the ultrasound to the silica gel regeneration. For example, at the regeneration air temperature of 55 or 65 °C, the TSEC for regenerating the silica gel with the 20-W ultrasound from the initial moisture ratio of 0.35 ($q_{ini} = 0.35$) to the terminal moisture ratio of 0.25 ($q_{tar} = 0.25$) is higher than that in the absence of ultrasound. It can be reflected as well from Figure 2.66 in which negative values of the ESR appear in the former stage of regeneration (e.g., the moisture ratio in silica gel decreases from the initial value to 0.2 or 0.25 or 0.3) under the conditions of 55 or 65 °C combined with the 20-W ultrasound. As shown in Figure 2.66, most of the ESRs brought by the power ultrasound tends to decrease with the increase of regeneration air temperature. This indicates that the effect of the power ultrasound on the improvement of silica gel regeneration will become less significant under the higher regeneration air temperature. The ESEC can be essentially considered as energy dissipation. The higher ESEC means the more energy loss. The results of ESEC versus the regeneration air temperature are presented in Figure 2.67. The regular pattern of change in the ESEC with the regeneration temperature is basically similar to that in the TSEC. The minimum ESEC happens at around 55 °C (in the regeneration air temperature) for the presence of 40 or 60-W ultrasound, and at around 45 °C for the 20-W ultrasonic applied. The results indicate that too low or too high drying

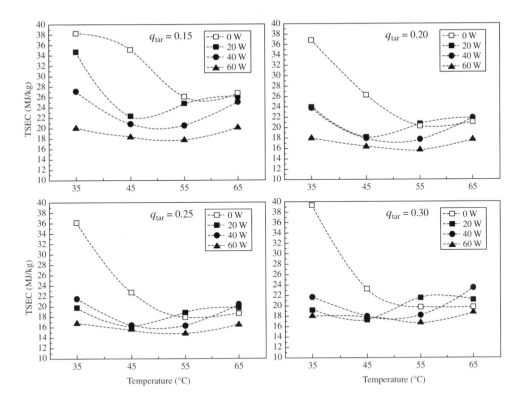

Figure 2.65   TSEC versus regeneration air temperature under different regeneration conditions

temperature is not beneficial for decreasing the ESEC. The reason may be that, as can be seen from the definition of the ESEC, it is a function of regeneration air temperature and the regeneration time. A higher regeneration temperature is not conducive for reducing the ESEC, but it can shorten the regeneration time and this helps to decrease the ESEC. Thus, there is a balance between the regeneration temperature and the regeneration time.

2.7.2.2   Different Acoustic Power Levels

Figure 2.68 shows the TSEC versus acoustic power level applied. As shown in Figure 2.68, when the regeneration process lasts till 0.15 in the moisture ratio (i.e., $q_{tar} = 0.15$), the TSEC decreases with the increase of the ultrasonic power. Whereas, for the other regeneration processes ending, respectively, at 0.2, 0.25, and 0.3 in the moisture ratio (i.e., $q_{tar} = 0.2, 0.25,$ and 0.3), the change trend of TSEC against the ultrasonic power differs from each other under different regeneration air temperatures. Taking the case "$q_{tar} = 0.2$," for example, the TSEC drops as the power increases from 0 to 20 W under the regeneration air temperature of 35 and 45 °C, and it is reverse for the other temperatures (i.e., 55 and 65 °C). The results indicate as well that there ought to be a threshold of power above which the ultrasound can bring about energy saving for the silica gel regeneration, and the threshold value increases with the rising of the regeneration air temperature. In this case, the threshold of ultrasonic power is found to

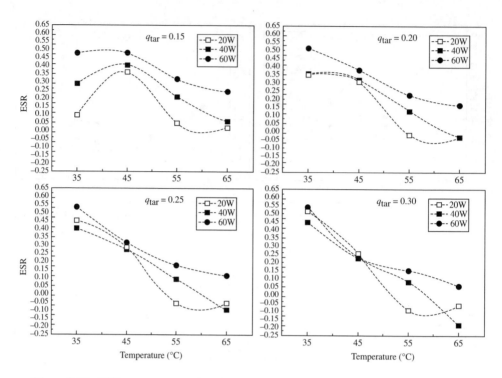

Figure 2.66   ESR versus regeneration air temperature under different regeneration conditions

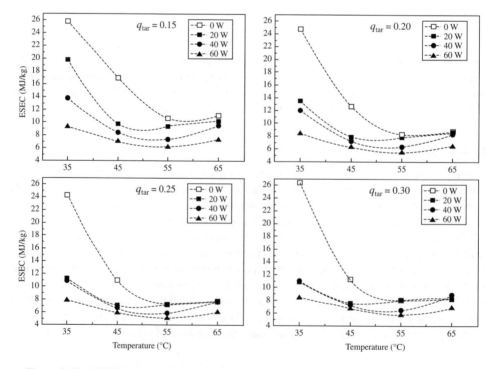

Figure 2.67   ESEC versus regeneration air temperature under different regeneration conditions

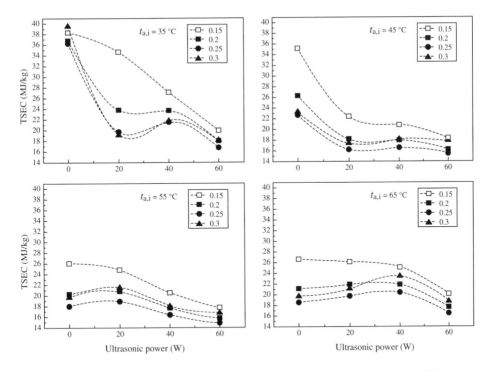

Figure 2.68 TSEC versus ultrasonic power level under different regeneration conditions

be less than 20 W at 35 and 45 °C in the regeneration temperature, and it exceeds 20 and 40 W, respectively, at 55 and 65 °C.

For the ESR versus the ultrasonic power (see Figure 2.69), regular patterns can only be found for the case of "$q_{tar} = 0.15$" in which the ESR increases with the increase of the acoustic power level. For the other cases (i.e., $q_{tar} = 0.2, 0.25, 0.3$), however, the change patterns of ESR versus ultrasonic power differ from one another under different regeneration temperatures. As shown in Figure 2.69, the 40-W ultrasound acquires the higher ESR than the 20-W ultrasonic at 55 °C, and the situation is opposite at 65 °C.

As shown in Figure 2.70, the ESEC decreases with the increase of the acoustic power in most cases. Among these cases, the regeneration air temperature of 55 °C combined with the 60-W ultrasound acquires the lowest ESEC (i.e., the minimum energy loss during regeneration).

2.7.2.3 Different Target Moisture Ratios

The target moisture ratio (q_{tar}) in silica gel is another important parameter that affects the energy consumption of regeneration. The changes of TSEC and ESEC versus the target MR are presented, respectively, in Figures 2.71 and 2.72. The results show that the lowest TSEC and ESEC both occur at the target moisture ratio of 0.25 ($q_{tar} = 0.25$) under different conditions. This indicates that the energy utilization efficiency is the highest for the regeneration process ending at about 0.25 in the target moisture ratio ($q_{tar} = 0.25$) which can be recommended as a control point for the ultrasound-assisted regeneration under such circumstances.

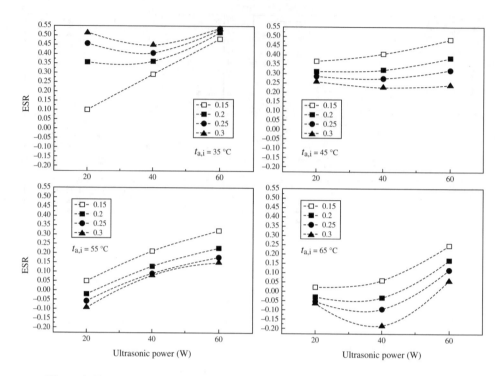

Figure 2.69   ESR versus ultrasonic power level under different regeneration conditions

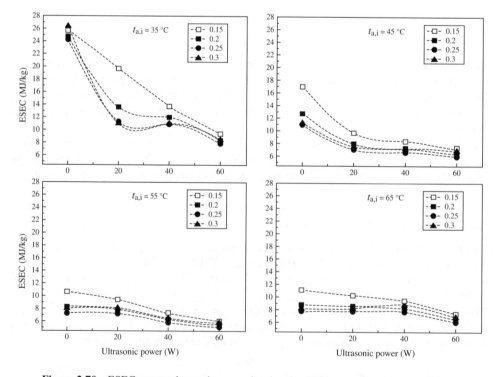

Figure 2.70   ESEC versus ultrasonic power level under different regeneration conditions

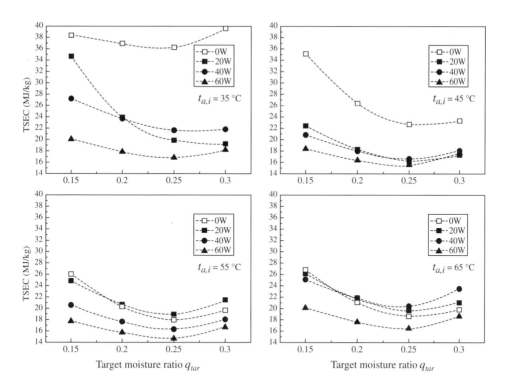

Figure 2.71   TSEC versus target moisture ratio under different regeneration conditions

2.7.3   Brief Summary

This section investigates the characteristics of energy consumption of the ultrasound-assisted regeneration with the energy consumption indices including the TSEC and the ESEC. By using the index of TSEC, the energy-saving characteristics of the new regeneration technology can be definitely described under different conditions based on which the optimum control scheme aiming at the minimum energy consumption can be made. The ESEC represents energy loss during the process of regeneration. It gives us a good idea of energy dissipation under different circumstances, and helps us to understand why the TSEC under one condition is higher than that under the others. The ESRs of the ultrasound-assisted regeneration have been compared under different conditions. Moreover, the influences of regeneration air temperature, ultrasonic power, and target moisture ratio on the total and the ESEC have been discussed in detail. The results manifest that the energy-saving characteristic of the ultrasound-assisted regeneration will be affected greatly by the regeneration conditions. Generally, there should be a different threshold of ultrasonic power above which the energy saving can be achieved under different circumstances, and one target moisture ratio (at which the regeneration process terminates) that will result in the lowest TSEC and ESEC for the regeneration.

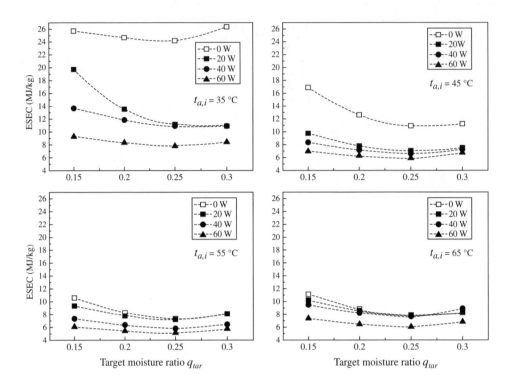

Figure 2.72 ESEC versus target moisture ratio under different regeneration conditions

2.8   Effects of Ultrasound-Assisted Regeneration on Desiccant System Performance

2.8.1   Study Objective and Method

The objective of this part is mainly to investigate the effects of the ultrasound-assisted regeneration on the performance of dehumidification system under a wide range of conditions. Meanwhile, the optimal regeneration time for different conditions is then discussed.

To begin with, some performance-related parameters are necessarily suggested for this study. These mainly include the moisture removal capacity (MRC), the additional moisture removal capacity (AMRC) brought about by ultrasound, and the dehumidification coefficient of performance (DCOP).

The MRC is the mass flow rate of moisture removed by the desiccant system:

$$MRC = G_a \left(w_{a,in} - w_{a,out} \right) \tau_{ads} / \left(\tau_{reg} + \tau_{ads} \right), \tag{2.102}$$

where G_a is mass flow rate of process air in kg/s; $w_{a,in}$ and $w_{a,out}$ are inlet and outlet air humidity (kg/(kg dry air)), respectively; τ_{ads} is time for desiccant moisture absorption in seconds (s), and τ_{reg} is time for desiccant regeneration in seconds (s).

The AMRC brought about by ultrasound is defined as:

$$AMRC = MRC_U - MRC_{NU}. \tag{2.103}$$

The DCOP represents the ratio of air enthalpy change during the dehumidification process to the energy consumption required for the regeneration process, which is written as:

$$DCOP = \left[m_{s,dry} \int_{q_{ini}}^{q_{tar}} H_{ads} dy \right] \Big/ \left(P_{total} \tau_{reg} \right), \tag{2.104}$$

where

$$P_{total} = P_u + G_a \left(h_{a,in} - h_{a,env} \right) / \eta_{heater} \text{ for the presence of ultrasound,} \tag{2.105a}$$

$$P_{total} = G_a \left(h_{a,in} - h_{a,env} \right) / \eta_{heater} \text{ for the absence of ultrasound.} \tag{2.105b}$$

In Equations (2.104), (2.105a), and (2.105b) $m_{s,dry}$ is mass of dry desiccant, kg; ΔH_{ads} is desorption/adsorption heat of water vapor from desiccant in J/kg; q_{ini} and q_{tar} stand, respectively, for the initial and the terminal moisture ratio in desiccant (kg/(kg dry sample)); P_{total} is total power for the regeneration in W; P_u is power for producing ultrasound in Watts; G_a is regeneration air flow rate in kg/s; $h_{a,in}$ is air enthalpy for the regeneration in J/kg; $h_{a,env}$ is ambient air enthalpy in J/kg; and η_{heater} is heating efficiency of heater.

The model developed earlier is employed to study the performances of the dehumidification system with the ultrasound-assisted regeneration. The simulation conditions are as below: the air temperature for the desiccant regeneration ranging from 50 to 90 °C; the air temperature and relative humidity for the desiccant adsorption ranging from 20 to 35 °C and from 70 to 95%, respectively; the air flow rate for the regeneration and adsorption process ranging from 0.5 to 4.5 m/s; the initial moisture ratio of desiccant for the regeneration process, which is equal to the target moisture ratio for the adsorption process, ranging from 0.10 to 0.20. The ultrasonic frequency and power are set as 23 kHz and 30 W, respectively. Taking into account the heat loss of air duct, the efficiency of the electric heating system, η_{heater}, can be approximately set as 0.8 during the calculation.

2.8.2 Results and Discussions

2.8.2.1 Effect of Operation Conditions on System Performance

Effect of Regeneration Temperature
The performances of dehumidification system with and without ultrasonic regeneration under different regeneration temperatures are presented in Figure 2.73. The legend "U" and "NU" denote regeneration with and without ultrasonic radiation, respectively. As shown in Figure 2.73a, the MRC of the desiccant system and the AMRC (brought about by ultrasound) all increase with the increase of the regeneration air temperature. This indicates that the power ultrasound can bring about an increased dehumidification capacity under a higher regeneration temperature. However, the increase rate in the AMRC is approaching zero after the regeneration temperature exceeds 80 °C.

The regeneration temperature will produce great influence on the DCOP of the desiccant system. As shown in Figure 2.73b, the *DCOP* first increases and then decreases with the increase of regeneration temperature. It indicates that for any specific conditions, there should be a best regeneration temperature under which the system will have the highest *DCOP*, and the best regeneration temperature (aiming at the highest *DCOP*) under the "U" case (about 62 °C) will be higher than that under the "NU" case (about 50 °C). In addition, it can be found as well

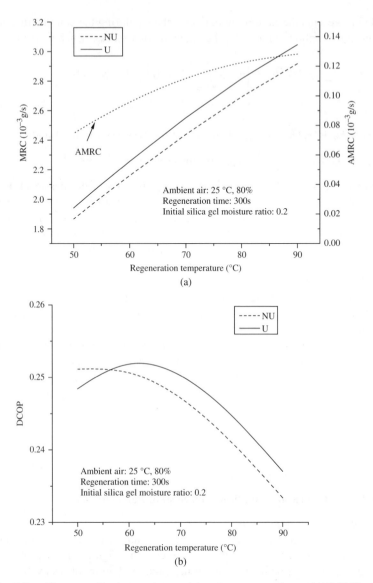

Figure 2.73 Effect of regeneration temperature on (a) moisture removal capacity (MRC) and (b) dehumidification coefficient of performance (DCOP)

from Figure 2.73b that it may not be energy saving by applying the ultrasound to the regeneration process if the regeneration temperature is lower than a critical point. Usually, the MRC and the AMRC brought by the ultrasonic effect should be considered together with the DCOP. Therefore, the regeneration temperature between 65 and 85 °C is recommended for this case.

Effect of Ambient Air Conditions
The effects of ambient air conditions (i.e., air temperature and relative humidity) on the *MRC* and *DCOP* of the system are shown in Figures 2.74 and 2.75, respectively. Basically, the *MRC*

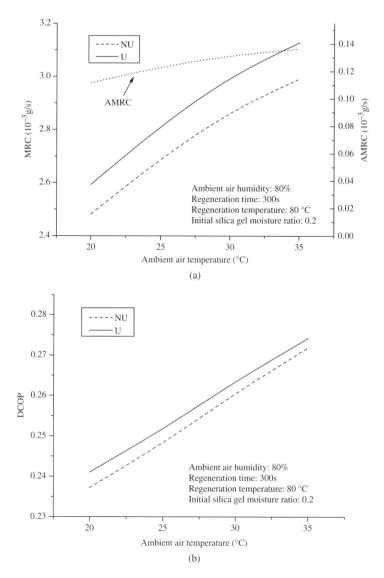

Figure 2.74 Effect of ambient air temperature on (a) moisture removal capacity (MRC) and (b) dehumidification coefficient of performance (DCOP)

and *DCOP* increase with the increase of the ambient air temperature. As we know, increasing the ambient air temperature means increasing the air humidity ratio when the air relative humidity is kept unchanged, and hence, promotes the adsorption process and causes the *MRC* to increase. Meanwhile, the increase of ambient air temperature can reduce the energy consumption for the regeneration. So, the higher ambient air temperature will result in the higher *DCOP* of the system. However, the *DCOP* is shown to decrease with the increase of the relative humidity of the ambient air (as shown in Figure 2.75b). This is because more energy will be consumed for the regeneration under the higher humidity ratio of ambient air.

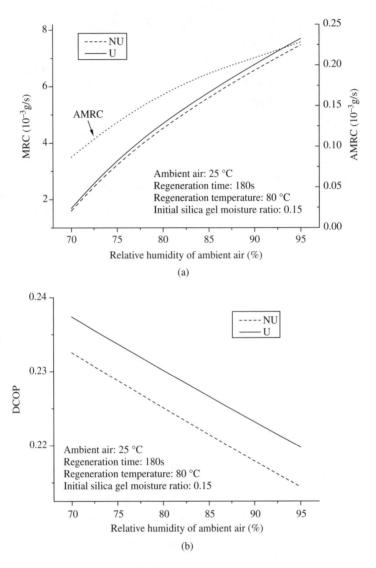

Figure 2.75 Effect of relative humidity of ambient air on (a) moisture removal capacity (*MRC*) and (b) dehumidification coefficient of performance (*DCOP*)

The curve of *AMRC* (brought about by ultrasound-assisted regeneration) in Figures 2.74a and 2.75a shows that the *AMRC* increases with the ambient air temperature and humidity increasing, and this indicates that the application of ultrasound to the regeneration process can bring about more improvement of dehumidification capacity of the system under the hotter and the more humid climate.

Effect of Air Flow Rate

The air flow rate will produce a great influence on the *MRC* and *DCOP* of the system. It can be seen from Figure 2.76 that the *MRC* has a remarkable rise with the air flow rate increasing, while it is reverse for the *DCOP*. Although the increase of air flow rate can enhance the heat and mass transfer and bring about a larger *MRC*, it will increase the energy consumption of the fan. So, whether or not the *DCOP* increases with the air flow rate increasing depends on the pay and the gain of the increased air flow rate.

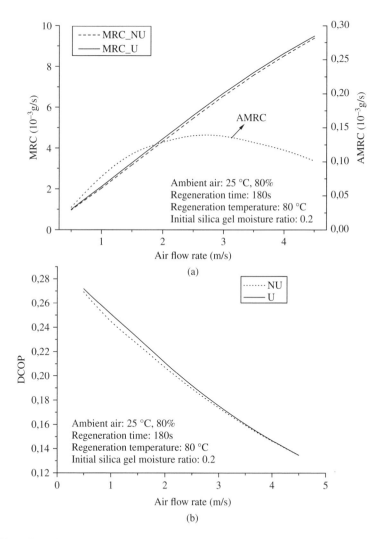

Figure 2.76 Effect of air flow rate on (a) moisture removal capacity (MRC) and (b) dehumidification coefficient of performance (DCOP)

As can also be seen from Figure 2.76a, the *AMRC* increases first and then drops with the increase of the air flow rate. It means that there should be a best air flow rate under which the ultrasonic-assisted regeneration can bring about the largest *AMRC*.

Effect of Initial Moisture Ratio of Silica Gel
Figure 2.77 gives the *MRC* and *DCOP* of the system under different initial moisture ratios of silica gel ranging from 0.11 to 0.2. The results show that the *MRC* decreases with the initial

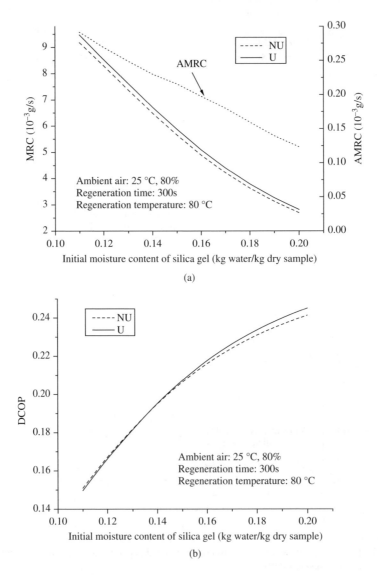

Figure 2.77 Effect of initial moisture of silica gel on (a) moisture removal capacity (MRC) and (b) dehumidification coefficient of performance (DCOP). (a) Variations of MRC and AMRC with (MR)$_o$ and (b) variations of DCOP with (MR)$_o$

moisture ratio increasing. This is because the time required for the adsorption process increases greatly as the initial moisture ratio for the regeneration increases. For the *DCOP*, the higher initial moisture ratio of silica gel is preferred since the dehumidification capacity of silica gel can be improved under such circumstance. The results of *AMRC* in Figure 2.77a manifests that the higher initial moisture ratio of desiccant will cause the effect of the ultrasound-assisted regeneration on the *MRC* to decline.

2.8.2.2 Optimization of Regeneration Time

The regeneration time is a key parameter affecting the performance of the system. The variations of *MRC*, *AMRC*, and *DCOP* against the regeneration time ranging from 60 to 600 seconds have been investigated under different conditions.

Figure 2.78 presents the variations of *MRC*, *AMRC*, and *DCOP* against the regeneration time for the regeneration air temperature of 70, 80, and 90 °C. As shown in Figure 2.78, the *AMRC* and *DCOP* first rises and then drops with the regeneration time. The regeneration time corresponding to the peak value of *AMRC* and *DCOP* is found to decrease with the regeneration temperature increasing. For the regeneration temperature of 70, 80, and 90 °C, the maximum *AMRC* occurs at about 450, 360, and 320 seconds in the regeneration time, respectively. And the regeneration time corresponding to the maximum *DCOP* is about 240, 210, and 200 seconds, respectively. It can be seen from Figure 2.78b that there exists a critical regeneration time after which the *DCOP* in the presence of ultrasonic radiation will be lower than that in the absence of ultrasonic radiation. This indicates that the regeneration time should not exceed the critical time when the ultrasound is applied to the regeneration for improving the *DCOP*.

Figure 2.79 shows the variations of *MRC*, *AMRC*, and *DCOP* against the regeneration time for different ambient air temperatures (20, 25, 30, and 35 °C). It can be seen that the higher *AMRC* is achieved and the longer regeneration time is required for obtaining the maximum *AMRC* under the higher ambient air temperature. In this case, the regeneration time corresponding to the maximum AMRC is about 360, 370, 390, and 420 seconds, for the ambient air temperature of 20, 25, 30, and 35 °C respectively. It seems that the regeneration time corresponding to the maximum DCOP is hardly affected by the ambient air temperature, which is identically about 200 seconds for all the ambient air temperatures. The critical regeneration time (which is the upper limit time for the ultrasound-assisted regeneration to improve the *DCOP*) decreases with the ambient air temperature increasing. As shown in Figure 2.79b, the critical regeneration time is estimated to be about 430, 420, 400, and 380 seconds, for the ambient air temperature of 20, 25, 30, and 35 °C respectively.

Figure 2.80 gives the variations of *MRC*, *AMRC*, and *DCOP* against the regeneration time for different relative humidity levels of ambient air (70, 80, 90%). From Figure 2.80a,b, we can see that the regeneration time corresponding to the maximum AMRC and DCOP increases slightly (from 260 to 280 seconds and from 150 to 160 seconds, respectively) as the relative humidity of the ambient air rises from 70 to 90%. The influence of the ambient air relative humidity on the critical regeneration time (which is the upper limit time for the ultrasound-assisted regeneration to improve the *DCOP*) is not very distinct. For this case, the critical regeneration time is about 300 seconds for all the relative humidity levels of ambient air.

From Figure 2.81a, it can be seen that the optimal regeneration time aiming at the maximum *AMRC* increases with the decrease of the regeneration air velocity, which is about 220, 270, 370, and 620 seconds, for the air velocity of 2.0, 1.5, 1.0, and 0.5 m/s, respectively. And a

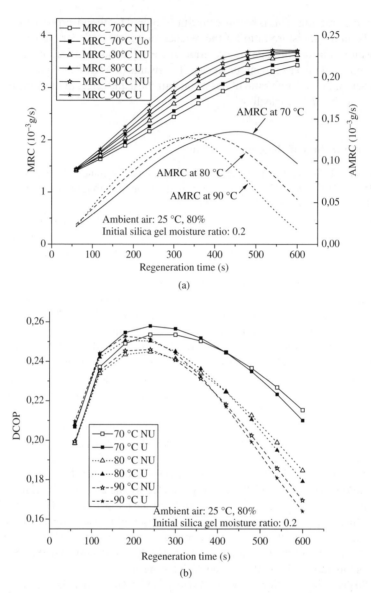

Figure 2.78 Variations of *MRC*, *AMRC*, and *DCOP* against regeneration time under different regeneration temperatures. (a) Variations of *MRC* and *AMRC* with time and (b) variations of *DCOP* with time

similar regular pattern can be found as well for the optimal regeneration time aiming at the maximum *DCOP*. As shown in Figure 2.81b, the maximum DCOP occurs at about 120, 150, 190, and 310 seconds, respectively, for the air velocity of 2.0, 1.5, 1.0, and 0.5 m/s. The critical regeneration time (which is the upper limit time for the ultrasound-assisted regeneration to improve the *DCOP*) will be different from each other for different regeneration air velocities.

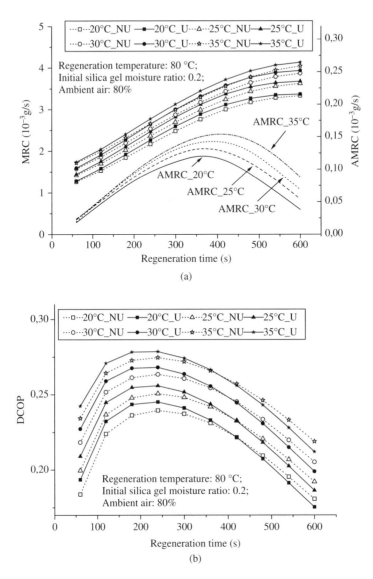

Figure 2.79   Variations of *MRC*, *AMRC*, and *DCOP* against regeneration time under different ambient air temperatures. (a) Variations of MRC and AMRC with time and (b) variations of DCOP with time

From Figure 2.81b, the critical regeneration time can be found to be about 250, 300, 350, and 250 seconds for the air velocity of 2.0, 1.5, 1.0, and 0.5 m/s, respectively.

The influences of the initial moisture ratio of desiccant for the regeneration process on the variations of MRC, AMRC, and DCOP against the regeneration time can be illustrated by Figure 2.82. The results show that the optimal regeneration time aiming at the maximum AMRC and the maximum DCOP increases with the initial moisture ratio of desiccant for regeneration increasing. Under the simulation conditions given in Figure 2.82, the regeneration

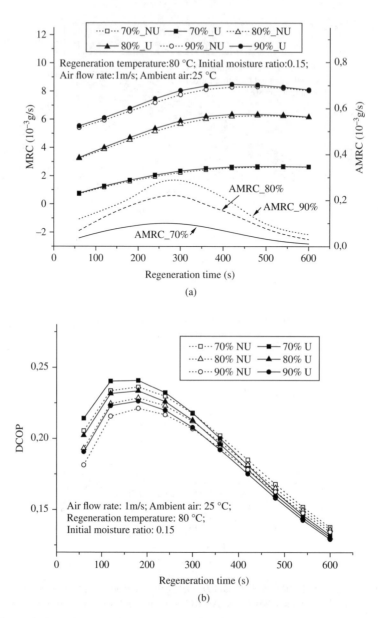

Figure 2.80 Variations of *MRC*, *AMRC*, and *DCOP* against regeneration time under different ambient air humidity levels. (a) Variations of MRC and AMRC with time and (b) variations of DCOP with time

time for the maximum AMRC increases from about 200 to 350 seconds when the initial moisture ratio of desiccant for regeneration rises from 0.12 to 0.18, and that for the maximum *DCOP* increases from about 120 to 180 seconds. From Figure 2.82b, it can be found that the critical regeneration time (which is the upper limit time for the ultrasound-assisted regeneration to improve the *DCOP*) increases with the increase of the initial moisture ratio of desiccant

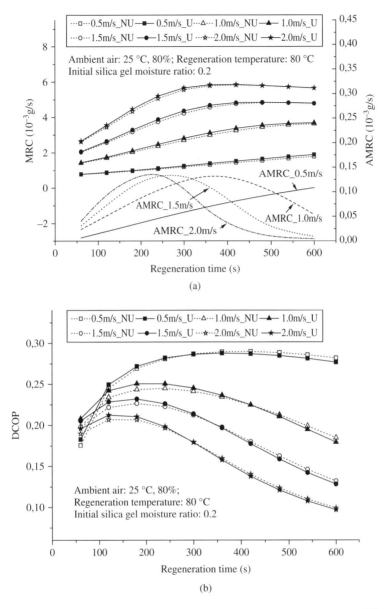

Figure 2.81 Variations of *MRC*, *AMRC*, and *DCOP* against regeneration time under different regeneration air velocities. (a) Variations of MRC and AMRC with time and (b) variations of DCOP with time

for the regeneration process. The critical regeneration time is about 230, 280, 310, and 360 seconds, respectively, for the initial desiccant moisture ratio of 0.12, 0.14, 0.16, and 0.18.

The *AMRC* reflects the contribution of ultrasound-assisted regeneration to the dehumidification ability of desiccant material, and we expect it to be as large as possible. At the same time, the *DCOP* must be taken into account from the perspective of energy saving. As indicated

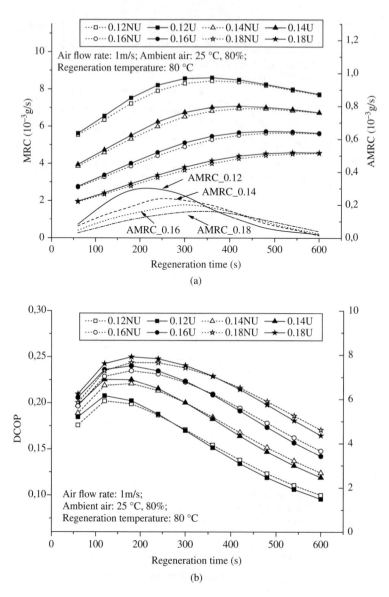

Figure 2.82   Variations of *MRC*, *AMRC*, and *DCOP* against regeneration time under different initial moisture ratios of desiccant for the regeneration process. (a) Variations of *MRC* and *AMRC* with time and (b) variations of *DCOP* with time

from Figures 2.78–2.82, the optimal regeneration time required for the maximum *AMRC* is not consistent with that for the maximum *DCOP*, and the former is longer than the latter for all the simulation conditions. So, there should be a compromised regeneration time under which a relatively higher *AMRC* and *DCOP* of the dehumidification system can be acquired simultaneously.

2.8.3   Brief Summary

The following points can be summarized based on the above results:

1. A higher regeneration temperature is conducive for increasing the AMRC brought about by the ultrasound-assisted regeneration, and there is an appropriate regeneration temperature to get the highest DCOP.
2. A higher ambient air temperature is conducive for increasing the AMRC and DCOP of the dehumidification system.
3. A higher ambient air humidity level will result in a higher AMRC and a lower DCOP of the dehumidification system.
4. The DCOP of the dehumidification system will decrease with increasing air flow rate. And there is an appropriate air flow rate to get the largest AMRC.
5. A higher initial moisture ratio of desiccant for the regeneration process is preferred for improving the DCOP, but it is unfavorable for increasing the AMRC.
6. The optimal regeneration time aiming at the maximum AMRC or DCOP decreases as the regeneration temperature or the air flow rate increases, and the situation is opposite as the ambient air temperature or humidity or the initial moisture content of silica gel increases.

References

[1] Kaeger, J. and Ruthven, D.M. (1992) *Diffusion in Zeolites and Other Microporous Solids*, John Wiley & Sons, Inc., New York.
[2] Edwards, D.K., Denny, V.E. and Mills, A.F. (1979) *Transfer Processes*, 2nd edn, Hemisphere/McGraw-Hill Book Company, Inc., New York.
[3] Pesaran, A.A. (1983) Moisture transport in silica gel particle beds. PhD Dissertation. School of Engineering and Applied Sciences, University of California, Los Angeles.
[4] Roque-Malherbe, R.M.A. (2007) *Adsorption and Diffusion in Nanoporous Materials*, CRC Press, Taylor & Francis Group, Boca Raton, FL.
[5] Sladek, K.J., Gilliland, E.R. and Baddour, R.F. (1974) Diffusion on surface. II. Correlation of diffusivities of physically and chemically adsorbed species. *Industrial & Engineering Chemistry Fundamentals*, **13** (2), 100–105.
[6] Pesaran, A.A. and Mills, A.F. (1987) Moisture transport in silica gel packed beds – II. Experimental study. *International Journal of Heat and Mass Transfer*, **30** (6), 1051–1060.
[7] Muralidhara, H.S. and Ensminger, D. (1986) Acoustic drying of green rice. *Drying Technology*, **4** (1), 137–143.
[8] Kudra, T. and Mujurndar, A.S. (2009) Advanced drying technologies: Chapter 13, in *Sonic Drying*, 2nd edn, CRC Press, Taylor & Francis, Oxford.
[9] Lee, C.P. and Wang, T.G. (1990) Outer acoustic streaming. *Journal of the Acoustical Society of America*, **88** (5), 2367–2375.
[10] Hyun, S. and Lee, D.R. (2005) Byoung-Gook Loh. Investigation of convective heat transfer augmentation using acoustic streaming generated by ultrasonic vibrations. *International Journal of Heat and Mass Transfer*, **48** (3–4), 703–718.
[11] Crank, J. (1979) *The Mathematics of Diffusion*, 2nd edn, Clarendon Press, Oxford.
[12] Nuh, D.N. and Brinkworth, B.J. (1997) Novel thin-layer model for crop drying. *Transactions of the American Society of Agricultural Engineers*, **40** (3), 659–669.
[13] Panades, G., Castro, D., Chiralt, A. *et al.* (2008) Mass transfer mechanisms occurring in osmotic dehydration of guava. *Journal of Food Engineering*, **87** (3), 386–390.
[14] Lu, H., Gong, X., Wang, C. *et al.* (2008) Effect of vibration on water transport through carbon nanotubes. *Chinese Physics Letter*, **25** (3), 1145–1148.
[15] Rege, S.U., Yang, R.T. and Cain, C.A. (1998) Desorption by ultrasound: phenol on activated carbon and polymeric resin. *AICHE Journal*, **44** (7), 1519–1528.

[16] Midilli, A. and Kucuk, H. (2003) Mathematical modeling of thin layer drying of pistachio by using solar energy. *Energy Conversion and Management*, **44** (7), 1111–1122.

[17] Sogi, D.S., Shivhare, U.S., Garg, S.K. and Bawa, A.S. (2003) Water sorption isotherm and drying characteristics of tomato seeds. *Biosystems Engineering*, **84** (3), 297–301.

[18] Marabi, A., Dilak, C., Shah, J. and Saguy, L.S. (2004) Kinetics of solids leaching during rehydration of particulate dry vegetables. *Journal of Food Science*, **69** (3), 91–96.

[19] Akpinar, E.K., Bicer, Y. and Cetinkaya, F. (2006) Modeling of thin layer drying of parsley leaves in a convective dryer and under open sun. *Journal of Food Engineering*, **75** (3), 308–315.

[20] Mcminn, W.A.M. (2006) Thin-layer modeling of the convective, microwave, microwave-convective and microwave-vacuum drying of lactose powder. *Journal of Food Engineering*, **72** (1), 113–123.

[21] Sacilik, K. (2007) Effect of drying methods on thin-layer drying characteristics of hull-less seed pumpkin. *Journal of Food Engineering*, **79** (1), 23–30.

[22] Khazaei, J. and Daneshmandi, S. (2007) Modeling of thin-layer drying kinetics of sesame seeds: mathematical and neural networks modeling. *International Agrophysics*, **21** (4), 335–348.

[23] Kocijan, J., Girard, A., Banko, B. and Murray-Smith, R. (2005) Dynamic systems identification with Gaussian processes. *Mathematical and Computer Modelling of Dynamical Systems*, **11** (4), 411–424.

[24] Pedreno-Molina, J.L., Monzo-Cabrera, J., Toledo-Moreo, A. and Sanchez-Hernandez, D. (2005) A novel predictive architecture for microwave-assisted drying processes based on neural network. *International Communications in Heat and Mass Transfer*, **32** (8), 1026–1033.

[25] Crighton, D.E., Dowling, A.P., Williams, J.E.F. *et al.* (1992) *Modern Methods in Analytical Acoustics*, Springer, New York.

[26] Kumar, T.S., Patle, M.K. and Sujith, R.I. (2007) Characteristics of acoustic standing waves in packed bed columns. *AICHE Journal*, **53** (2), 297–304.

[27] Morse, R.W. (1952) Acoustic propagation in granular media. *The Journal of the Acoustical Society of America*, **24** (6), 696–700.

[28] Kinsler, L.E., Frey, A.R., Coppens, A.B. and Sanders, J.V. (1999) *Fundamentals of Acoustics*, 4th edn, John Wiley & Sons, Inc., New York.

[29] Zwikker, C. and Kosten, C.W. (1949) *Sound Absorbing Materials*, Elsevier Publishing Company, New York.

[30] Biot, M.A. (1956) Theory of propagation of elastic waves in a fluid: saturated porous solid. II. Higher frequency range. *The Journal of the Acoustical Society of America*, **28** (2), 179–191.

[31] Buck, A.L. (1981) New equations for computing vapor pressure and enhancement factor. *Journal of Applied Meteorology*, **20** (12), 1527–1532.

[32] Xu L.H., Yan J. *Fundamental of Thermal Engineering and Industrial*. Kiln and Metallurgical Industry Press, Beijing, 2006 (in Chinese).

[33] Lin, S. and Zhang, F. (2000) Measurement of ultrasonic power and electro-acoustic efficiency of high power transducers. *Ultrasonics*, **37** (8), 549–554.

[34] Awad, M.M., Ramzy, A., Hamed, A.M. and Bekheit, M.M. (2008) Theoretical and experimental investigation on the radial flow desiccant dehumidification bed. *Applied Thermal Engineering*, **28** (1), 75–85.

[35] Chang, K.S., Wang, H.C. and Chung, T.W. (2004) Effect of regeneration conditions on the adsorption dehumidification process in packed silica gel beds. *Applied Thermal Engineering*, **24** (5–6), 735–742.

[36] Herrera, C.A., Levy, E.K. and Ochs, J. (2002) Characteristics of acoustic standing waves in fluidized beds. *AICHE Journal*, **48** (3), 503–513.

[37] Jakevičius, L. and Demcenko, A. (2008) Ultrasound attenuation dependence on air temperature in closed chambers. *Ultrasound*, **63** (1), 18–22.

[38] Sujith, R. and Zinn, B. (2000) A theoretical investigation of enhancement of mass transfer from a packed bed using acoustic oscillations. *The Canadian Journal of Chemical Engineering*, **78** (6), 1145–1150.

[39] García-Pérez J.V., Rossello C., Carcel J.A., Fuente S de la, Mule A. Effect of air temperature on convective drying assisted by high power ultrasound. *Defect and Diffusion Forum*, 2006, 258–260 (10): 563–574.

[40] Sharma, G. and Prasad, S. (2006) Specific energy consumption in microwave drying of garlic cloves. *Energy*, **31** (12), 1921–1926.

[41] Baker, C. and Mckenzie, K. (2005) Energy consumption of industrial spray dryers. *Drying Technology*, **23** (1), 365–386.

3

Ultrasound-Assisted Regeneration for a New Honeycomb Desiccant Material

3.1 Brief Introduction

In rotating and honeycomb-form dry desiccant systems, the state-of-the-art contactor is a silica gel-impregnated rotor consisting mainly of fibrous glass paper with a silicate binder. The cylindrical rotating wheel is obtained through rolling up sheets of a supporting material coated with an adsorbent substance in order to get a large number of parallel channels with typically a sinusoidal or triangular cross-sectional geometry, as shown in Figure 3.1. The main materials used as support are usually paper, aluminum, synthetic fibers, or plastic and common adsorbents are silica gel and activated alumina. Two air streams pass through the cross-sectional area of the device: the process air, which is dehumidified and heated, and the regeneration air, which removes water from the adsorbent material. The two streams are always arranged as counter current flows.

In this chapter, the effect of ultrasonic radiation on a new macro-porous desiccant (a kind of paper-basis material coated with nano-particles silica gel) based honeycomb matrix is investigated. According to the information provided by the manufacturer, the new honeycomb desiccant has a water adsorption of more than 60% at the ambient relative humidity of 100%, a pore diameter in the range of 40 to about 200 Å (angstrom) and pore volume from about 0.40 to about 0.80 cm^3/g. The preferred regeneration temperature ranges from 50 to 70 °C. The objectives of this chapter mainly include: (i) to study the regeneration characteristics of the new honeycomb type desiccant under different regeneration conditions with ultrasonic radiation; (ii) to make a quantitative analysis on the contributions of the "vibration effect" and the "heating effect" of the high-intensity ultrasound to the regeneration enhancement; and (iii) to establish a computational model to predict the regeneration efficiency of the honeycomb-type desiccant in the presence of ultrasonic radiation.

Ultrasonic Technology for Desiccant Regeneration, First Edition. Ye Yao and Shiqing Liu.
© 2014 Shanghai Jiao Tong University Press. All rights reserved. Published 2014 by John Wiley & Sons Singapore Pte Ltd.

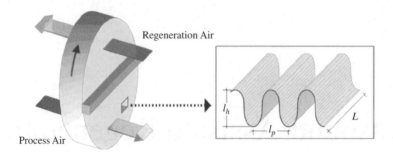

Figure 3.1   Honeycomb desiccant wheel and channel schemes

3.2   Experimental Study

3.2.1   Experimental System

The schematic diagram and photo of experimental system are presented in Figures 3.2, 3.3 and 3.4, respectively. The basic information of the instruments is listed in Table 3.1. As seen from Figure 3.2, the experiments of regeneration with and without ultrasonic radiation can be performed simultaneously with the experimental system, meanwhile, the air conditions before the two desiccant beds can be easily kept consistent. The inlet air conditions (temperature and humidity) for the regeneration experiments can be acquired through adjusting the electric heater and the humidifier. The air flow rate in the two branch ducts can be kept consistent through adjusting the air valve. The ducts in the experimental system are thermally insulated to reduce the ambient disturbances and maintain the stable regeneration conditions. In addition, the distance between the T-bend of duct and the air flow sensor is long enough to ensure a laminar air flow passing through the air flow sensor and improve the measurement precision (Figure 3.4).

3.2.2   Raw Material and Experimental Conditions

The size of samples is $65 \times 60 \times 60$ mm, and the thickness of the monolayer material is 0.16 mm. Two kinds of sample with different pitch sizes, that is, SS-type and M-type (see Table 3.2), are investigated experimentally in this study.

The experimental regeneration conditions are listed as below:

1. Initial moisture ratio of the trial desiccant sample for the regeneration experiments is identically given as 0.35 kg/(kg dry sample);
2. Inlet air conditions include three temperatures (i.e., 40, 45, and 50 °C) combined with two humidity ratios (i.e., 0.008 and 0.012 kg/(kg DA)) and three air flow velocities (i.e., 0.5, 0.7, and 0.9 m/s);
3. Frequency and indicator power of ultrasound employed for this experimental study are 23 kHz and 40 W (or 60 W), respectively;
4. Regeneration time identically lasts for 10 minutes.

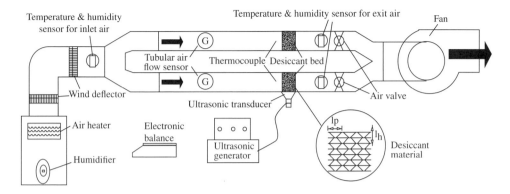

Figure 3.2   Schematic diagram of the experimental system [1]

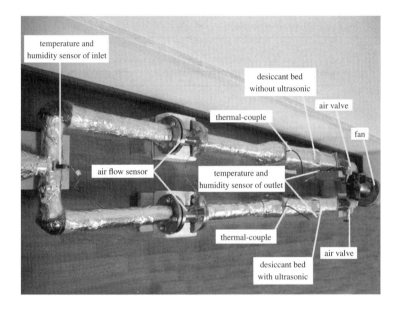

Figure 3.3   Photo of the experimental system

Ultrasonic producer   Ultrasonic transducer

Figure 3.4   Photo of the experimental system and relevant equipments. (a) Ultrasonic producer and (b) ultrasonic transducer

Table 3.1 Instruments and equipments in the experimental study

Item	Number	Basic information	Type/manufacturer
Air flow sensor	2	Measurement precision: ±0.3% of the measured data	ETFM-20/ Shanghai Cao Yi Ltd, China
Temperature and humidity sensor	3	Measurement precision: ±0.8 in relative humidity (%); ±0.1 °C in temperature	HHC2-S/ OTRONIC Ltd, Sweden
Electronic balance	1	Measurement precision: ±0.1 g in absolute error	BS1500M /Shanghai Yousheng Ltd
Ultrasonic producer	1	Ultrasonic frequency: 0–100 MHz, ultrasonic power: 0–300 W	UGD/Taheda Hi-tech Company, China
Ultrasonic transducer	3	Compatible frequency: 19 kHz Maximum power input: 500 W	UGD/Taheda Hi-tech Company, China
Oscilloscope	1	Measurement precision: 2 mV	DS1052E/Pu Yuan Ltd, China
Centrifugal fan	1	Rated power: 800 W Wind pressure: 700 Pa	SYDF/Shanghai Yingda Fan Company

Table 3.2 Basic properties of material sample

Structural parameter	SS-type	M-type
Height of air passage L_h (mm)	0.9	1.7
Width of air passage l_p (mm)	2.2	3.4
Specific surface area per unit volume (m^2/m^3)	5100	2800
Specific heat (kJ/(kg·K))	2.0	2.0

To guarantee the same initial moisture ratio in the desiccant sample for all the experiments of regeneration, the desiccant samples were firstly dried by an electric oven with 90 °C for 2.5 hours, and then were humidified to the expected moisture ratio (0.35) before the experiments.

3.2.3 Analysis Parameters

To estimate the influence of ultrasound on the regeneration process, the additional moisture removed (AMR) and the enhancement percentage of regeneration (EP) brought about by ultrasound over a period of time of regeneration (τ) are given by Equations (3.1) and (3.2), respectively:

$$AMR(\tau) = \int_0^\tau RR_U d\tau - \int_0^\tau RR_{NU} d\tau, \tag{3.1}$$

$$EP = \frac{AMR_\tau}{\displaystyle\int_0^\tau RR_{NU} d\tau} \times 100\%, \tag{3.2}$$

where superscript "U" and "NU" denote the regeneration with and without ultrasound, respectively; *RR* is regeneration rate (i.e., moisture desorption rate) in kg/s, which can be calculated by Equation (3.3).

$$RR = G_a \left(w_{a,out} - w_{a,in} \right), \qquad (3.3)$$

where G_a is mass flow rate of regeneration air in kg/s and $w_{a,in}$ and $w_{a,out}$ are air humidity (in kg *water*/(kg dry air)) entering and leaving the desiccant bed, respectively.

The maximum energy efficiency of ultrasound (MEEU) for any specific regeneration conditions is used to determine the optimal ultrasonic time during the regeneration, which is defined as below:

$$MEEU = \frac{MAMR \times H_{des}}{P_U \cdot \tau} \times 100\%, \qquad (3.4)$$

where, P_U is ultrasonic power (in W) input for the regeneration; τ is regeneration time in s; H_{des} is moisture desorption heat from the desiccant material in J/kg; and *MAMR* is the maximum additional moisture removal (in kg) brought about by ultrasound during the regeneration, which is expressed by,

$$MAMR = \max \left[AMR \left(\tau \right) \right]. \qquad (3.5)$$

The average energy efficiency (AEE) is used to investigate the energy-saving features of the ultrasound-assisted regeneration for the new type material, which is defined as below:

$$AEE = \frac{\left(\int_0^{\tau} RR d\tau \right) \times H_{des}}{P_{total} \cdot \tau} \times 100\%. \qquad (3.6)$$

For the absence of ultrasound,

$$P_{total} = G_a c_a \left(t_{a,reg} - t_{a,env} \right) / \eta_{heater} \qquad (3.7)$$

and for the presence of ultrasound,

$$P_{total} = G_a c_a \left(t_{a,reg} - t_{a,env} \right) / \eta_{heater} + P_U, \qquad (3.8)$$

where P_{total} is total energy power input for the regeneration in W; G_a is regeneration air flow rate in kg/s; c_a is specific heat of air in J/(kg·°C); and $t_{a,reg}$ and $t_{a,env}$ are regeneration air temperature and ambient air temperature (in °C), respectively. η_{heater} is heating efficiency of the air heater, normally 0.8–0.9.

According to the theory of error analysis, the experimental relative error of regeneration rate (ERERR) can be written as:

$$ERERR = \frac{\Delta RR}{RR} \times 100\% = \sqrt{ \left(\frac{\Delta G_a}{G_a} \right)^2 + \left(\frac{\Delta w_{a,out}}{w_{a,out} - w_{a,in}} \right)^2 + \left(\frac{\Delta w_{a,in}}{w_{a,out} - w_{a,in}} \right)^2 } \times 100\%,$$
$$(3.9)$$

where "Δ" stands for the absolute error.

3.2.4   Experimental Results

3.2.4.1   Equilibrium Regeneration

To investigate the effect of ultrasound on the moisture desorption of desiccant material during the whole regeneration process, the equilibrium regeneration has been experimentally

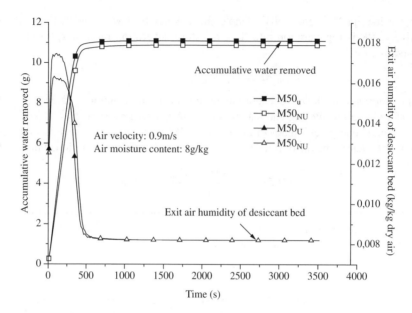

Figure 3.5   Equilibrium regeneration of M-type material under certain regeneration conditions

performed. The results are presented in Figure 3.5, where the legend "$M50_U$" stands for the regeneration of M-type material under the regeneration temperature of 50 °C plus ultrasound, and "$M50_{NU}$" for that without ultrasound. The results show that it takes about 600 seconds for the desiccant material to arrive at the quasi-equilibrium state under the regeneration air conditions of 50 °C, 0.9 m/s, and 8.0 g/(kg dry air). It can be found as well that after the quasi-equilibrium state, the ultrasonic will produce little effect on the regeneration. Therefore, in the following sections, only the experimental results of 600-second regeneration with and without ultrasound are to be discussed.

3.2.4.2   Experimental Results of Exit Air Humidity and Regeneration Rate

Figure 3.6 shows the experimental results of the exit air humidity during the regeneration of M- and SS-type material with and without ultrasound under different regeneration air temperatures (the other conditions: UP = 40 W; $w_{a,i}$ = 8.0 g/(kg dry air); u_a = 0.7 m/s). Based on the experimental exit air humidity and the other conditions, the regeneration rates (RRs) of the two types of material under different regeneration conditions with and without ultrasound are obtained, respectively, as shown in Figure 3.7. The symbol "M" and "SS" in the legend represent the M-type and the SS-type material, respectively; the number (e.g., 40, 45, and so on) in the legend represents the regeneration air temperature (e.g., 40, 45 °C, and so on); and the subscript "with" and "without" stand for the case with and without ultrasonic radiation.

   It can been seen from Figures 3.6 and 3.7 that the variation curves of the exit air humidity and that of the regeneration rate all experience the following three stages: (i) the rapidly rising

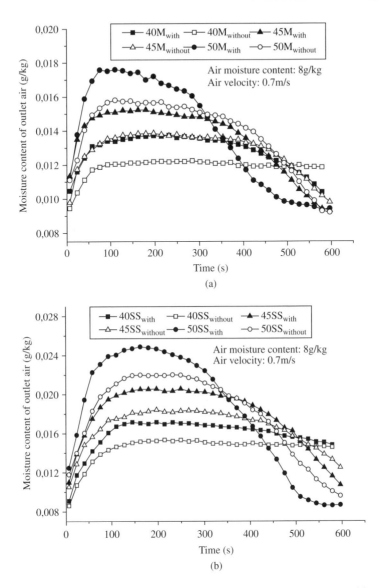

Figure 3.6 Experimental results of exit air humidity during the regeneration of M- and SS-type material with and without ultrasound under different inlet air temperatures (UP = 40 W; $w_{a,i}$ = 8.0 k/(kg dry air); u_a = 0/7 m/s). (a) M-type material and (b) SS-type material

stage; (ii) the relatively stable stage; and (iii) the rapidly dropping stage. And the conditions more beneficial for improving the desiccant regeneration (i.e., the higher inlet air temperature, the lower inlet air humidity ratio, and the bigger air flow rate.) will reduce the time of all the stages. Comparing the regeneration rate (*RR*) curves with and without ultrasound under the same regeneration conditions, it is found that the crossover point of the two curves usually

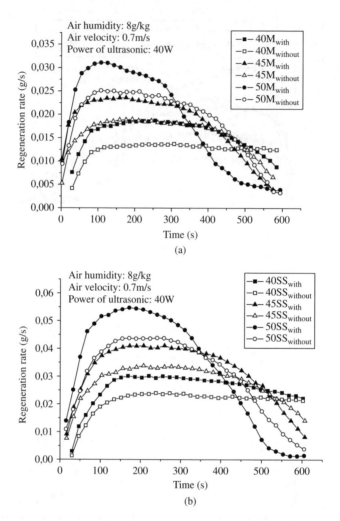

Figure 3.7 Comparisons of experimental R_r of M- and SS-type material with and without ultrasound under different inlet air temperatures (UP = 40 W; $w_{a,i}$ = 8.0 k/(kg dry air); u_a = 0/7 m/s). (a) M-type material and (b) SS-type material

occurs in the rapidly dropping stage of the *RR* curve with ultrasound. Moreover, the time of occurrence of the crossover point will be obviously shorter under the higher regeneration air temperatures. So is the same case for the lower regeneration air humidity ratios and the larger airflow rates. It indicates that the positive effect of ultrasound on the regeneration rate will largely depend on the regeneration air conditions and the regeneration time.

Figure 3.8 shows the experimental regeneration rate of SS-type material under different ultrasonic power levels (other regeneration conditions: $t_{a,i}$ = 45 °C; $w_{a,i}$ = 8.0 g/kg; u_a = 0.5 m/s). From Figure 3.8, it can be seen that the regeneration rate (*RR*) changes little as the ultrasonic power rises from 40 to 60 W. It indicates that there should be a threshold of ultrasonic

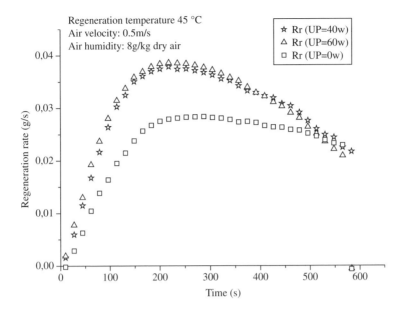

Figure 3.8   Experimental *RR* of SS-type material under different ultrasonic power levels

power above which the regeneration rate benefits from the increased power. The influence of the ultrasonic power on the regeneration rate will be further discussed afterwards with model simulations.

3.2.4.3   Enhancement of Regeneration Brought about by Ultrasound

The EP of regeneration brought about by ultrasound for different regeneration time and conditions can be obtained according to the experimental data. The results are shown in Figure 3.9 where the symbol "M0.5" and "SS0.5" stand for the regeneration of the M- and the SS-type material under the air flow rate of 0.5 m/s, respectively; "50H" stands for 50 °C and 12.0 g/(kg dry air) in the regeneration air temperature and humidity, respectively; and so on.

From Figure 3.9, we can see that:

1. The enhancement of desiccant regeneration brought about by the ultrasound will be affected by the regeneration time. In this case, the EP brought about by the ultrasound for the 5-minute regeneration is obviously higher than that for the 10-minute regeneration.
2. The enhancement will be affected as well by the regeneration conditions. As shown in Figure 3.9, the EP decreases with the increase of air flow rate. However, no regular patterns can be found about the influence of regeneration air temperature and humidity on the enhancement.
3. The structure of material will produce influence on the enhancement brought about by the ultrasound. It shows that the EPs for the M-type material are mostly higher than those for the SS-type one under the same regeneration conditions.

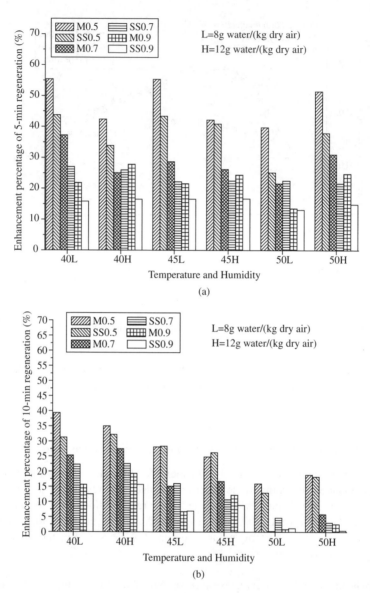

Figure 3.9 Comparisons of EP of regeneration brought about by ultrasound under different regeneration conditions. (a) 5-minute regeneration and (b) 10-minute regeneration

3.2.4.4 Energy Efficiency of Ultrasound for the Regeneration

Figures 3.10 and 3.11 present the variations of AMR (brought about by the ultrasound) versus the regeneration time under different regeneration conditions for the M- and the SS-type material, respectively. From these figures, we can see that:

1. The AMR increases firstly to a peak value, and then decreases, which indicates that there should be a best regeneration time for which the maximum additional moisture removal (MAMR) is acquired for any specific conditions of regeneration.

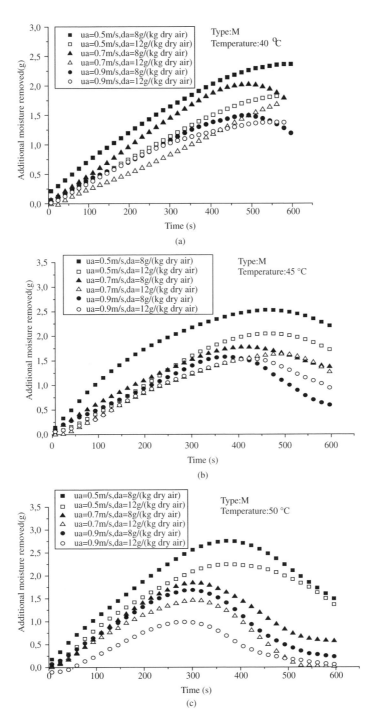

Figure 3.10 Variations of AMR for the M-type material under different regeneration conditions. (a) Regeneration air temperature: 40 °C, (b) regeneration air temperature: 45 °C, and (c) regeneration air temperature: 50 °C

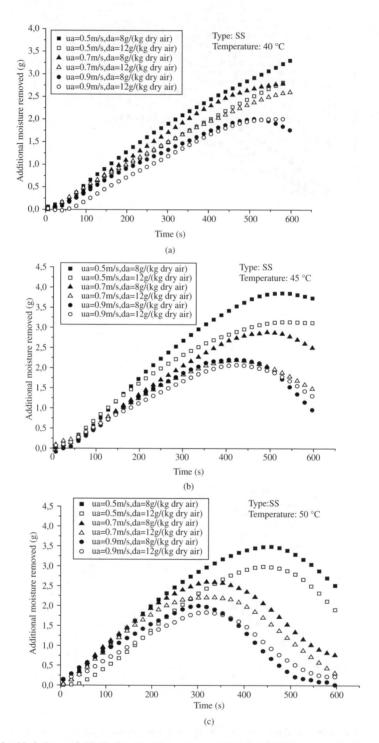

Figure 3.11 Variations of AMR for the SS-type material under different regeneration conditions. (a) Regeneration air temperature: 40 °C, (b) regeneration air temperature: 45 °C, and (c) regeneration air temperature: 50 °C

2. The AMR increases with the decrease of the airflow rate or regeneration air humidity ratio. It indicates that the effect of ultrasound on the enhancement of the desiccant regeneration will be more significant under lower airflow rate or lower air humidity.
3. The AMR for the SS type is larger than that for the M-type material under the same regeneration conditions, and it takes more time for the SS-type material to reach the MAMR.

The MAMR for different regeneration conditions can be obtained based on the curves of AMR, as shown in Figure 3.12. It can be found that the MAMR for the SS-type material is obviously higher than that for the M-type one for any specific conditions. Meanwhile, it will be affected by the regeneration conditions, that is, the regeneration air temperature and humidity as well as the airflow rate. The results show that the lower air humidity or the lower airflow rate results in the larger MAMR. However, there are no distinct regular patterns of the MAMR with the regeneration air temperature. In this case, the MAMR at 45 and 50 °C are mostly higher than that at 40 °C.

The MEEU under different regeneration conditions is shown in Figure 3.13. Basically, the MEEU for the SS-type material is higher than that for the M-type one. It can attain as high as 45% for the SS-type material under the conditions of 50 °C in the regeneration air temperature, 0.008 kg/(kg dry air) in the regeneration air humidity and 0.5 m/s in the airflow rate.

3.2.4.5   Average Energy Efficiency (AEE)

The results of AEE of the 10-minute regeneration under different regeneration conditions are shown in Figure 3.14. Overall, the AEE decreases with the increase of the regeneration air temperature, the regeneration air humidity or the airflow rate. Although the AEE of

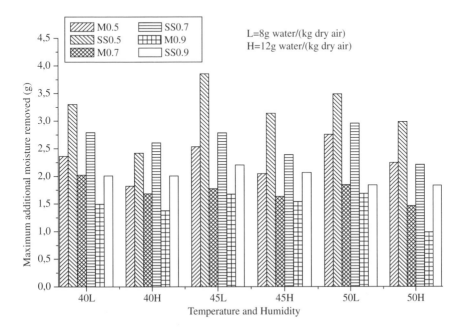

Figure 3.12   The MAMR brought about by ultrasound under different regeneration conditions

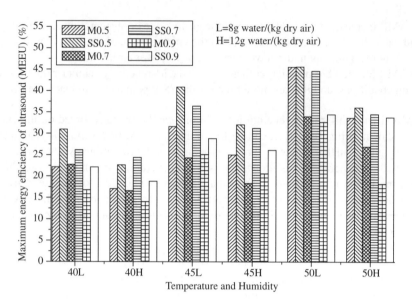

Figure 3.13 The MEEU brought about by ultrasound under different regeneration conditions

regeneration can be improved by applying the ultrasound, it is far from what we expected. The reason is that the acoustic conductivity of the desiccant materials used in this study is not very good, and this results in a rapid energy decay of ultrasound during its propagation in the medium. In a later section, the ultrasonic energy attenuation in the honeycomb-type desiccant will be further investigated.

3.2.4.6 Experimental Error Analysis

The *ERERRs* for the M-type material are shown in Figure 3.15 [1]. The "U" and "NU" denote the regeneration in the presence and absence of ultrasound, respectively. The "0.5," "0.7," and "0.9" in the legend describe different regeneration air velocities (m/s). The error analysis manifests that the experimental errors of the regeneration rate are normally lower than 15% in most of the regeneration time except for the beginning of the regeneration when the experimental errors are a little higher (15–25%).

3.2.5 Energy Attenuation and Absorptivity of Ultrasound in the Material

3.2.5.1 Acoustic Energy Attenuation Test

The schematic diagram for testing the acoustic energy attenuation in the desiccant material is given by Figure 3.16. Driven by an ultrasonic producer, the ultrasonic transmitter produces 23-kHz ultrasound with a steady intensity. The acoustic intensity can be sensed by an ultrasonic receiver and measured by an oscillograph recording system in which the sound intensity is expressed in the form of voltage.

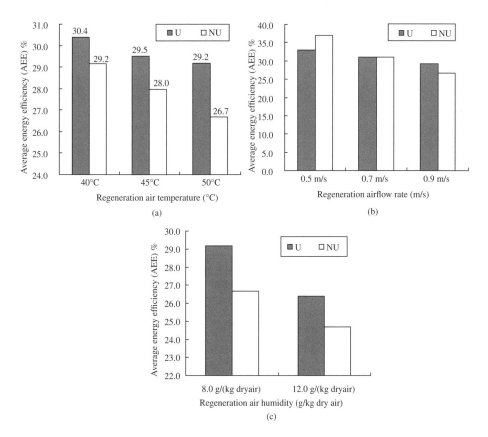

Figure 3.14 Average energy efficiency (AEE) under different regeneration conditions (10-minute regeneration). (a) Different regeneration air temperatures, (b) different regeneration airflow rates, and (c) different regeneration air humidity ratios

To investigate the influence of moisture ratio in honeycomb-type desiccant on the acoustic energy attenuation, the test has been performed for the SS-type material with different moisture ratios (i.e., q = 0.15, 0.25, and 0.35). Figure 3.17 gives the test results, which manifests that the energy attenuation of ultrasound in the medium is almost unaffected by the moisture ratio in the material.

Usually, the dimensionless acoustic energy in the material can be expressed by

$$P_u(x)/P_u(0) = \exp(-\xi \cdot x), \tag{3.10}$$

where $P_u(0)$ is ultrasonic power (in W) or sound pressure (in Pa) at the sound source, and $P_u(x)$ is that at the position being x (in cm) away from the sound source; ξ is attenuation coefficient of sound in medium.

The dimensionless acoustic energy in the M-type and the SS-type desiccant material are shown in Figure 3.18, and the ultrasonic attenuation coefficient, ξ, in the M-type and the SS-type material can be recognized as 2.2 and 1.6, respectively.

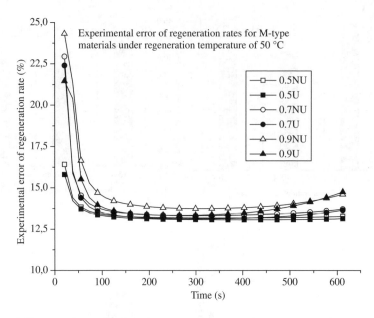

Figure 3.15   Experimental errors of regeneration rates during the regeneration process [1]

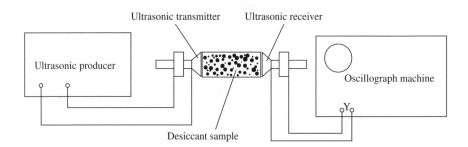

Figure 3.16   Schematic for acoustic energy attenuation test

3.2.5.2   Acoustic Energy Absorptivity Test

As known from Equation (2.74) in the earlier chapter, the key parameter for determining the ultrasonic absorptivity (a_u) is the temperature rise of medium caused by the ultrasonic radiation. So, the acoustic energy absorptivity of the honeycomb-type desiccant material can be obtained through observing the temperature rise of the material when it is subjected to the ultrasonic radiation for a specific time. Several temperature sensors (thermocouples) are employed for the acoustic energy absorptivity test of the honeycomb-type desiccant material, and the detailed locations of the temperature sensors in the material are shown in Figure 3.19.

Figure 3.20 presents the variations of temperatures at different locations in the SS-type material when subjected to the ultrasonic radiation starting from the 128th second. The ultrasonic

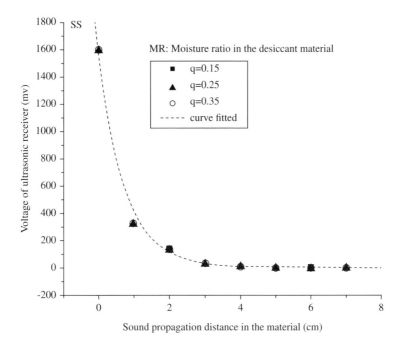

Figure 3.17   Influence of moisture ratio on ultrasonic energy attention in the honeycomb material

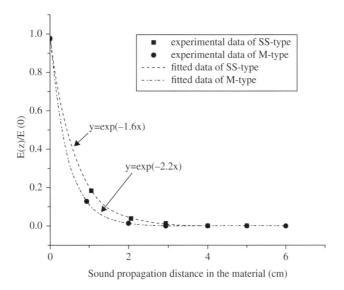

Figure 3.18   Dimensionless acoustic energy in the SS- and M-type honeycomb material

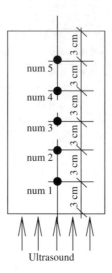

Figure 3.19   Locations of temperature sensors in the sample for acoustic energy absorptivity test

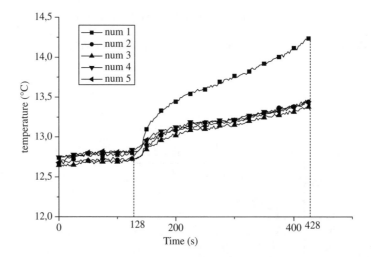

Figure 3.20   Temperature variations in the SS-type material when subjected to ultrasonic radiation at 128 seconds

radiation lasts totally for 5 minutes. It can be seen that the temperature at the location of num.1 rises much more rapidly than that at the other locations. The acoustic energy absorptivity is calculated in terms of the average change value of the five local temperatures and the other parameters including the mass of the test sample, the specific heat of the desiccant material and the ultrasonic power applied. The acoustic energy absorptivity for the M- and the SS-type desiccant material in this study are identified as about 0.022 and 0.032, respectively.

3.3 Theoretical Model for Honeycomb-Type Desiccant Regeneration

3.3.1 Basic Assumptions

To assist a better use of ultrasound in the honeycomb-type desiccant regeneration, it is necessary to develop a theoretical model that is capable of predicting the effect of ultrasound on the regeneration process under different conditions. The fundamental theory of heat and mass transfer model is employed as well, and the following assumptions are made for the model development:

1. There exists an air layer on the surface of material that has an equilibrium state with the regeneration air.
2. Moisture transfer is externally controlled, and the heat and mass transfer resistance inside the material is negligible.
3. Heat and mass transfer in the parallel direction is negligible.

3.3.2 Governing Equations

Likewise, the mass and energy balance equations for the regeneration air and desiccant materials can be expressed, respectively, as below:

$$\rho_a u_a(x)\left(-\frac{\partial w_a}{\partial x}\right) = (1-\varepsilon)\,\rho_s\frac{\partial q_s}{\partial \tau}, \tag{3.11}$$

$$-(1-\varepsilon)\,\rho_s\frac{\partial q_s}{\partial \tau} = K_m S_V\left(w_s{}^* - w_a\right), \tag{3.12}$$

$$\rho_a c_{p,a} u_a(x)\frac{\partial t_a}{\partial x} = H_m S_V\left(t_s - t_a\right) + K_m S_V c_{p,v}\left(w_s{}^* - w_a\right)\left(t_s - t_a\right), \tag{3.13}$$

$$(1-\varepsilon)\,\rho_s c_s\frac{\partial t_s}{\partial \tau} = H_m S_V\left(t_a - t_s\right) - H_{des} K_m S_V\left(w_s{}^* - w_a\right) + \frac{a_u \eta_T P_u}{(1-\varepsilon)\,V}. \tag{3.14}$$

In Equations (3.11)–(3.14), x is distance from the windward side of material in m; τ is time in s; ρ is density in kg/m^3; u is regeneration air velocity in m/s; q is moisture ratio in the medium; K_m is convective mass transfer coefficient in kg/(m^2·s); H_m is coefficient of convective heat transfer in W/(m^2·°C); a_u is coefficient of ultrasonic absorption by medium; w_a is humidity of regeneration air in kg/(kg dry air); w_s^* is humidity of air layer on the surface of solid desiccant in kg/(kg dry air); S_V is volumetric surface area of the honeycomb-type material in m^2/m^3; t is temperature in °C; c_p is constant pressure specific heat in J/(kg·°C); P_u is ultrasonic power in W; V is volume of desiccant in m^3; H_{des} is moisture desorption heat from the desiccant material in J/kg; η_T is working efficiency of ultrasonic transducer; and ε is void fraction of the desiccant material. The subscript "a," "s," and "v" denote regeneration air, solid desiccant, and water vapor, respectively.

The boundary and initial conditions should be given for solving the above equations. These include initial moisture content in the desiccant $q_s(x,0) = q_{s0}$, initial temperature of desiccant $t_s(x,0) = t_{s,0}$, inlet air humidity of desiccant bed $w_a(0,\tau) = w_{a,0}$, and inlet air temperature of desiccant bed $t_a(0,\tau) = t_{a,0}$.

3.3.3   Determination of Key Parameters

The convective mass transfer coefficient, K_m (in kg/(m^2·s)), is defined as [1]:

$$K_m = \frac{RR}{S_V V \left(w_s^* - w_a\right)}, \tag{3.15}$$

where RR is regeneration rate in kg/s.

The relationship between the coefficient of convective mass transfer (K_m) and the coefficient of convective heat transfer (H_m) can be expressed by Equation (3.16) by assuming Lewis number to be one.

$$K_m = \frac{H_m D_v}{\lambda_a}, \tag{3.16}$$

where D_v is mass diffusivity coefficient of water vapor in the air, 2.4×10^{-5} m^2/s. λ_a is thermal conductivity of air, 0.023 W/(m·°C).

By using the method illustrated in the previous chapter, the convective heat transfer coefficient, H_m, can be obtained through experiments as below [2]:

$$H_m = 0.683 \rho_a u_a c_{p,a} \text{Re}^{-0.51}, \tag{3.17}$$

where Re is the Reynolds number, $\text{Re} = \rho_a \bar{u}_a d_e / \mu_a$. The equivalent diameter of a single air passage in the material, d_e, can be calculated by Holman [3],

$$d_e = \frac{4 \times \left(0.5 L_{\text{height}} L_{\text{width}}\right)}{L_{\text{width}} + \sqrt{4 L_{\text{height}}^2 + L_{\text{width}}^2}}, \tag{3.18}$$

where L_{height} and L_{width} are the height (in m) and the width (in m) of a single air passage in the honeycomb desiccant material, respectively.

The average velocity of air ($\bar{u}_a$) is the composed result of regeneration air flow rate (u_a) passing through the honeycomb-type desiccant and the vibration speed induced by ultrasound (u_s), which is written as:

$$\bar{u}_a = \sqrt{u_a^2 + u_s^2}. \tag{3.19}$$

Since there is a significant energy attenuation in the honeycomb-type desiccant, the vibration velocity induced by ultrasound (u_s) will change greatly along the direction of sound propagation (y), and it can be calculated by

$$u_s(y) = p_u(y) / Z_s, \tag{3.20}$$

where y is distance from the sound source. p_u is sound pressure in Pa; Z_s is acoustic impedance of ultrasound in medium in N·s/m^5, which is defined as:

$$Z_s = \rho_s c_o, \tag{3.21}$$

where ρ_s is density of desiccant material in kg/m^3 and c_o is sound propagation velocity in the material in m/s.

The sound pressure at the sound source ($y = 0$) is given as [1]:

$$p_u(0) = \sqrt{\left(P_u \eta_T / S\right) \cdot Z_s}, \tag{3.22}$$

where P_u is power required for producing ultrasound in W and S is area of ultrasonic radiation in m^2. Thus, $p_u(z)$ can be calculated by using Equation (3.10).

The absolute humidity of air layer on the surface of material (w_s^*: kg/(kg dry air)) can be obtained based on its corresponding relative humidity (φ^*:%) which can be expressed as a function of moisture ratio in the desiccant (q_s) given as below [4]:

$$\varphi^* = s_1 t_s q_s^2 + s_2 t_s q_s + s_3 q_s^4 + s_4 q_s^3 + s_5 q_s^2 + s_6 q_s, \tag{3.23}$$

where t_s is temperature of material in °C and s_1, s_2, s_3, s_4, s_5, and s_6 are empirical coefficients, which have been obtained as 1.34926, 2.4988, 577.3534, -191.3, -596.76, and 374.1, respectively, by using the experimental method illustrated in the previous chapter.

3.3.4   Model Validation

The model is validated by experiments in terms of the regeneration rate. The comparisons of regeneration rate between the theoretical results and the experimental data under different conditions are shown in Figure 3.21 for the M-type desiccant and Figure 3.22 for the SS-type desiccant, respectively. The subscript "L" and "H" stand for 0.008 and 0.012 kg/(kg dry air) in the regeneration air humidity, respectively; and "U" and "NU" denote the regeneration process with and without ultrasonic radiation, respectively. From these figures, several points can be summarized as below:

1. All the regeneration rates under different regeneration conditions share a similar process, that is, increase to a peak value in the beginning and drop afterwards. As mentioned earlier, the regeneration rate will be affected by the temperature and moisture content of desiccant which change with the regeneration time. At the beginning of regeneration, the material's temperature increases due to the heat transfer from the regeneration air to the desiccant, and this causes the regeneration rate to rise. But the moisture content in the desiccant drops with the regeneration carrying on, which results in the decrease of regeneration rate. When the moisture ratio drops to a critical point, the influence of material's temperature on the regeneration rate becomes less dominant than that of moisture content in the desiccant. As a result, the regeneration rate begins to drop.
2. In the former stage of regeneration, the curves of regeneration rate for the presence of ultrasound are all above that without ultrasound. However, the situation is opposite in the latter stage. This is because the regeneration rate with ultrasound is faster than that without ultrasound in the beginning stage, which makes the moisture content of the desiccant in the presence of ultrasonic radiation drop more rapidly and arrive at the critical point (after which the regeneration rate starts to decline) earlier than that without ultrasound.
3. There occurs a crossover point between the regeneration rate curves for the presence and the absence of ultrasound under any specific regeneration conditions. The time of occurrence decreases with the increase of the regeneration air temperature or the airflow rate and the decrease of the regeneration air humidity. The time for the crossover point can be taken as the valid effect time of ultrasound in the regeneration. From the results, it can be seen that valid effect time of ultrasound will be shorter under the higher regeneration air temperature or the lower regeneration air humidity.

4. The theoretical results are very close to the experimental data including the trend of variation with the regeneration time. As shown in Figure 3.23, the relative errors of theoretic results compared with the experimental data in terms of the total amount of moisture desorption in the 10-minute regeneration are all lower than 7.0% for these cases. Therefore, we can safely conclude that the theoretical model employed here has a good accuracy in predicting the dehydration kinetics of the honeycomb-type desiccant during the ultrasound-assisted regeneration.

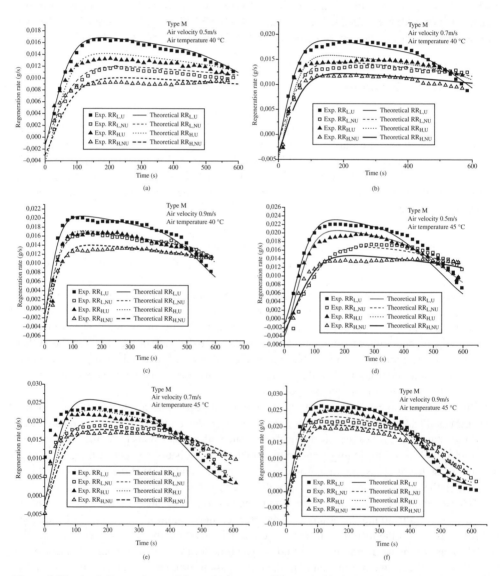

Figure 3.21 (a–i) Comparisons of theoretical regeneration rate with the experimental one for M-type material under different regeneration conditions

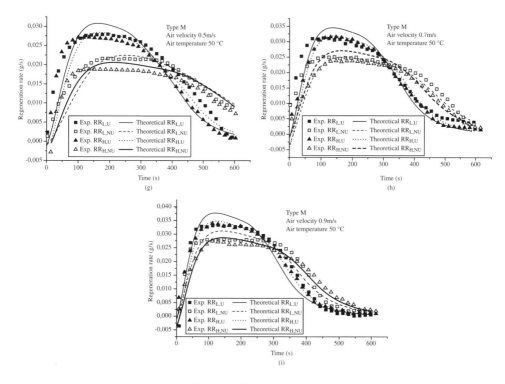

Figure 3.21 (*continued*)

3.4 Model Simulations and Analysis

3.4.1 Parametric Study

3.4.1.1 Different Regeneration Air Temperatures

Taking the SS-type material, for example, the five regeneration temperatures (i.e., 35, 40, 45, 50, and 55 °C) are investigated under the other regeneration conditions as below: 0.35 in the initial moisture ratio of the desiccant, 0.5 m/s in the regeneration air flow velocity passing through the honeycomb-type material, 8.0 g/(kg dry air) in the regeneration air humidity. The ultrasound used for assisting the regeneration is of 40 W in power and 19 kHz in frequency. The results of the regeneration rate for these regeneration temperatures are presented in Figure 3.24. It can be seen from Figure 3.24 that the regeneration rate increases with the increase of the regeneration air temperature in the former stage of regeneration (about 500 seconds), and the maximum difference between the regeneration rates in the presence ultrasound and that in the absence of ultrasonic radiation increases as well with the regeneration air temperature. Besides, the slope of regeneration rate against the time becomes larger under the higher regeneration air temperature, and it takes almost the same time for the regeneration rate to arrive at the peak value for different regeneration air temperatures.

The influences of regeneration air temperature on the MEEU in terms of different regeneration time (i.e., 240, 480, 720, and 960 seconds) are investigated under the specific regeneration

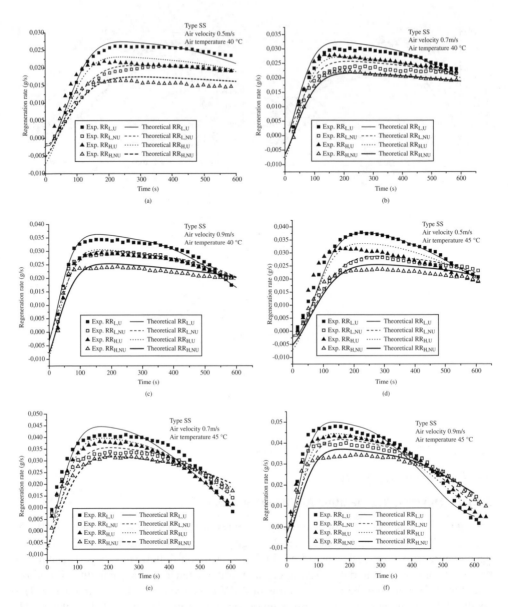

Figure 3.22 (a–i) Comparisons of theoretical regeneration rate with the experimental one for SS-type material under different regeneration conditions

conditions, as shown in Figure 3.25. It can be seen that the MEEU for the 240-second regeneration is obviously higher than that for the other longer-time regenerations (e.g., 480, 720, and 960 seconds). Meanwhile, the change patterns of MEEU with the regeneration temperature will be affected by the regeneration time. For the 240-second regeneration, the MEEU increases with the rising of regeneration temperature, while it is the opposite as the regeneration time is 960 seconds.

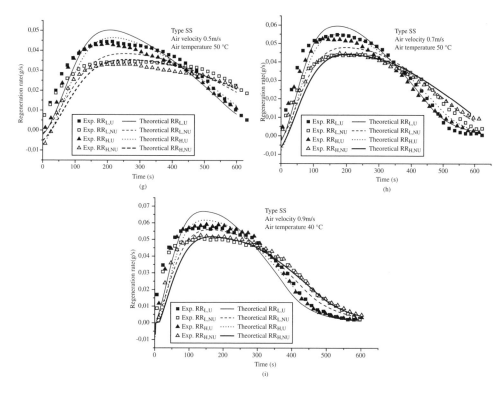

Figure 3.22   (*continued*)

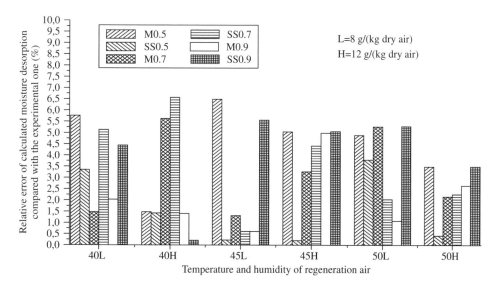

Figure 3.23   The relative errors of theoretic results compared with the experimental data of total amount of moisture desorption in the 10-minute regeneration

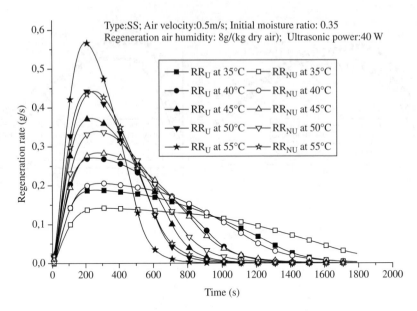

Figure 3.24 Theoretical regeneration rates of SS-type material under different regeneration air temperatures

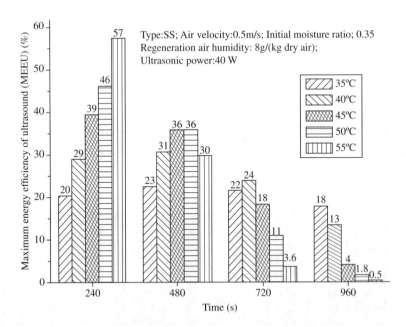

Figure 3.25 Theoretical maximum energy efficiency of ultrasound under different regeneration air temperatures

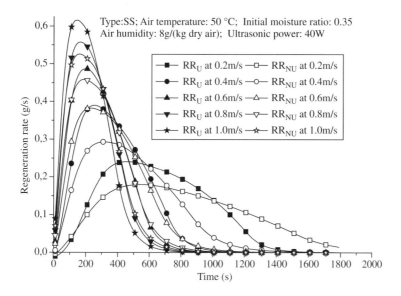

Figure 3.26　Theoretical regeneration rates of SS-type material under different air flow velocities

3.4.1.2　Different Regeneration Air Flow Rates

The regeneration rates under different regeneration air flow velocities (0.2, 0.4, 0.6, 0.8, and 1.0 m/s) are simulated as shown in Figure 3.26. It can be seen that in the rising stage, the larger regeneration air flow velocity result in the larger regeneration rate and the steeper slope of regeneration rate against the time, and the time for the regeneration rate to attain the peak value reduces as the air flow velocity increases.

The influences of the regeneration air flow velocity on the MEEU in terms of different regeneration time (i.e., 240, 480, 720, and 960 seconds) are investigated under the specific regeneration conditions as shown in Figure 3.27. It can be found that there should be a best air flow rate under which the largest MEEU can be acquired for any specific regeneration conditions and regeneration time. Among these cases, the best air flow rate is around 0.6 and 0.4 m/s for the 240 and 480-second regeneration, respectively. And when the regeneration time exceeds 720 seconds, the MEEU decreases with the increase of the air flow velocity.

3.4.1.3　Different Ultrasonic Power Levels

Figure 3.28 shows the regeneration rates of the SS-type desiccant under different ultrasonic power levels (i.e. 20, 30, 40, 50, and 60 W). Evidently, the higher ultrasonic power will surely bring about the larger regeneration rate of desiccant. However, the influence of ultrasonic power on the regeneration rate is much smaller than that of the regeneration air temperature or the regeneration air flow velocity. Meanwhile, the higher ultrasonic power may result in the reduction of the energy efficiency. As seen from Figure 3.29, the MEEU drops with the ultrasonic power increasing.

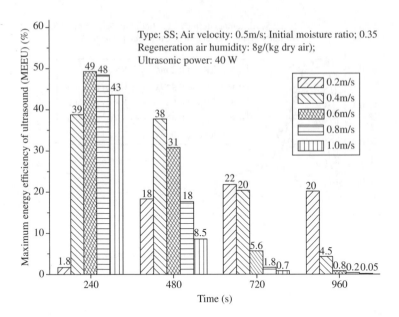

Figure 3.27   Theoretical maximum energy efficiency of ultrasound under different air flow velocities

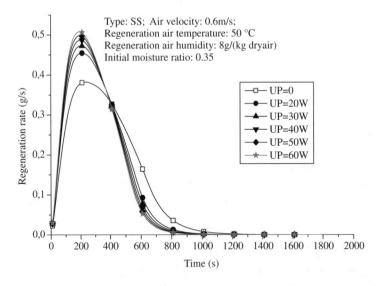

Figure 3.28   Theoretical regeneration rates of SS-type material under different ultrasonic power levels

3.4.1.4   Different Regeneration Air Humidity Ratios

Figure 3.30 presents the regeneration rates of the SS-type desiccant under different regeneration air humidity ratios (i.e., 6.0, 8.0, 10.0, 12.0, and 14.0 g/(kg dry air)). It manifests that the lower regeneration air humidity ratio will be conducive to improving the regeneration rate

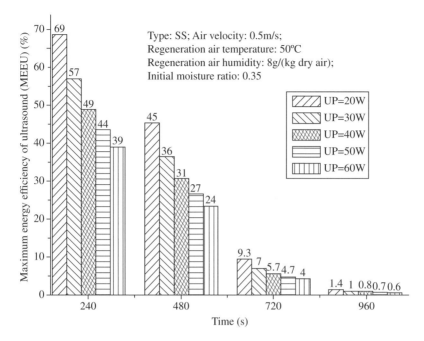

Figure 3.29 Theoretical maximum energy efficiency of ultrasound under different ultrasonic power levels

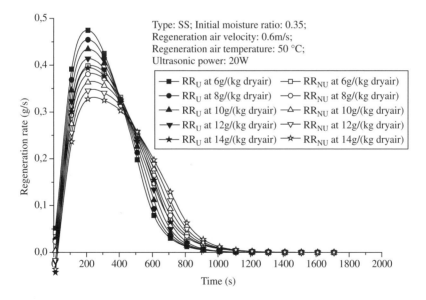

Figure 3.30 Theoretical regeneration rates of SS-type material under different regeneration air humidity ratios

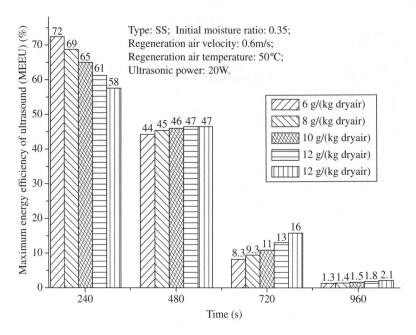

Figure 3.31 Theoretical maximum energy efficiency of ultrasound under different regeneration air humidity ratios

in the former stage of regeneration (about 400 seconds). In addition, the change rate of the regeneration rate against the time is larger under the lower air humidity ratio.

The influences of regeneration air humidity on the MEEU in the regeneration are shown in Figure 3.31 from which we can see that the lower regeneration air humidity is favorable for improving the MEEU for the 240-second regeneration. But with the regeneration time increasing, the situation will be the opposite.

3.4.1.5 Different Moisture Ratios in Desiccant

The initial moisture ratio is another important factor that affects the regeneration rate of desiccant. As seen from Figure 3.32, the larger initial moisture ratio of desiccant will bring about the bigger peak value of the regeneration rate during the regeneration, and it will take more time to arrive at the peak value of the regeneration rate.

The MEEU under different initial moisture ratios of desiccant are presented in Figure 3.33. From Figure 3.33, we can see that the MEEU increases with the increase of the initial moisture ratio of desiccant in most cases. This indicates that a larger initial moisture ratio of desiccant is preferred for the ultrasound-assisted regeneration.

3.4.1.6 Different Desiccant Materials

The material structure will affect the ultrasound-assisted regeneration process and the MEEU in the regeneration. The three types of desiccants, that is, the SS-type, the S-type, and the

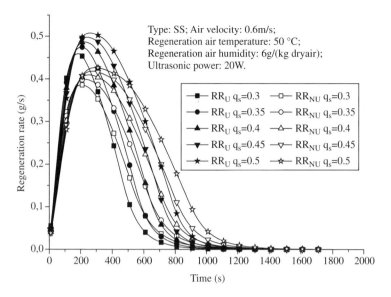

Figure 3.32   Theoretical regeneration rates of SS-type material under different initial moisture ratios of desiccant

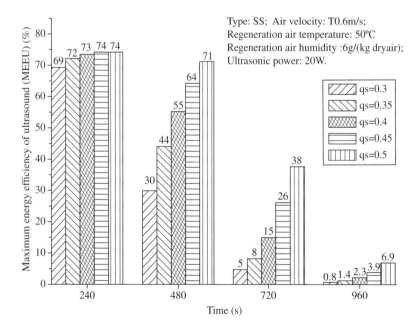

Figure 3.33   Theoretical maximum energy efficiency of ultrasound under different initial moisture ratios of desiccant

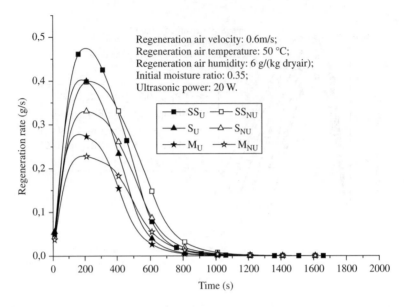

Figure 3.34   Comparisons of regeneration rates among different types of desiccants

M-type, are compared in terms of the regeneration rate and the MEEU under the specific regeneration conditions. The results are given in Figures 3.34 and 3.35, respectively. Among these types of desiccants, the SS-type is the most suitable for the ultrasound-assisted regeneration from the perspective of energy efficiency of regeneration.

3.4.1.7   Brief Summary

From the above results, the following points can be summarized:

1. A higher regeneration air temperature or a larger moisture ratio of desiccant will be conducive to improving the MEEU for the regeneration.
2. There ought to be a best air flow rate for any specific regeneration conditions under which the highest MEEU can be acquired.
3. The ultrasonic power employed for the regeneration should be selected aiming at the highest MEEU.
4. The smaller pore size of honeycomb-type desiccant is more favorable for ultrasound-assisted regeneration.

3.4.2   Quantitative Contributions of Ultrasonic Effects to the Regeneration of Honeycomb-Type Desiccant

The contributions of individual effect of ultrasound (i.e., the thermal effect and the mechanical vibration effect) to the regeneration of honeycomb-type desiccants are to be investigated under different regeneration conditions, and the analysis method illustrated in the previous chapter is

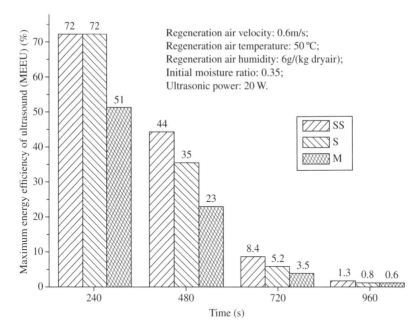

Figure 3.35 Theoretical maximum energy efficiency of ultrasound for the regeneration of different-type desiccants

again employed for this investigation. The results are presented in Figures 3.36 and 3.37, which manifests that the "thermal effect" of ultrasound contributes much less to the regeneration of honeycomb-type desiccants compared with the "mechanical vibration effect" of ultrasound. From Figure 3.36, it can be found that the contribution of "thermal effect" of ultrasound to the regeneration has a declining trend with the rising of regeneration air temperature. The regeneration air humidity may also affect the thermal effect of ultrasound to the regeneration. Under the lower air flow rate (e.g., 0.5 m/s), the contribution of ultrasonic thermal effect to the regeneration increases with the increase of regeneration air humidity. However, it is reversed under a higher air flow rate like 0.9 m/s.

As can be inferred from Figure 3.37, the higher regeneration air temperature will be conducive to enhancing the mechanical vibration effect of ultrasound to the regeneration. However, no regular pattern can be found about the influence of the other conditions, like the air humidity, the airflow rate, and the type of material, on the mechanical vibration effect of ultrasound to the regeneration.

The quantitative analysis on the ultrasonic energy flow in the regeneration process is shown in Figure 3.38 according to the analysis above. It is noted that the "vibration effect" and "heating effect" of ultrasound on the regeneration occur simultaneously and are coupled with each other, which may produce a synergistic effect. This has been proved in the previous chapter.

Known from the above analysis, the contributions of ultrasonic effects to the regeneration are closely related to the regeneration conditions. In the real applications, what we care about most is the energy utilization efficiency. Therefore, the best conditions for ultrasound-assisted regeneration should be chosen aiming at the maximum energy efficiency of the desiccant system.

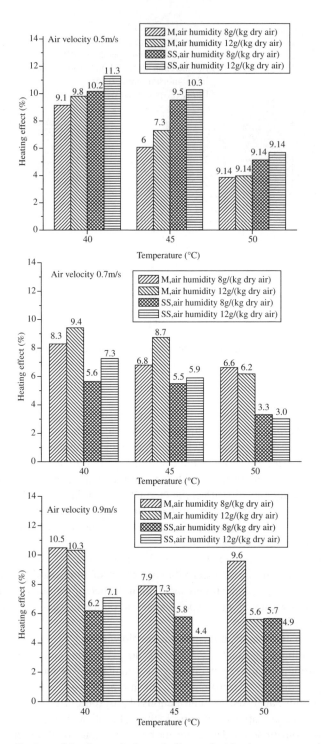

Figure 3.36 Contributions of the ultrasonic thermal effect to the 10-minute regeneration under different regeneration air temperatures and humidity ratios

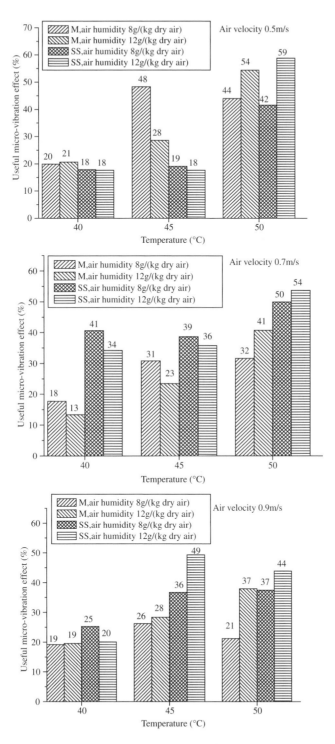

Figure 3.37 Contributions of the ultrasonic mechanical effect to the 10-minute regeneration under different regeneration temperatures and humidity ratios

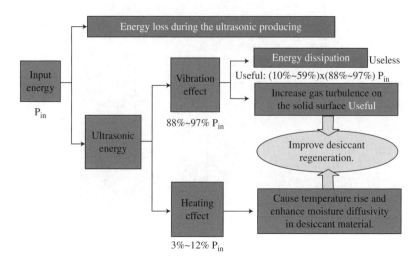

Figure 3.38   Quantitative analysis on the ultrasonic energy flow in the 10-minute regeneration process

3.5   Summary

In this chapter, the effects of ultrasound on the regeneration of the new honeycomb-type desiccants have been investigated. To begin with, the experiments of regeneration in the presence and absence of ultrasound have been performed under a series of conditions. And the enhancement of regeneration brought about by ultrasound and the energy efficiency of ultrasound for the regeneration are discussed based on the experimental data. Then, acoustic energy attenuation and absorptivity of ultrasound in the honeycomb-type desiccant samples have been tested, and a theoretical model (which has been validated by a large amount of experiments) is developed according to the fundamental theory of heat and mass transfer and ultrasonic propagation. Lastly, a wide range of conditions for ultrasound-assisted regeneration are simulated, and the MEEU for regeneration under different conditions has been analyzed based on the model results. Meanwhile, the contributions of ultrasonic "mechanical vibration effect" and "thermal effect" to the regeneration of honeycomb-type desiccants have been quantitatively analyzed.

To improve the effect of ultrasound on regeneration, the structure and basis material of the honeycomb desiccants should be reconfigured in favor of ultrasonic propagation. Although a smaller pore size of honeycomb-type desiccant is more favorable for ultrasound-assisted regeneration, it will inevitably increase regeneration air flow resistance. So, there is a balance between them.

References

[1] Yang, K. and Yao, Y. (2013) Investigation on the performance of applying ultrasonic to the regeneration of a new honeycomb desiccant material. *International Journal of Thermal Sciences*, **72** (10), 159–171.
[2] Zhang, G., Zhang, Y.F. and Fang, L. (2008) Theoretical study on simultaneous water and VOCs adsorption and desorption in a silica gel rotor. *Indoor Air*, **18** (1), 37–43.
[3] Holman, J.P. (2005) *Heat Transfer*, Mechanical Industry Press, Beijing.
[4] Puig, A., Perez-Munuera, I., Carcel, J.A. *et al.* (2008) Theoretical and experimental investigation on the radial flow desiccant dehumidification bed. *Applied Thermal Engineering*, **28** (1), 75–85.

4

Ultrasound-Atomizing Regeneration for Liquid Desiccants

4.1 Overview

4.1.1 Principles and Features of the Liquid-Desiccant Dehumidification

The Liquid desiccant dehumidification method uses a hygroscopic salt solution as an absorbent. Examples of such solutions are lithium bromide solution, a solution of lithium chloride, calcium chloride solution, and so on. A salt solution and the air to be treated are in direct contact as a result of which there exists water vapor partial pressure difference between the surface of the hygroscopic salt solution and the air to be treated due to moisture absorption. This drives the water and moisture in the air to pass through the absorbent solution, enabling the processing of air humidity. When the water vapor partial pressure of the surface of the desiccant solution is lower than that of the air, the moisture in the air moves into the solution and the air is dehumidified; on the contrary, the water in the solution will move into the air, thus generating a more concentrated solution. The absorbed moisture will reduce the dehumidifying ability of the desiccant solution. To recycle the desiccant, the solution must be regenerated after the air dehumidification process.

The liquid desiccants have the following advantages as compared with the solid ones:

1. *Higher energy efficiency*. The liquid desiccants are easier to be cooled, and the isothermal dehumidification process can be acquired, which helps to reduce the irreversible energy loss and improve the energy efficiency.
2. *More flexible form*. Since the liquid is of fluidity, the desiccant regenerator and the air dehumidifier can be designed separately. This makes it possible for the waste heat far away from the air-handling units to be utilized for the desiccant regeneration.
3. *Easier for the air humidity control*. This can be achieved by regulating the circulation flow rate of desiccant solution in order to control the humidity of air to be treated;
4. *Storage convenience*. The desiccant solution has a large capacity of energy storage due to its high adsorption heat, which is conducive to utilizing the intermittent renewable energy (such as solar) for the air dehumidification throughout the day.

Ultrasonic Technology for Desiccant Regeneration, First Edition. Ye Yao and Shiqing Liu.

4.1.2   Thermo-Physical Properties of Liquid Desiccant Materials

The dehumidification and regeneration processes of the desiccant solution are closely related to the water vapor partial pressure on the surface of the solution, and the physical and chemical properties of the desiccant solution will affect greatly the dehumidification and the regeneration process as well as the energy performance of the whole desiccant system.

Generally, the desiccant solution should meet the following requirements:

1. *Excellent hygroscopic capacity.* A higher hygroscopic capacity of the desiccant solution is preferred because it helps to reduce the amount of usage of the desiccant and the size of equipments for a specific dehumidification load.
2. *Lower regeneration temperature.* A lower regeneration temperature is conducive to utilizing the low-grade energy for the desiccant regeneration and improving the dehumidification system's energy performance.
3. *Larger specific heat capacity of the solution.* The dehumidifying process causes the release of latent heat of the solution and results in the temperature rise of the desiccant solution, which is not good for the air dehumidification. A larger specific heat capacity of the solution can reduce such kind of influence and keep the dehumidification performance more stable.
4. *Lower crystallization temperature of the solution.* The higher concentration of the desiccant solution is conducive to improving the water absorption capacity, but it is prone to crystallization and causes the blocking of pipe. The crystallization temperature limits the maximum concentration of the desiccant that can be used.
5. *Lower density and viscosity.* A lower density and viscosity of the desiccant is preferred because they will reduce the energy consumption of the pumps which are responsible for the solution transportation.
6. *Noncorrosive and nontoxic.* During air dehumidification, a small amount of desiccant solution will enter into the air since the solution is usually in direct contact with the air. In order to guarantee the safety of the processed air, the solution must be noncorrosive and nontoxic.
7. The solution should be inexpensive and easy to obtain.

Desiccant solutions currently available are divided into two main categories: the organic and the inorganic. The organic solutions mainly include triethylene glycol (TEG) and diethylene glycol, and the inorganic mainly include lithium bromide, lithium chloride, calcium chloride solution, and so on. The TEG is the first liquid desiccant to be used as a dehumidifying agent [1], but its viscosity is relatively high, and the system is prone to stagnation during the circulation. Adhesion on the surface of the air-conditioning system affects the stability of the system operation. Also, the diethylene glycol, the TEG, and the other volatile organic substances may potentially cause harm to people's health. The above shortcomings limit the applications of the TEG in the liquid desiccant system. Thus, the metal halide salt solution (e.g., lithium bromide and lithium chloride) is employed to replace the TEG. These salt solutions meet most of the requirements of the desiccants mentioned earlier including nonvolatile, nontoxic, and low price. In addition, the salt solutions have been proved to be anti-bacteria. Although they have a certain degree of corrosion, the use of corrosion resistant materials such as plastic can prevent salt solution from corroding pipes and equipments.

Currently, the desiccant salt solutions have been increasingly focused by people. Many local and foreign research institutions and scholars [2–7] have carried out detailed tests on lithium bromide (LiBr), lithium chloride (LiCl), and calcium chloride ($CaCl_2$) solutions properties such as desiccant surface vapor pressure, crystallization curve, specific heat, density, viscosity, and other parameters. They have put forward the results of changes in solution performance parameters with the variations in temperature and concentration.

4.1.2.1 Surface Vapor Pressure

Surface vapor pressure data of the solution need to be obtained through experimental tests. Normally, the surface vapor pressure equations are firstly formed with respect to the measured data under different temperatures and concentrations of the solution. Afterwards, they are used to calculate the surface vapor pressure of the solution for any specific states of the solution.

Table 4.1 shows the research information about the surface vapor pressure of lithium bromide, lithium chloride, and calcium chloride salt solutions, and the detailed formulas are given in the Appendix.

4.1.2.2 Crystallization Curve

Crystallization must be avoided because it may block the solution pipeline. Under the higher concentration or a lower temperature, the solution is prone to crystallize. So, the crystallization curve is very crucial for the usage of the desiccant solution. Some scholars have studied the crystallization of some desiccant solutions including the lithium bromide, the lithium chloride, and the calcium chloride. The relevant studies are listed in Table 4.2, and the complete crystallization curves and formulas are given in the Appendix.

Table 4.1 Studies on the surface vapor pressure of salt solutions

Type of solution	Researchers and publication year	Methods	Scope of data
Lithium bromide	Uemura, 1964 [2]	Experimental test	0–70% by mass, 0–180 °C
	McNeely, 1979 [3]	Experimental test	0–70% by mass, 0–180 °C
	Patil, 1990 [4]	Experimental test and formula fitting	0% – salting point, 30–70 °C
	Chua, 2000 [5]	Formula fitting based on the others' data	0–75% by mass, 0–190 °C
	Yuan, 2005 [6]	Formula fitting based on the theory of Gibbs free energy	All concentrations (pure water to crystal line), 5–250 °C
Lithium chloride	Patil, 1990 [4]	Experimental test and formula fitting	0% – salting point, 30–70 °C
	2003; ASHRAE Handbook [8]	Providing the surface vapor pressure curves	0–45% by mass, 15–45 °C
	Conde, 2004 [7]	Formula fitting	0–60% by mass, 0–100 °C
Calcium chloride	Patil, 1990 [4]	Experimental test and formula fitting	0% – salting point, 30–70 °C
	Ertas, 1992 [9]	Experimental test	20% by mass, 25–65 °C
	Conde, 2004 [7]	Formula fitting	0–70% by mass, 0–100 °C

Table 4.2   Studies on the crystallization curve of salt solutions

Type of solution	Researchers and publication year	Main results
Lithium bromide	Domestic lithium bromide aqueous solution property graph set, 1976 [10]	Crystallization curves
	Patek, 2006 [11]	The complete fitting formula
Lithium chloride	Conde, 2004 [7]	Fitting formula and crystallization curves
Calcium chloride	Conde, 2004 [7]	Fitting formula and crystallization curves

Table 4.3   Studies on the specific heat capacity of salt solutions

Type of solution	Researchers and publication year	Methods	Scope of data
Lithium bromide	Lower, 1960 [12]	Experimental test	0–60% by mass, 0–130 °C
	Rockenfeller, 1994 [13]	Experimental test	45–65% by mass, 60–130 °C
	Chua, 2000 [5]	Formula fitting	0–75% by mass, 0–190 °C
	Yuan, 2005 [6]	Formula fitting	All concentrations (pure water to crystal line), 5–250 °C
Lithium chloride	Conde, 2004 [7]	Formula fitting	0–50% by mass, −10 to 80 °C
Calcium chloride	Conde, 2004 [7]	Formula fitting	0–55% by mass, −15 to 90 °C

4.1.2.3   Specific Heat Capacity

Dehumidification process along with phase change process causes water vapor to condense into liquid water, and this will release latent heat and cause the desiccant solution temperature to rise. The degree of temperature rise largely depends on the specific heat capacity of the solution. Table 4.3 shows the research information about the specific heat capacity calculation formula of salt solutions, and the corresponding formulas are given in the Appendix.

4.1.2.4   Density and Viscosity

Desiccant salt solution density and viscosity are the important parameters affecting the energy consumption of the solution pumps in the system. This is because the pressure head due to the gravity is directly related to the density, and the drag force due to the friction is directly related to the viscosity. Table 4.4 lists the research information about the density and viscosity calculation formula of salt solutions which are given in the Appendix as well.

The main purpose of the study on the thermo-physical properties of the desiccant solutions is to choose the most suitable desiccant for a liquid desiccant system. Zhao and Shi [15] selected desiccant solutions for solar energy liquid desiccant air-conditioning system through comparing the solution surface vapor pressure, corrosiveness, and the cost (price) of different salt solutions (i.e., LiBr, LiCl, and $CaCl_2$). The results show that the LiCl solution is the most suitable for the solar liquid desiccant dehumidification air conditioning system.

Table 4.4 Studies on the density and viscosity of salt solutions

Type of solution	Researchers and publication year	Methods	Scope of data
Lithium bromide	Wimby, 1994 [14]	Experimental test	10–60% by mass, 20–90 °C
	Chua, 2000 [5]	Formula fitting	0–70% by mass, 0–200 °C
	Yuan, 2005 [6]	Formula fitting	All concentrations (pure water to crystal line), 5–250 °C
Lithium chloride	Conde, 2004 [7]	Formula fitting	0–56% by mass
Calcium chloride	Conde, 2004 [7]	Formula fitting	0–60% by mass

Table 4.5 Weighting factor and scoring criteria for the desiccant solutions

Items	Weighting factor	Scoring criteria
Safety	1.0	Lethal dose (LD 50)
Corrosivity	0.8	Corrosion rate
Mass transfer potential	0.8	Equilibrium water vapor pressure
Heat of mixing	0.6	Energy consumption per kilogram water adsorption
Price	0.5	US dollar per kilogram solution
Heat transfer potential	0.5	Heat conductivity
Flow property	0.3	Viscosity

Studak and Perterson [16] conducted a preliminary evaluation of several liquid desiccants including LiBr, LiCl, $CaCl_2$, and TEG solution. They proposed the weight factor of each characteristic of the desiccant solution according to the importance. Then, the partial score of each characteristic for any specific solution was estimated by experts. Finally, the final scores of the desiccants could be obtained based on the weight factor and the partial score. The one with the highest final score is the best desiccant. The weight factor and the corresponding scoring criteria for evaluating the desiccant are listed in Table 4.5. The study by Studak and Perterson [16] shows that the $CaCl_2$ solution gets the highest final score among these liquid desiccants because its price is much lower than the others.

Yi [17] studied the performances of two kinds of liquid desiccants (i.e., LiBr and LiCl solutions) by using the weight analysis method with the weight factor scoring criteria proposed by Studak and Perterson [16]. The results indicated that the overall performance difference between the LiBr solution and the LiCl solution is small, and the LiCl solution has a slight advantage.

The lithium bromide, lithium chloride, and calcium chloride salt solutions have their own advantages and disadvantages: for example, the lithium bromide and the lithium chloride have good desiccant dehumidification ability but their prices are relatively high, and the situation is the opposite for the calcium chloride solution. Thus, the mixed solutions were put forward and studied by some researchers. Ertas *et al.* [9] proposed a mixed desiccant solution by using the lithium chloride and the calcium chloride. According to the study, the mass ratio of the lithium chloride to the calcium chloride is favorably as 1 : 1 for the mixed solution. Ahmed and Gandhidasan [18] also performed an experimental study on the thermo-physical properties of the mixed LiCl-CaCl$_2$ solution including its water vapor partial pressure, density, viscosity, and specific heat capacity. The results show that the mixed solution as desiccant can achieve the best combination of dehumidification and economic performance.

4.1.3   Research Status of Solution Regenerators

The main components of a liquid desiccant system are the dehumidifier and regenerator. Desiccants absorb (or release) moisture because of the difference in vapor pressure between the surface of the desiccant and the surrounding air. The air dehumidification process occurs when the vapor pressure on the desiccant surface is lower than that of the air. The moisture diffusion from the air to the desiccant causes a dilution of the desiccant which must be regenerated (i.e., concentrated) to return back to its original state. The regeneration of desiccant is usually performed by heating the desiccant and bringing it in contact with an air stream. The process equipment utilized for the liquid-gas contacting is generally falling film, spraying, or packed towers.

Different regenerator designs have been studied by researchers. Jain *et al.* [19] studied experimentally a falling film plate regenerator. The results were compared with the predictions from theoretical model. Scalabrin and Scaltri [20] analyzed a spray tower in which a stream of air comes into direct contact with the weak lithium chloride solution sprinkled over an internally-heated tube bank. Chung *et al.* [21] designed a U-shaped spray tower to prevent carry over and developed a mass transfer correlation for the air stripping process using the TEG. The packed tower configurations have received more attention due to the fact that they can provide large heat and mass transfer area between the air and desiccant. As early as 1980, Factor and Grossman [22] compared the experimental data and the theoretical results of a packed regenerator using LiBr and pre-heated air. Afterwards, Etras *et al.* [23] investigated the influence of some key variables on a packed regenerator performance. Potniz and Lenz [24] tested a packed regenerator using random polypropylene and structured packings. The experimental study showed that the evaporation rate was 130–300% greater in the randomly packed bed than that in the structured packing bed. Martin and Goswami [25] used TEG as the liquid desiccant and packed bed as the regenerator to conduct an experimental study, and the work presented effects of the inlet parameters of the air and the desiccant on the humidity effectiveness of the regenerator. Oberg and Goswami [26] developed two novel performance correlations for the effectiveness of a packed dehumidifier/regenerator. A comparison to the experimental data showed that the correlations presented correctly predict the influence of the design variables on the performance within 15% error. Fumo and Goswami [27] assessed the regeneration process under the effect of the design variables of a packed regenerator using the LiCl solution. Gandhidasan [28] investigated the influence of the heating source on the evaporation rate of a packed bed regenerator. Elsarrag [29] made a performance study of the simultaneous heat and mass transfer between the air and the desiccant in a structured packed-stripping tower where the cellulose rigid medium pads are used as the structured packing. The effects of the air flow rate, the liquid flow rate, the air humidity, the desiccant temperature, and concentration have been investigated on the evaporation rate and the humidity effectiveness of the column. Longo and Gasparella [30] performed an experimental study on the structured and the random packed regenerator using LiBr/H$_2$O and a new solution KCOOH/H$_2$O and studied the effect of the mass flow rate ratio of the desiccant to the air on the tower efficiency. Yin and Zhang [31] compared the performance of the adiabatic (Figure 4.1a) and the internally heated (Figure 4.1b) regenerators. The results indicated that the internally heated regenerator could not only increase the regenerate rate, but also exhibited higher energy utilization efficiency. Different from the adiabatic regenerator, the internally heated regenerator can provide comparable regeneration efficiency and regeneration rate (RR) at low desiccant flow rates, so it is a good alternative to avoid the carryover of desiccant droplets.

Although the packed towers facilitate more mass transfer by providing a larger contact area, the air pressure drop through the packing stuff is often high. Moreover, an increased solution mass flow rate is required for ensuring the right wetting ratio of the packed bed which is a crucial factor affecting the performance of the packed regenerator. The method of ultrasonic atomizing (which is mainly produced by the special effect of cavitation of power ultrasound occurring in a liquid environment) for the solution regeneration may become a promising way to develop a kind of new regenerator in which the solution is atomized into droplets of 40–50 μm instead of flowing along the surface of the packed material. The new structure configuration will greatly enlarge the air-desiccant contacting area and reduce the air flow resistance, and hence, the performance and regeneration efficiency can be significantly improved. In this section, a new solution regenerator with ultrasonic atomizing is to be proposed and tested for the desiccant solution regeneration. Meanwhile, a model is developed to compare the performance of the newly developed regenerator with that of the conventional one.

4.2 Theoretical Analysis

4.2.1 Mass Transfer Coefficients for the Droplets

The proposed mechanism for the process of mass transfer between the air and solution droplet is shown in Figure 4.2 [32]. This consists of the following steps:

1. Diffusion from the air to the air-liquid interface;
2. Adsorption and mass accommodation at the interface;
3. Dissolution (absorption) and diffusion into the bulk solution.

The product of the mass transfer coefficient and the overall driving force gives the mass transfer flux (K_m: kg/(m^2 s)) of moisture molecule between the gas phase and the liquid droplet.

$$K_m = u_w \left(C_w - C_w^* \right), \tag{4.1}$$

where u_w is moisture transfer velocity in m/s; C_w is moisture concentration (kg/m^3) in the gaseous phase (mainstream air), C_w^* is the moisture concentration (kg/m^3) in equilibrium with the liquid phase.

The moisture transfer velocity (u_w) can be defined as

$$u_w = \frac{Sh \cdot D_{w\text{-}a}}{d_s}, \tag{4.2}$$

where u_w is moisture transfer velocity in m/s; Sh is the Sherwood number; D_{w-a} is diffusion coefficient (m^2/s) of water vapor in the air; and d_s is diameter of droplet in m.

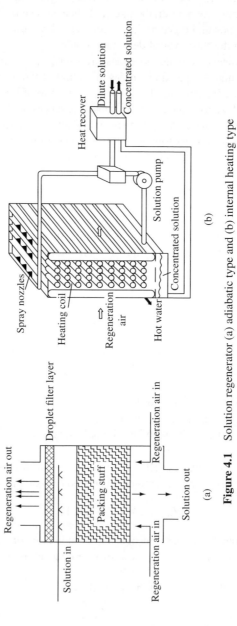

Figure 4.1   Solution regenerator (a) adiabatic type and (b) internal heating type

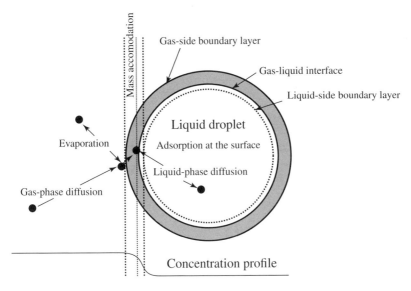

Figure 4.2 Schematic diagram for the process of mass transfer between the air and solution droplet [32]

The Sherwood number (Sh) is the key parameter to determine the mass transfer coefficient. According to Fick's law, the mass transfer flux (K_m: kg/(m^2 s)) can be written as:

$$K_m = -D_{w-a}\frac{dC_w}{dx} = \frac{G_w}{4\pi r_s^2}, \tag{4.3}$$

where C_w is the mass concentration of water vapor in the air, kg/m^3; G_w is moisture flow rate in kg/s; x is distance from the center of droplet in m; and r_s is radius of droplet in m.

By using the method of separation of variables, Equation (4.3) can be transformed as:

$$-D_{w-a}dC_w = \frac{G_w}{4\pi}\frac{dx}{r_s^2}. \tag{4.4}$$

Integrating from the particle surface to infinity, we can obtain:

$$-D_{w-a}\int_{r_s}^{\infty} dC_w = \frac{G_w}{4\pi}\int_{r_s}^{\infty}\frac{dx}{r_s^2}. \tag{4.5}$$

Then,

$$-D_{w-a}\left(C_{w,\infty} - C_{w,r}\right) = \frac{G_w}{4\pi}\frac{1}{r_s}, \tag{4.6}$$

$$G_w = -4\pi r_s D_{w-a}\left(C_{w,\infty} - C_{w,r}\right), \tag{4.7}$$

where the subscript "∞" stands for the mainstream air, and the subscript "r" for the air on the surface of the droplet.

Combining Equations (4.3) and (4.7), the mass transfer flux (K_m: kg/(m^2 s)) can be written as

$$K_m \big|_{r_s} = -\frac{D_{\text{w-a}}}{r_s} \left(C_{\text{w},\infty} - C_{\text{w,r}}\right) = -u_{\text{w}} \left(C_{\text{w},\infty} - C_{\text{w,r}}\right). \tag{4.8}$$

Thus,

$$u_{\text{w}} = \frac{D_{\text{w-a}}}{r_s}. \tag{4.9}$$

According to the definition of moisture transfer velocity (u_{w}) in Equation (4.2), the Sherwood number (Sh) can be identified as 2.

Considering the effect of the free stream turbulence on the sphere (or droplet) mass transfer, the Sherwood (Sh) numbers can be expressed by Equation (4.10) for the case of low Re ($Re < 5$) [33],

$$Sh = 2 + 0.6Re^{1/2}Sc^{1/3}, \tag{4.10}$$

where Re is the Reynolds number, $Re = \rho_a \left(u_f - u_a\right) d_s / \mu_a$; Sc is the Schmidt number, $Sc = \mu_a / \left(\rho_a D_{\text{w-a}}\right)$; u_f and u_a are falling speed of droplet and velocity of freestream air in m/s, respectively; and μ_a is the viscosity of air in Pa·s.

According to the analogy theory of heat and mass transfer, the heat transfer coefficient (H_m:W/(m^2·°C)) of droplet can be expressed by Equation (4.11).

$$H_m = \frac{Nu \cdot \lambda_a}{d_s}. \tag{4.11}$$

The Nusselt number (Nu) can be calculated by

$$Nu = 2 + 0.6Pr^{1/3} \cdot Re^{1/2}, \tag{4.12}$$

where Pr is the Prandtl number, $Pr = c_{p,a}\mu_a / \lambda_a$ and λ_a is heat conductivity of air, W/(m·°C).

Figure 4.3 presents the effect of droplet size and falling velocity on the mass transfer coefficient of droplet exposed to an airstream at 40 °C. The water molecular diffusivity in the air ($D_{\text{w-a}}$) is approximated as 2.88×10^{-5} m^2/s, and the air kinematic viscosity (v_a) as 17.65×10^{-6} m^2/s in the calculation. As illustrated in Figure 4.3a, the mass transfer coefficient of droplet increases greatly with the decrease of the droplet size, especially within the range of 0–50 μm; and it increases with the increase of falling velocity relative to the airstream. It can be seen as well from Figure 4.3b that the influence of droplet size on the mass transfer coefficient tends to be less significant as the size of droplet increases, and so does that of falling velocity on the mass transfer coefficient under the higher falling velocity of droplet.

The mass transfer accompanying evaporation from a droplet surface requires heat transfer from the gas phase to maintain a constant rate of evaporation. During the unsteady period, the droplet cools, leading to a reduction of the vapor pressure of the droplet constituents at the droplet surface and a corresponding decrease in the rate of mass transfer. After this transient period, a limiting steady state is reached, characterized by evaporation at a constant droplet temperature under which the heat flux to the droplet from the gas-phase balances the heat lost due to the evaporation, the droplet temperature is steady, and the mass transfer rate is constant. Maxwell has solved the steady-state conservation equations of mass and energy in the gas phase for the evaporation into a stagnant gas in order to explain the phenomenon of

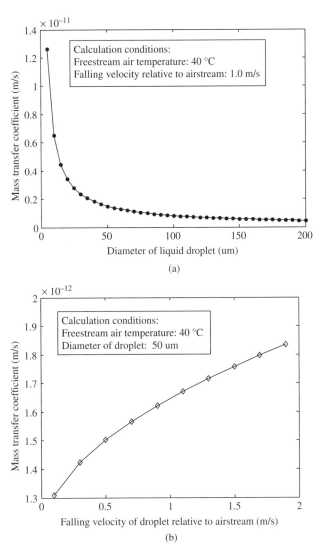

Figure 4.3 Influence of droplet size and falling velocity on mass transfer coefficient of droplet. (a) Mass transfer coefficient versus droplet size and (b) mass transfer coefficient versus falling speed

the wet-bulb temperature [34]. This derivation assumes that equilibrium should be established for the gas and liquid at the interface and the convection effects be negligible.

The steady treatment is only appropriate for the isothermal evaporation in which the vapor pressures of the components are approximately constant. The rate of mass transfer is independent of time and determined by the concentration gradient of the evaporating component from the droplet surface to an "infinite" distance, the diffusion constant of the evaporating component in the gas phase and the droplet surface-to-volume ratio. Although the steady state theory can bring convenience for modeling, it may cause a relatively larger error in predicting the vapor pressures of the droplet constituents in some cases like the rapid evaporation of highly volatile components in which the temperature depression of the surface must be explicitly included in the heat and mass transfer analysis to account for the time-dependent evaporation rate. By contrast, the quasi-steady state theory provides a better description of the physical processes occurring than the steady state theory while reducing the complexity of the unsteady state model, and hence, it may be a good compromise between the steady and the unsteady-state approaches. The quasi-steady treatment of droplet evaporation considers that, at any particular instant in time, the concentration and temperature profiles in the gas-phase surrounding the particle, and the mass and energy fluxes can be treated as if they were in steady state. Any changes in concentration or temperature profile or flux can be expressed as arising from a change in the boundary conditions at the droplet surface or at a large distance from the droplet. This can be assumed to be a good approximation if the time for establishing a stable concentration or temperature profile in the surrounding gas phase is much shorter than the time-scale over which the physical state of the system (droplet radius, temperature, and composition) is changing.

Newbold and Amundson [35] derived a quasi-steady model for the evaporation of a multi-component droplet in a stagnant gas. The model describes the transport of different vapors through an inert gas around a single isolated and stationary aerosol droplet. The transport equations for the gas-phase are first solved, and the expressions for the mass and heat flux from the droplet can be obtained. Differential equations describing the change rate in the droplet radius, the concentration, and the temperature are then derived based on the law of mass and energy conservation. The droplet surface and the boundary conditions are assumed to change slowly with time, allowing the evaporation to be treated within the quasi-steady approximation. In addition, the gas–liquid interface is assumed to be at equilibrium state, which inherently means that the interfacial transfer between the liquid and gas phases is much faster compared to the gas diffusion away from the droplet. This is a valid assumption for the systems considered here (i.e., liquid desiccant regeneration by ultrasonic atomizing) where all the droplets fall within the continuum region. Meanwhile, the mean free path of the gas-phase molecules ($L_{g,molecule}$) at the pressures presented in this work is much shorter at the pressures than the droplet radius (r_s), and the Knudsen number, K_n, defined by Equation (4.13), is considerably less than 1.

$$K_n = \frac{L_{g,molecule}}{r_s}.$$

(4.13)

Under the normal atmospheric pressure of 1.01×10^5 Pa and 50 μm in the droplet radius, the Knudsen number is estimated as 1.36×10^{-3}, and the gas-phase molecules ($L_{g,molecule}$) is considered as 6.8×10^{-8} m according to Ref. [36]. Thus, the evaporation is limited by the gas diffusion.

To assess the validity of the quasi-steady treatment, Seinfeld and Pandis [37] demonstrate that the time-scale for gas-phase diffusion to establish a steady-state concentration profile of a species around the particle (τ_s) can be estimated from the gas diffusion constant for the species (D_g) and the droplet radius (r_s) as below:

$$\tau_s = \frac{r_s^2}{4D_g}.$$

(4.14)

For a droplet size of 50 μm in radius under the normal atmospheric pressure of 1.01×10^5 Pa, the time-scale for water evaporation is estimated about 2.4×10^{-5} seconds (the moisture diffusivity in the air is 2.6×10^{-5} m^2/s). Thus, for such small droplets, the gas phase compositional profile around the droplet responds to changes at the particle surface very rapidly, and the use of quasi-steady model can be justified.

If the thermal expansion and volume change due to mixing are neglected, the volume change of the droplet corresponds to the volume of liquid vaporized. The rate of change of droplet volume can be expressed as:

$$\frac{dV_s}{d\tau} = -4\pi r_s^2 \sum_{i=1}^{n} N_i V_{\text{mole},i}, \tag{4.15}$$

where V_s is the volume of droplet in m^3; N_i is the molar flux (in mol/(m^2·s)) of the ith component from the droplet surface; $V_{\text{mole},i}$ is the molar volume (m^3/mol) of the ith component; and n denotes total number of components in a multi-component droplet evaporating into the air.

By using Equation (4.15), the change rate of droplet radius with the time can be derived as below:

$$\frac{dr_s}{d\tau} = -\sum_{i=1}^{n} N_i V_{\text{mole},i}. \tag{4.16}$$

From considering the mass balance on the component i, the change rate of the mole fraction of the component i in the droplet is given by:

$$\frac{dx_i}{d\tau} = \frac{3\overline{V}_{\text{mole}}}{r_s} \sum_{i=1}^{n} N_i \left(x_i - N_i \Big/ \sum_{i=1}^{n} N_i \right), \tag{4.17}$$

where, $\overline{V}_{\text{mole}}$ is the mean molar volume defined as $\overline{V}_{\text{mole}} = \sum_{i=1}^{n} x_i V_{\text{mole},i}$.

The energy balance, ignoring the radiation effects, can be used to derive an expression for the change in the droplet temperature (T_{rs}) with time:

$$\frac{dT_{rs}}{d\tau} = \frac{3\overline{V}_{\text{mole}}}{r_s c_{\text{a,mole}}} \left\{ \frac{\lambda_a \vartheta}{\left(e^\vartheta - 1\right) r_s} \left(T_\infty - T_{rs} \right) - \sum_{i=1}^{n} N_i \left[\Delta h_{i,\text{vap}} + \Delta c_{i,\text{mole}} \left(T_{rs} - T_{\text{ini}} \right) \right] \right\}, \tag{4.18}$$

where ϑ is defined as:

$$\vartheta = \frac{r_s}{\lambda_a} \sum_{i=1}^{n} N_i c_{i,\text{mole}}. \tag{4.19}$$

T_∞ is the temperature at an infinite distance from the droplet, T_{rs} is the temperature at the droplet surface, and T_{ini} is the initial droplet temperature (in K); $c_{i,\text{mole}}$ is the molar heat capacities (in J/(mol·°C)) of the ith component of droplet, $\Delta c_{i,\text{mole}}$ is the difference of molar heat capacities (in J/(mol·°C)) of the ith component between the gas and the liquid phases, and $c_{\text{a,mole}}$ is the molar heat capacity (in J/(mol·°C)) of the bulk air; λ_a is the air thermal conductivity in W/(m·°C); and $\Delta h_{i,\text{vap}}$ is the enthalpy of vaporization (in J/kg) for the ith component of the droplet.

For the system considered in this study (i.e., regeneration of liquid desiccant with ultrasonic atomizing), only the moisture in the droplet evaporates into the air. So, N_i and $V_{\text{mole},i}$ in Equations (4.15)–(4.19) are actually the molar flux and molar volume of water vapor, respectively.

To determine the molar flux of component i (N_i) from the droplet surface, the following assumptions must be made: the binary diffusivities are replaced by an effective binary diffusivity of component i with respect to the mixture; and the quasi-steady approximation is made. This leads to the following expression [38]:

$$N_i = \frac{C_{total} D_a}{r_s p_{total}} \ln(\chi) \left[\frac{x_i p_{vap,i}(T_s) \chi^{D_a/D_{im}} - p_{i,\infty}}{\chi^{D_a/D_{im}} - 1} \right], \tag{4.20}$$

where C_{total} is total gas concentration; D_a and D_{im} are the diffusion coefficient (m²/s) of the dry air and the ith vaporizable component in the droplet; x_i is the mole fraction of the ith component in the droplet; and $p_{vap,i}(T_s)$ is the vapor pressure of pure component i at the droplet surface temperature. p_{total} is the total gas pressure and $p_{i,\infty}$ is the partial pressure of component i at an infinite distance from the droplet; χ is defined as,

$$\chi = x_{a,\infty}/x_{a,r_s}, \tag{4.21}$$

where $x_{a,\infty}$ and x_{a,r_s} are mole fraction of dry air in the bulk gas (i.e., moist air) at an infinite distance from the droplet and that at the gas–liquid interface of droplet.

Equations (4.16)–(4.18) are the basis for describing the evolution in the radius, the composition and the surface temperature of the multi-component droplet with time. The equations can be integrated by using a fourth order Runge-Kutta routine with a variable step size enabling this series of equations to be solved numerically.

Figure 4.4 illustrates the calculated influence of the total air pressure on the variations of size, temperature, and molar concentration of LiCl-H₂O droplet (composed of 50% LiCl by molar) initially 25 and 50 µm in radius and 70 °C in solution temperature at an evaporation time of 2 ms under the ambient air temperature of 30 °C and humidity of 70%. The diffusion coefficient of air and water vapor are both taken as 2.6×10^{-5} m²/s in the calculation. In these predictions, a change in the ambient air pressure causes a change in the mass and heat transfer rate, that is, the droplet size, the temperature, and the mole fraction of salt in the droplet all change, and the change range of all these parameters becomes greater under the lower ambient pressure. In addition, the smaller droplet size will bring about a larger change rate of these parameters. As shown in Figure 4.4, the change percentages of the droplet radius, the temperature, and the salt molar concentration under the initial radius of 25 µm are about 1/5 of those under the initial radius of 50 µm.

Figure 4.5 shows the calculated influence of the initial droplet temperature on the variations of the size, the temperature, and the molar concentration of LiCl-H₂O droplet (composed of 50% LiCl by molar) initially 25 and 50 µm in radius and 100 kPa in ambient pressure at an evaporation time of 2 ms under the ambient air temperature of 30 °C and humidity of 70%. It can be easily understood that a higher droplet temperature will cause the higher rate of heat and mass transfer between the salt droplet and the ambient air, and hence, the change rate of the droplet parameters will become increasingly significant as the initial droplet temperature rises. The degree of influence of droplet temperature on the change rate of these droplet parameters will be affected as well by the droplet size. As seen from Figure 4.5, the change percentages of the droplet radius, the temperature, and the salt molar concentration under the initial radius of 25 µm are about 1/7 of those under the initial radius of 50 µm.

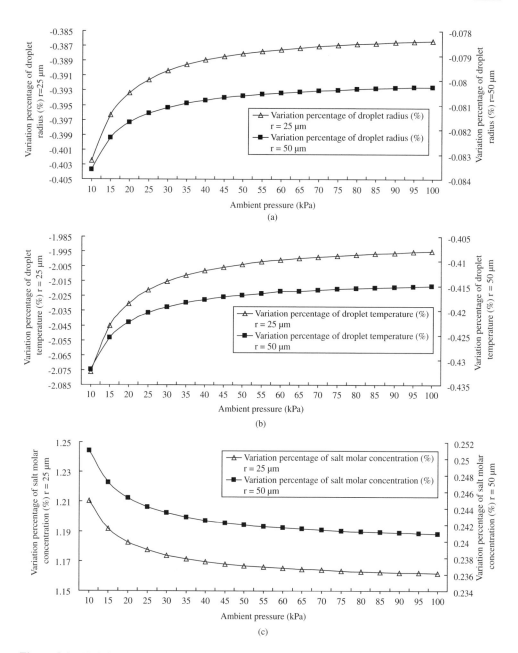

Figure 4.4 Variations of size, temperature, and molar concentration of LiCl-H$_2$O droplet at an evaporation time of 2 ms under different ambient pressures. (a) Variations of droplet size versus the ambient pressure, (b) variations of droplet temperature versus the ambient pressure, and (c) variations of salt molar concentration versus the ambient pressure

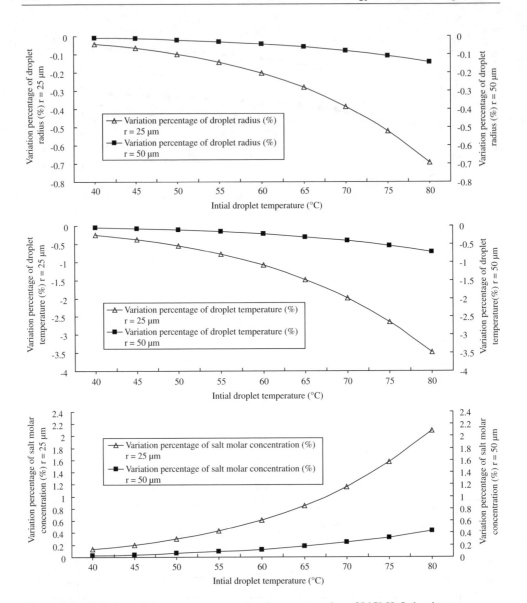

Figure 4.5 Variations of size, temperature, and molar concentration of LiCl-H$_2$O droplet at an evaporation time of 2 ms under different initial droplet temperatures

4.2.2 Atomized Size of Droplet by Ultrasonic Atomizing

The diameter of droplet (d_s: in m) ultrasonically atomized may be theoretically calculated by Yasuda *et al.* [39]:

$$d_s = 2.8\left[\sigma_s/\left(\rho_s f^2\right)\right]^{1/3}\left(\mu_s/\mu_w\right)^{-0.18},\qquad(4.22)$$

where σ_s is surface tension of droplet in N/m; ρ_s is solution density in kg/m^3; and μ_s and μ_w are the dynamic viscosity (in Pa·s) of bulk solution and water, respectively.

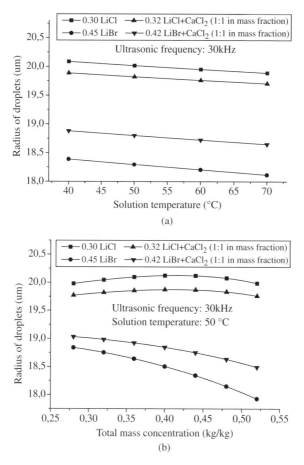

Figure 4.6 Influence of solution temperature and salt mass concentration on the atomized size of droplet. (a) Droplet size versus solution temperature and (b) droplet size versus total mass concentration

Since the temperature and the total mass concentration of solution affect its surface tension and viscosity, the atomizing droplet size will be affected as well by the temperature and the total mass concentration of solution. As shown in Figure 4.6a, the droplet size by ultrasonic atomizing decreases with the increase of the solution temperature. However, the regular patterns about the influence of the total mass concentration of desiccant solution on the droplet size by ultrasonic atomizing may depend on the type of desiccant solution. As seen from Figure 4.6b, the size of atomizing droplet decreases with the increase of the total mass concentration of solution for the LiBr and LiBrCaCl$_2$ desiccant solution, and it is very different for the LiCl and LiClCaCl$_2$ desiccant solution in which the atomizing size hardly changes with the total mass concentration of solution. The ultrasonic frequency is another parameter that makes great influence on the atomized droplet size. In Figure 4.7, we can see that the size of atomized droplet decreases by nearly 50% when the ultrasonic frequency applied rises from 20 to 50 kHz.

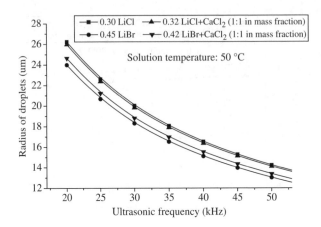

Figure 4.7   Influence of ultrasonic frequency on the atomized size of droplet

4.2.3   Droplet Distribution Characteristics and Measurement Techniques

4.2.3.1   Droplet Distribution Characteristics

The atomization of liquid droplets through the nozzle generally consists of a group of particles ranging in droplet size. In order to describe and assess the atomization of liquid droplet quality and presentation of its atomization characteristics, it requires representation of both particle diameter size as well as an expression of quantity or quality of different diameters. The expression is the so-called droplet size distribution. Currently, there has not been an expression able to theoretically describe in detail the distribution of liquid particles. So, the empirical formulas are often used for the expression of droplet distribution.

Generally, we have the following four forms for the description of droplet size distribution [40]:

1. *Number integral distribution,* which is defined as the ratio of the number of droplets whose diameters are bigger (or smaller) than a given value to the total number of droplets.
2. *Mass integral distribution,* which is defined as the ratio of the mass of droplets whose diameters are bigger (or smaller) than a given value to the total mass of droplets.
3. *Number differential distribution* (denoted as $\Delta N\,(\Delta d)\,/N_{total}$), which is defined as the ratio of number increment of droplet within the diameter range of $\left[d_s - \Delta d_s/2\right] < d_s < \left[d_s + \Delta d_s/2\right]$ $(\Delta N\,(\Delta d))$ to the total number of droplets (N_{total}).
4. *Mass differential distribution* (denoted as $\Delta m\,(\Delta d)\,/m_{total}$), which is defined as the ratio of mass increment of droplet within the diameter range of $\left[d_s - \Delta d_s/2\right] < d_s < \left[d_s + \Delta d_s/2\right]$ $(\Delta m\,(\Delta d))$ to the total mass of droplets (m_{total}).

The typical curve of these parameter indicators is shown in Figure 4.8. It can be seen from Figure 4.8 that the typical size distribution curve has a single peak shape which can be expressed by a uniform mathematical function as below:

$$\frac{\Delta N}{\Delta d} = c_1 \cdot d^{n_1} \cdot \exp\left(-c_2 \cdot d^{n_2}\right), \tag{4.23}$$

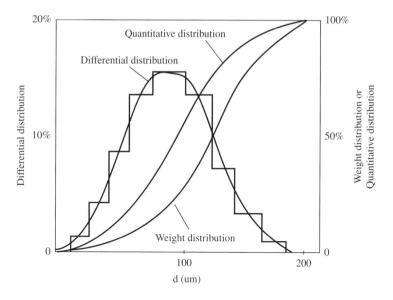

Figure 4.8 Size distribution of atomized droplet

where ΔN stands for the number of droplets (or a distribution of drops) whose diameters lie between d and $d + \Delta d$; c_1, c_2, n_1 and n_2 are empirical constants which can be obtained through experimental data with the graphical method.

One of the most commonly used model (known as the *Rosin-Rammler* expression) for the mass integral distribution of droplets (m_{dis}) can be written as

$$m_{\text{dis}} = \left[1 - \exp\left(-\left(d_i/\overline{d} \right)^n \right) d_i \right] \times 100\%, \tag{4.24}$$

where m_{dis} is the mass percentage of droplets whose diameters are smaller than d_i; $\overline{d}$ is the characteristic diameter of droplets; and n is exponential distribution parameter. $\overline{d}$ and n can be determined through experimental data.

The average droplet size is taken as a hypothetical uniformly sized atomization, which is often employed to simplify the modeling of heat and mass transfer for the spray drying. The requirements of average droplet size differ with applications. The most commonly used average droplet diameters are the mass mean diameter (MMD) and Sauter mean diameter (SMD). The mass mean diameter is defined as above or below 50% of droplet mass of this diameter. The SMD is defined as:

$$SMD = \frac{\sum S_i \times d_i}{\sum S_i} = \frac{\sum N_i \times d_i^3}{\sum N_i \times d_i^2}, \tag{4.25}$$

where d_i stands for different sizes of droplets in diameter; S_i is the summation of droplet surface area corresponding to the diameter of d_i; and N_i is the number of droplet whose diameter is d_i.

4.2.3.2   Particle Size Measurement with the Technique of Laser Diffraction

The technique of laser diffraction is based on the principle that particles passing through a laser beam will scatter light at an angle that is directly related to their size. As the particle size decreases, the observed scattering angle increases logarithmically. The observed scattering intensity is also dependent on particle sizes and diminishes, to a good approximation, in relation to the particle's cross-sectional area. Large particles therefore scatter light at narrow angles with high intensity, whereas small particles scatter at wider angles but with low intensity.

The primary measurement that has to be carried out within a laser diffraction system is the capture of the light scattering data from the particles under study. A typical system consists of:

1. laser to provide a source of coherent, intense light of fixed wavelength;
2. sample presentation system to ensure that the material under test passes through the laser beam as a homogeneous stream of particles in a known, reproducible state of dispersion;
3. series of detectors which are used to measure the light pattern produced over a wide range of angles.

The size range accessible during the measurement is directly related to the angular range of the scattering measurement. Modern instruments make measurements from around 0.02° to 135°. A logarithmic detector sequence, where the detectors are grouped closely together at small angles and more widely spaced at wide angles, yields the optimum sensitivity. Finally, the detector sequence is generally set up such that equal volumes of particles of different sizes produce a similar measured signal. This requires the size of the detectors to be increased as the measured scattering angle increases. The particle size distribution can be obtained by using the Mie Theory or the Fraunhofer Approximation method.

4.2.4   Vapor Pressure of Liquid Desiccant Mixture

The vapor pressure of solution is the critical parameter that affects the working performance of the liquid desiccant. The computation models for the vapor pressure of a single-salt solution (e.g., LiCl solution) have been given in the Appendix. The poor performance of $CaCl_2$ solution is due to its high vapor pressure and its performance depends on the conditions of inlet air. While the LiCl solution has a lower vapor pressure and better stability, its cost is much higher. Hence in the practical applications, the desiccant solutions with two or more salts are often concerned. By combining salts, such as LiCl and $CaCl_2$, improved characteristics can be expected as well as a considerable reduction in cost. For a better use of the mixed desiccant solution, the models for accurate estimation of any given property for each of these classes of mixture are necessarily developed. Equations of both theoretical and empirical nature can be found in the literature for predicting the properties of the mixtures [9, 18, 41, 42]. Some of these equations are shown to be derived from only a few simple, but very general, mixing rules [43], which is given as:

$$\varpi = \varpi_a x_a + \varpi_b x_b, \tag{4.26}$$

where ϖ, ϖ_a, and ϖ_b denote any specific property (e.g., vapor pressure) of the mixed solution and each salt component in the mixture, respectively; x_a and x_b are mass fraction of each salt component in the mixture, which are related as:

$$x_a + x_b = 1. \tag{4.27}$$

Equation (4.26) requires that one only needs to know the value of the given property for each component and their concentrations. The concentration terms may be either mass fractions or volume fractions, depending upon the kind of property concerned.

In the great majority of mixtures, the simplest mixture rules are not capable of accurately predicting the properties of a binary mixture. To make accurate predictions, some kinds of additional information may be applied to an equation appropriate to the type of mixture. Thus, the mixing-rule model for the properties of the mixtures can usually be rewritten as:

$$\varpi = \varpi_a x_a + \varpi_b x_b + c_{ab} x_a x_b, \tag{4.28}$$

where c_{ab} is the interaction parameter that is determined through experiments.

As indicated by Equation (4.28), the effectiveness of the mixing-rule model depends largely on the experimental data, and it is difficult to be extended to wide conditions. Thus, many independent theories have been developed for studying electrolyte solutions [41, 44, 45] in the sense of chemical field. According to the Rault rule, the water vapor pressure of the solution can be obtained by calculating the activity of the water. Activity is always corresponding to the activity coefficient and the concentration (including mole fraction and mass fraction). A lot of theories have been developed to calculate the activity coefficient of the electrolyte solutions, single or mixed, such as the Debye-Huckel theory, the Pitzer theory, and the Local Composition Model. Most of them are based on the statistical thermodynamics principles. Since the NRTL (nonrandom two-liquid) equation of Local Composition Model [41] has a relatively good precision and its parameters are easy to obtain, it has been applied to calculate the activity coefficient (which reflects the water vapor pressure of the solution) of the mixed desiccant solution in our study.

The vapor-liquid equilibrium relation for water in an electrolyte solution can be simplified as [42]:

$$A_w = p_{w,s}/p_{w,w}, \tag{4.29}$$

$$A_w = x_w f_c, \tag{4.30}$$

where A_w is activity of water (no units); $p_{w,s}$ and $p_{w,w}$ are the water vapor pressure of solution and pure water, respectively.

The water vapor pressure of pure water, $p_{w,w}$, can be calculated by using the Antoine equation

$$\log_{10} p_{w,w} = A - \frac{B}{C + t}, \tag{4.31}$$

where the temperature t is in degrees Celsius, and the water vapor pressure of pure water, $p_{w,w}$, is in mmHg (1 mmHg = 133.32 Pa). The constants in Equation (4.31) are given as below: A = 8.07131, B = 1730.63, and C = 233.426 for the temperature ranging from 1 to 100°C.

The vapor pressure of a solution can be calculated only if the activity of water A_w has been determined in advance. At the same time, A_w is the product of mole concentration and activity coefficient f_c (which is corresponding to the mole fraction).

The activity coefficient f_c can be calculated by NRTL equation based on the Local Composition Model. The expression of NRTL for calculating the activity coefficient of the solvent could be divided into two parts: one part is known as the long-range interaction contribution and the other part is called the short-range interaction contribution [7]. The vapor pressure is affected mostly by the temperature and the concentration, and so does the activity coefficient. Assuming the temperature is constant, the long-range interaction can be described as:

$$\ln f_{c,LRC} = \left(\frac{1000}{M_w}\right)\frac{2A_\phi I_x^{3/2}}{1 + \chi I_x^{1/2}}, \tag{4.32}$$

where $f_{c,LRC}$ is long-range contribution; I_x is ionic strength in mole fraction scale, $I_x = 0.5\sum x_i z_i^2$; x_i is mole fraction of any species (e.g., ions) in the desiccant mixture, and z corresponds to the absolute value of ionic charge; M_w is water molecular weight, $M_w = 18$ kg/kmol; A_ϕ is Debye-Huckel constant for the osmotic coefficient, $A_\phi = 0.3910$; and χ is the closest approach parameter of the Pitzer-Debye-Huckel equation, $\chi = 14.9$.

For short-range interaction, the equation of activity coefficient calculation for water (one of the solvent) in a multi-component electrolyte could be expressed as:

$$\ln f_{w,SRC} = \frac{\sum_j X_j G_{jw} E_{jw}}{\sum_k X_k G_{kw}} + \sum_{s'} \frac{X_{s'} G_{ws'}}{\sum_k X_k G_{ks'}}\left[E_{ws'} - \frac{\sum_k X_k G_{ks'} E_{ks'}}{\sum_k X_k G_{ks'}}\right]$$
$$+ \sum_c \sum_{a'} \frac{X_{a'}}{\sum_{a''} X_{a''}}\frac{X_c G_{wc,a'c}}{\sum_k X_k G_{kc,a'c}}\left[E_{wc,a'c} - \frac{\sum_k X_k G_{kc,a'c} E_{kc,a'c}}{\sum_k X_k G_{kc,a'c}}\right]$$
$$+ \sum_a \sum_{c'} \frac{X_{c'}}{\sum_{c''} X_{c''}}\frac{X_a G_{wa,c'a}}{\sum_k X_k G_{ka,c'a}}\left[E_{wa,c'a} - \frac{\sum_k X_k G_{ka,c'a} E_{ka,c'a}}{\sum_k X_k G_{ka,c'a}}\right]. \tag{4.33}$$

The relationship between the parameters could be defined as:

$$E_{ji} = \left(g_{ji} - g_{ii}\right)/RT; E_{ji,ki} = (g_{ji} - g_{ki})/RT;$$
$$G_{ji} = \exp(-\alpha_{ji}E_{ji}); \quad G_{ji,ki} = \exp(-\alpha_{ji,ki}E_{ji,ki}).$$
$$G_{as} = G_{cs}; \quad G_{cw} = \sum_a X_a G_{ca,w}/\sum_{a'} X_{a'}; \quad G_{aw} = \sum_c X_c G_{ca,w}/\sum_{c'} X_{c'}.$$
$$\alpha_{as} = \alpha_{cs} = \alpha_{ca,s}; \quad \alpha_{sc,ac} = \alpha_{sa,ca} = \alpha_{s,ca}; \quad \alpha_{ca,s} = \alpha_{s,ca}.$$
$$E_{as} = E_{cs} = E_{ca,s}; \quad E_{sc,ac} = E_{sa,ca} = E_{s,ca}; \quad E_{ca,c'} = -E_{c',ca}; \quad E_{ca,ca'} = -E_{c',ca}.$$

In the above equations, $X_j = C_j x_j$; $C_j = |z_j|$ in the ion case and $C_j = 1$ in the molecule case. α is NRTL nonrandomness factor, $\alpha = 0.2$; g is Gibbs energy of molecules; E is NRTL binary interaction energy parameter; R is gas constant; and T is temperature in K. The subscripts: a, a' and a'' stand for anion; c, c' and c'' for cation; $i, j,$ and k for any species including anion, cation, and molecule; s and s' for solvent; w for water; $ji, ki,$ and ii for the interaction among all species; as for the interaction between the anion and the solvent; cs for the interaction between the cation and the solvent; cw for the interaction between the cation and the water; kw for the interaction between the water and any species (including anion, cation, and molecule); ws' for the interaction between the water and the solvent; ks' for the interaction between the solvent and any species (including anion, cation, and molecule); "ca, s" for the interaction between the electrolyte (ca) and the solvent; "sc, ac" for the interaction between the electrolyte (ac) and the solution with cation; "sa, ca" for the interaction between the electrolyte (ca) and the solution with anion; "s, ca" and "ca, s" for the interaction between the electrolyte (ca) and the solution; "$wc, a'c$" means the interaction between the cation (c) and the water is equivalent to that between the electrolyte ($a'c$) and the water; "$wa, c'a$" means the interaction between the anion (a) and the water is equivalent to that between the electrolyte ($c'a$) and the water; "$kc, a'c$" means the interaction between the electrolyte kc and $a'c$ when k is anion, and it means the interaction between the cation and the electrolyte ($a'c$) when k is cation; and "$ka, c'a$" means the interaction between the electrolyte ka and $c'a$ when k is cation, and it means the interaction between the anion and the electrolyte ($c'a$) when k is anion.

The whole interaction expression is:

$$\ln f_c = \ln f_{c,LRC} + \ln f_{c,SRC}. \tag{4.34}$$

The activity coefficient could be obtained with Equation (4.34) and later used to deduce the activity and the vapor pressure at a certain temperature with Equations (4.29) and (4.30).

In the case of mixed LiCl-CaCl$_2$ solution, the form of long-range interaction can be obtained as,

$$\ln f_{c,LRC} = 5.289 \frac{\left(x_{LiCl} + 3x_{CaCl_2}\right)^{3/2}}{1 + 14.9\left(x_{LiCl} + 3x_{CaCl_2}\right)^{1/2}}. \tag{4.35}$$

The form of short-range interaction can be obtained as

$$
\begin{aligned}
\ln f_{c,SRC} =\ & \frac{2x_{LiCl}G_{LiCl,w}E_{LiCl,w} + 4x_{CaCl_2}G_{CaCl_2,w}E_{CaCl_2,w}}{2G_{LiCl,w}x_{LiCl} + 4G_{CaCl_2,w}x_{CaCl_2} + x_w} \\
&+ (-x_w)\frac{2x_{LiCl}G_{LiCl,w}E_{LiCl,w} + 4x_{CaCl_2}G_{CaCl_2,w}E_{CaCl_2,w}}{\left(2G_{LiCl,w}x_{LiCl} + 4G_{CaCl_2,w}x_{CaCl_2} + x_w\right)^2} \\
&+ \left(x_{LiCl}G_{w,LiCl}\right)\frac{E_{w,LiCl}x_{LiCl} + 2x_{CaCl_2}\left[G_{CaCl_2,LiCl}\left(E_{w,LiCl} - E_{CaCl_2,LiCl}\right) + E_{w,LiCl}\right]}{\left[x_{CaCl_2}\left(2G_{CaCl_2,LiCl} + 2\right) + x_{LiCl} + x_wG_{w,LiCl}\right]^2} \\
&+ \left(2x_{CaCl_2}G_{w,CaCl_2}\right)\frac{2E_{w,CaCl_2}x_{CaCl_2} + x_{LiCl}\left[G_{LiCl,CaCl_2}\left(E_{w,CaCl_2} - E_{LiCl,CaCl_2}\right) + E_{w,CaCl_2}\right]}{\left[\left(G_{LiCl,CaCl_2} + 1\right)x_{LiBr} + 2x_{CaCl_2} + x_wG_{w,CaCl_2}\right]^2} \\
&+ \left(x_{LiCl}G_{w,LiCl}\right)\frac{E_{w,LiCl}x_{LiCl} + 2x_{CaCl_2}G_{CaCl_2,LiCl}\left(E_{w,LiCl} - E_{CaCl_2,LiCl}\right)}{\left(x_{LiCl} + 2x_{CaCl_2}G_{CaCl_2,LiCl} + x_wG_{w,LiCl}\right)^2} \\
&+ \left(2x_{CaCl_2}G_{w,CaCl_2}\right)\frac{2E_{w,CaCl_2}x_{CaCl_2} + x_{LiCl}G_{LiCl,CaCl_2}\left(E_{w,CaCl_2} - E_{LiCl,CaCl_2}\right)}{\left(x_{LiCl}G_{LiCl,CaCl_2} + 2x_{CaCl_2} + x_wG_{w,CaCl_2}\right)^2}. \tag{4.36}
\end{aligned}
$$

In the above equation, there exits the relationship:

$$x_w = 1 - 2x_{LiCl} - 3x_{CaCl_2}. \tag{4.37}$$

Table 4.6   NRTL binary interaction energy parameters for the activity coefficient calculation (25 °C)

$E_{LiCl,w}$	$E_{w,LiCl}$	$E_{CaCl_2,w}$	$E_{w,CaCl_2}$	$E_{LiCl,CaCl_2}$	$E_{CaCl_2,LiCl}$
−5.1737	10.1242	−5.2549	10.5126	0	0

Table 4.6 gives the NRTL binary interaction energy for the activity coefficient calculation at 25 °C (represented by $E\left(25°C\right)$) [41], and the NRTL binary interaction energy under the other temperatures ($E\left(t\right)$) may be calculated by the following formula assuming that the Gibbs energy of molecules would not vary with temperature:

$$E\left(t\right) = \frac{298.15}{t + 273.15} E\left(25°C\right). \tag{4.38}$$

Figure 4.9 shows the calculated partial pressure of water vapor over LiCl-CaCl$_2$ mixture solution at different mass concentrations (20, 30, and 40%) by the mixing-rule model and the local composition model and compared with the measured results given by Etras *et al.* [23]. In the calculation by the mixing-rule model, the interaction parameter is taken as the product of the weight fraction of the desiccants, that is, $c_{ab} = 1.0$ in Equation (4.28). As can be seen from Figure 4.9, the NRTL local composition model has a better agreement with the measured data by Etras *et al.* than that of the mixing-rule model.

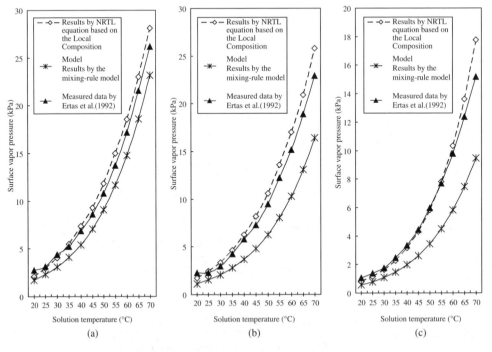

Figure 4.9   Comparisons of vapor pressure of LiCl-CaCl$_2$ mixture (1 : 1 in mass fraction) solution at different mass concentrations. (a) Total mass concentration at 20%, (b) total mass concentration at 30%, and (c) total mass concentration at 40%

4.3   Theoretical Modeling for the Ultrasound-Atomizing Regenerator

4.3.1   Assumptions

The schematic diagram of working principle for the regenerator is presented in Figure 4.10. A steady-state and one-dimensional model is employed to study the performance of the regenerator with ultrasonic atomizing. Several assumptions are given as below for the model development:

1. The regeneration process is considered as adiabatic.
2. The droplets are uniformly distributed in the regenerator and the regeneration air flows parallel to the droplets in the regenerator.
3. Heat conduction and mass transfer are neglected inside the droplet.
4. The drops are spherical and deformations are not considered.
5. The size of droplets is given on average and coalescence and collisions among droplets are not considered.

4.3.2   Basic Equations

The mass and energy conservation relations of the desiccant droplets and the regeneration air can be obtained as below:

$$G_a \mathbf{d} h_a + \mathbf{d}\left(G_s h_s\right) = 0, \tag{4.39}$$

$$G_a \mathbf{d} w_a + \mathbf{d} G_s = 0, \tag{4.40}$$

$$\mathbf{d}\left(G_s x_s\right) = 0, \tag{4.41}$$

where G_a and G_s are mass flow rate (in kg/s) of the air and the desiccant solution, respectively; h_a and h_s are enthalpy of air and solution (in kJ/kg), respectively; and x_s is the mass concentration of the desiccant solution.

Meanwhile, the energy and mass transfer between the regeneration air and the desiccant droplets can be expressed by Equations (4.42) and (4.43).

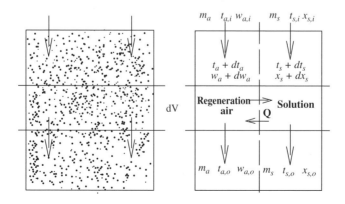

Figure 4.10   Schematic diagram for the regenerator modeling

$$G_a dh_a = H_m S_V \left(t_s - t_a\right) dV + h_{w,ts} G_a dw_a, \tag{4.42}$$

$$G_a dw_a = u_m S_V \left(\rho_{a,ts} w_a{}^* - \rho_{a,ta} w_a\right) \cdot dV, \tag{4.43}$$

where t_a, w_a, and h_a are temperature (in °C), humidity (in kg/(kg dry air)) and enthalpy (in J/kg) of regeneration air, respectively; $h_{w,ts}$ is enthalpy (in J/kg) of water vapor at the solution temperature; $\rho_{a,ta}$ and $\rho_{a,ts}$ are air density (in kg/m^3) at the regeneration air temperature and the solution temperature, respectively; S_V is specific area of contact between the regeneration air and solute droplets in m^2/m^3; H_m and u_m are convective heat coefficient (in W/(m^2·°C)) and moisture transfer velocity (in m/s), respectively; and t_s and $w_a{}^*$ are solution temperature (in °C) and humidity (in kg/(kg dry air)) of air in equilibrium with the desiccant solution, respectively.

In addition, the boundary conditions for the regeneration air and desiccant solution are given as below for the calculation:

$$t_a|_{L=0} = t_{a,in}, \quad w_a|_{L=0} = w_{a,in}, \quad t_s|_{L=0} = t_{s,in}, \quad x_s|_{L=0} = x_{s,in}, \quad G_s|_{L=0} = G_{s,in}.$$

All the governing equations are numerically computed by using a backward-finite-difference algorithm. They can be discretized as follows:

$$\left(h_a^i - h_a^{i-1}\right) + \gamma x_{s,in} \left(h_s^i/x_s^i - h_s^{i-1}/x_s^{i-1}\right) = 0, \tag{4.44}$$

$$\left(w_a^i - w_a^{i-1}\right) + \gamma x_{s,in} \left(1/x_s^i - 1/x_s^{i-1}\right) = 0, \tag{4.45}$$

$$G_s^i x_s^i = G_{s,in} x_{s,in}, \tag{4.46}$$

$$\frac{h_a^i - h_a^{i-1}}{\Delta l} = \frac{NTU \cdot Le}{L} \left[\left(h_a^{*i} - h_a^i\right) + \left(h_{w,ts}/Le - r_0\right)\left(\rho_{a,ts} w_a{}^* - \rho_{a,ta} w_a\right)\right], \tag{4.47}$$

$$\frac{w_a^i - w_a^{i-1}}{\Delta l} = \frac{NTU}{L} \left(\rho_{a,ts} w_a{}^* - \rho_{a,ta} w_a\right), \tag{4.48}$$

where $h_a{}^*$ is the air enthalpy (in J/kg) in equilibrium with the desiccant solution; r_0 is the latent heat of vaporization of water at 0 °C, J/kg; L is the height of the regenerator, m; NTU is a dimensionless parameter, $NTU = K_m S_V V/G_a$; Le is the Lewis number, $Le = \alpha_h / \left[\alpha_m \left(c_{p,a} + w_a c_{p,w}\right)\right]$; and γ is the mass flow rate ratio of liquid desiccant to air, $\gamma = G_{s,in}/G_a$. The superscript "i" and "$i-1$" stand for the adjacent positions in the regenerator. In the following calculation, total 1000 grids (i.e., $\Delta l = L/1000$) are segmented along the height of the regenerator.

4.3.3 Determination of Key Parameters

4.3.3.1 Falling Velocity of Droplet (u_f)

Assuming uniform movement, the falling velocity of droplets (u_f) can be gotten through solving Equation (4.49).

$$\frac{1}{6}\rho_s \pi d_s^3 g = \frac{1}{6}\rho_a \pi d_s^3 g + \frac{1}{8}C_f \rho_a \pi d_s^2 u_f^2, \tag{4.49}$$

where C_f is drag coefficient, which is a function of the droplet velocity.

Under the low Re ($Re < 5$), the drag coefficient (C_f) can be calculated by the relations given by Seinfeld [46]:

$$C_f = (24/Re) \left[1 + 3Re/16 + 9Re^2 \ln(2Re)/160 \right]. \tag{4.50}$$

4.3.3.2 Specific Contact Area between the Air and Desiccant Droplets (S_V)

The specific contact area between the air and desiccant droplets, S_V (in m^2/m^3), in the regenerator with ultrasonic atomizing, can be calculated by

$$S_V = \frac{6G_{s,in}}{\rho_{s,in} d_s S_{reg} u_{f-a}}, \tag{4.51}$$

where, $G_{s,in}$ is atomizing rate in kg/s; $\rho_{s,in}$ is density of inlet desiccant solution in kg/m^3; S_{reg} is cross-sectional area of regenerator in m^2; and u_{f-a} is velocity of droplet relative to the regeneration air in m/s.

4.3.3.3 Size of Atomizing Droplet (d_s)

The size of atomizing droplet (d_s) is one of the most important parameters that produce great influence on the results of desiccant solution regeneration. Theoretically, the size of atomizing droplet can be estimated by using Equation (4.22). However, it may cause a big error due to the complex uncertainties in the actual situation. So, an appropriate correction is very necessary with respect to the actual measurement.

4.3.3.4 Air Humidity Equilibrium with Desiccant Solution ($w_a{}^*$)

Humidity (in kg/(kg dry air)) of air in equilibrium with the desiccant solution $w_a{}^*$ can be calculated by:

$$w_a{}^* = 0.622 p_{w,s}/\left(B - p_{w,s}\right), \tag{4.52}$$

where B is atmospheric pressure in Pa and $p_{w,s}$ is water vapor pressure of desiccant solution in Pa. The calculation model for the vapor pressure of desiccant solution has been illustrated in the earlier section.

4.3.4 Model Validation

4.3.4.1 Raw Materials

The liquid desiccants adopted in this experimental study include LiCl solution and mixed LiCl-CaCl$_2$ solution. The mass concentration ratio of the two desiccants in the mixed solution is identically as 1 : 1.

4.3.4.2 Experimental Systems

Atomized Droplet Size Measurement
Figure 4.11 shows the schematic diagram of the experimental setup for the laser diffraction-based measurements. To obtain the droplet size information, the commercial instrument based Fraunhofer diffraction was adopted. It is comprised of two parts: the

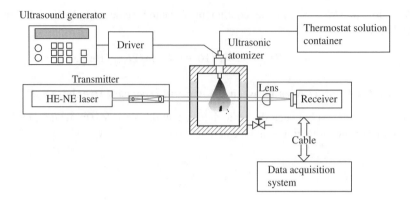

Figure 4.11 Experimental system for the atomized droplet size measurement

transmitter and receiver. The transmitter includes a continuous wave helium–neon laser (wavelength: 632.8 nm, power: 2 mW) and a beam collimator. After expanding and collimating into a beam of 8 mm diameter, the laser transmits the spray, and then the parallel lights are focused into a spot and the diffracted lights form a Fraunhofer diffraction pattern through the Fourier transformation lens set in the receiver. The intensity distribution of the diffracted lights at each ring of Fraunhofer pattern is captured by the annular sensors, and transferred to a computer for the size distribution analysis. In the data analysis system, a calibration algorithm is included to minimize the effects of multiple scattering on the measuring reliability. Because Fraunhofer diffraction pattern does not contain the spatial information of measured objects, the laser diffraction-based method can achieve merely the line-of-sight measurement of an average droplet size. For spatially resolved information, a deconvolution scheme is necessary.

An ultrasonic atomizer with the frequency of 30 kHz is employed for atomizing the desiccant solution, and the atomized droplet sizes of two kinds of desiccant solution, for example, the LiCl solution with the mass concentration of 0.30 and the LiCl-CaCl$_2$ mixture (1 : 1 in the mass fraction) solution with the total mass concentration of 0.32, have been measured under different solution temperatures including 40, 50, 60, and 70 °C.

The SMD, which is an average droplet diameter containing the information about the volume to surface ratio of the entire measured droplets (see Equation (4.25)), is used to analyze the atomized droplet size.

Solution Regeneration with Ultrasonic Atomizing

As shown in Figure 4.12, the experimental system mainly contains an atomizing chamber, an ultrasonic atomizer, a thermostatic solution tank, a fan, and wire mesh demisters. The hot-and-weak desiccant solution is atomized by an ultrasonic atomizer with the frequency of 30 kHz, and sprayed into the atomizing chamber in which the regeneration air passes through the atomizing chamber from top to bottom. The heat and mass transfer occurs between the droplets of weak solution and the regeneration air during the droplets falling. Most of these droplets fall into the strong solution tank below the atomizing chamber. The droplets entrained by the air flow will be intercepted by wire mesh demisters at the outlet of the chamber. Several temperature-and-humidity sensors (measurement error: ±0.8 in relative humidity (%), ±0.1 °C in temperature) are used for observing the inlet and outlet air conditions of the chamber, and

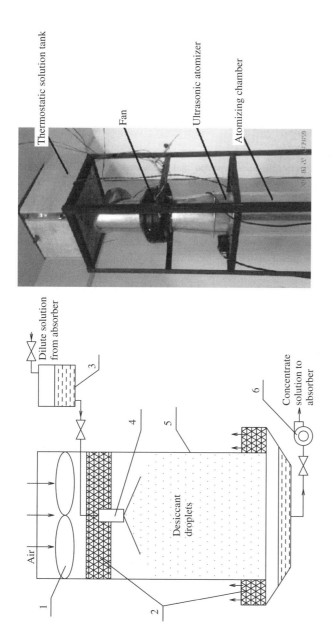

Figure 4.12 (a) Schematic and (b) field photo for liquid desiccant regenerator with ultrasonic atomizer

1: Fan; 2: Wire mesh demister; 3: Thermostatic solution tank;
4: Ultrasonic atomizer; 5: Atomizing chamber; 6: Solutions pump.

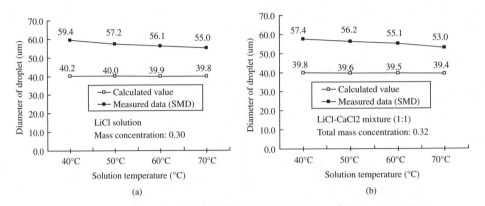

Figure 4.13 Comparisons between the calculated and measured ultrasonic atomized droplet size ($UF = 30\,$kHz). (a) For LiCl solution and (b) for LiCl-CaCl$_2$ solution

an anemometer (measurement error: $\pm5\%$ of the measured data) for the regeneration air flow rate. Several thermocouples (measurement error: $\pm0.2\,°$C) are used to measure the temperatures of inlet and outlet solution of the regenerator. The solution mass flow rate is obtained by measuring the mass flow of solution (measurement error: $\pm0.1\,$g) in a specific time period. The mass concentration of solution is obtained by measuring the mass (measurement error: $\pm0.01\,$g) and volume (measurement error: $\pm0.1\,$ml) of the solution sample, and the sample tests are made for every 1 minute during the regeneration.

4.3.4.3 Model Validation

A series of experiments under different experimental conditions have been performed to validate the proposed model. Figure 4.13 gives the experimental and calculated results on the ultrasonic atomized droplet diameter under different solution temperatures. The results show that the calculated droplet sizes got by Equation (4.25) are all much smaller than the experimental ones. In the following model simulations, the experimental droplet diameter will be adopted.

Reliable sets of experiments for the regeneration of liquid desiccant with the ultrasonic atomizing have been performed to examine the effectiveness of the model. Typically, the LiCl solution and LiCl-CaCl$_2$ (1 : 1 by mass) mixture solution are chosen as the liquid desiccant for the experiments. The states of the exit regeneration air (i.e., temperature and humidity) and the exit solution (i.e., temperature and salt mass concentration) are measured with respect to different experimental conditions. The experimental results for the two types of liquid desiccants are listed in Table 4.7 (LiCl solution) and Table 4.8 (LiCl-CaCl$_2$ mixture solution), respectively. Here, t, w, x, and G are temperature ($°$C), humidity (g/(kg dry air)), mass concentration and mass flow rate (kg/(m^2·s)), respectively; and the subscript "a," "s," "in," and "out" stand for the regeneration air, the desiccant solution, inlet, and outlet, respectively.

By using the experimental conditions including the inlet states of regeneration air and desiccant solution as well as their mass flow rates, the exit states of regeneration air and desiccant

Table 4.7 Experimental results for ultrasonic atomizing regeneration under different regeneration conditions (LiCl solution, ultrasonic frequency: 30 kHz; ultrasonic power: 60 W)

Experiment. number No.	Experimental conditions						Experimental results			
	$x_{s,in}$	$t_{a,in}$ (°C)	$w_{a,in}$ (g/(kg dry air))	G_a (kg/ (m²·s))	$t_{s,in}$ (°C)	G_s (kg/ (m²·s))	$t_{a,out}$ (°C)	$w_{a,out}$ (g/(kg dry air))	$t_{s,out}$ (°C)	$x_{s,out}$
1	0.2420	26.5	14.4	0.1344	51.4	0.0521	33.8	19.7	34.7	0.2454
2	0.2420	28.6	16.3	0.1302	52.3	0.0521	35.4	21.6	36.2	0.2453
3	0.2420	30.5	17.6	0.1287	49.8	0.0521	35.9	22.1	36.5	0.2448
4	0.2420	32.5	15.9	0.1387	51.9	0.0521	36.2	22.0	36.7	0.2459
5	0.2420	34.7	18.5	0.1321	52.6	0.0521	37.9	24.4	38.3	0.2457
6	0.2420	36.8	19.6	0.1265	50.3	0.0521	38.4	24.9	38.5	0.2453
7	0.2631	28.7	15.4	0.1298	62.2	0.0521	38.3	22.9	39.4	0.2683
8	0.2631	30.4	16.6	0.1376	61.9	0.0521	39.1	23.9	40.1	0.2682
9	0.2631	31.8	17.7	0.1345	63.2	0.0521	40.2	25.3	41.1	0.2684
10	0.2631	32.4	18.1	0.1304	59.6	0.0521	39.7	24.7	40.5	0.2676
11	0.2631	34.6	19.8	0.1268	61.2	0.0521	41.2	26.7	41.9	0.2679
12	0.2631	36.9	21.3	0.1332	62.9	0.0521	42.7	28.8	43.3	0.2683
13	0.2811	28.2	16.4	0.1305	70.5	0.0521	41.3	24.9	42.8	0.2874
14	0.2811	29.2	17.3	0.1382	71.8	0.0521	42.2	26.1	43.7	0.2876
15	0.2811	32.7	18.6	0.1332	73.2	0.0521	43.8	28.1	45.1	0.2882
16	0.2811	34.6	19.3	0.1331	72.4	0.0521	44.4	28.8	45.5	0.2882
17	0.2811	35.2	21.5	0.1289	71.0	0.0521	44.9	30.0	46.1	0.2874
18	0.2811	37.3	23.6	0.1342	70.2	0.0521	45.9	31.7	46.9	0.2871
19	0.3022	25.8	15.3	0.1304	80.5	0.0521	43.7	25.3	45.8	0.3102
20	0.3022	26.9	16.4	0.1283	79.5	0.0521	44.2	26.0	46.2	0.3099
21	0.3022	28.3	17.3	0.1295	82.5	0.0521	45.5	27.8	47.5	0.3106
22	0.3022	30.6	18.5	0.1353	80.6	0.0521	46.1	28.5	47.9	0.3102
23	0.3022	32.4	19.5	0.1311	82.7	0.0521	47.3	30.3	49.0	0.3108
24	0.3022	34.2	21.7	0.1332	78.8	0.0521	47.7	31.0	49.2	0.3096

solution can be obtained as well by the present model. The comparisons between the model results and the experimental data for the LiCl solution and the LiCl-CaCl$_2$ mixture solution are presented in Figures 4.14 and 4.15, respectively. Basically, there is an excellent agreement between the calculated results and the experimental data on the outlet temperature and salt mass concentration of the desiccant solution, and the deviations are all within 3%. However, the differences between the model and the experimental values for the outlet air temperature and humidity are relatively larger, and the maximum discrepancy is shown to be 5 and 8%, respectively.

The uncertainties of the experimental data can be estimated by the theory of experimental error analysis that has been illustrated in the earlier chapter (see "Section 2.4.5"). The experimental relative error (ERE) is identified as 0.03% for the temperature of air and solution (°C), 5.5% for the humidity of air (g/(kg dry air)), and 0.05% for the salt mass concentration.

Table 4.8 Experimental results for ultrasonic atomizing regeneration under different regeneration conditions (LiCl-CaCl$_2$ (1 : 1 by mass fraction) solution, ultrasonic frequency: 30 kHz; ultrasonic power: 60 W)

Experiment number	Experimental conditions						Experimental results			
	$x_{s,in}$	$t_{a,in}$ (°C)	$w_{a,in}$ (g/(kg dry air))	G_a (kg/ (m²·s))	$t_{s,in}$ (°C)	G_s (kg/ (m²·s))	$t_{a,out}$ (°C)	$w_{a,out}$ (g/(kg dry air))	$t_{s,out}$ (°C)	$x_{s,out}$
1	0.2451	27.5	15.2	0.1304	50.4	0.0521	33.6	20.7	34.3	0.2486
2	0.2451	28.4	16.3	0.1332	51.7	0.0521	34.6	21.9	35.2	0.2487
3	0.2451	29.5	16.9	0.1297	51.4	0.0521	35.1	22.4	35.6	0.2486
4	0.2451	31.5	17.4	0.1317	51.3	0.0521	35.7	23.1	36.2	0.2488
5	0.2451	32.7	18.2	0.1353	52.2	0.0521	36.5	24.2	36.9	0.2489
6	0.2451	33.8	19.3	0.1285	50.9	0.0521	36.9	24.8	37.3	0.2486
7	0.2622	28.6	15.8	0.1288	61.2	0.0521	37.2	23.6	38.1	0.2676
8	0.2622	29.4	16.2	0.1355	61.5	0.0521	37.6	24.2	38.5	0.2677
9	0.2622	31.2	17.5	0.1325	60.2	0.0521	38.2	25.0	39.0	0.2674
10	0.2622	32.8	18.6	0.1314	61.6	0.0521	39.3	26.5	40.1	0.2677
11	0.2622	34.2	19.2	0.1298	61.8	0.0521	39.9	27.3	40.6	0.2678
12	0.2622	36.4	21.9	0.1302	59.9	0.0521	41.0	29.1	41.5	0.2671
13	0.2841	28.7	16.7	0.1307	70.9	0.0521	40.5	26.0	41.8	0.2911
14	0.2841	29.4	17.9	0.1352	72.9	0.0521	41.5	27.6	42.8	0.2914
15	0.2841	31.5	19.0	0.1322	70.2	0.0521	41.8	28.0	43.0	0.2909
16	0.2841	33.1	19.8	0.1311	71.4	0.0521	42.8	29.3	43.8	0.2912
17	0.2841	34.3	21.7	0.1299	71.3	0.0521	43.6	30.9	44.7	0.2910
18	0.2841	36.3	22.6	0.1322	70.6	0.0521	44.3	31.7	45.2	0.2910
19	0.3062	25.8	15.3	0.1314	81.5	0.0521	42.6	26.4	44.4	0.3152
20	0.3062	26.9	16.4	0.1293	79.8	0.0521	42.9	26.9	44.6	0.3147
21	0.3062	28.3	17.3	0.1293	81.5	0.0521	43.9	28.3	45.6	0.352
22	0.3062	30.6	18.5	0.1313	80.3	0.0521	44.6	29.3	46.2	0.3150
23	0.3062	32.4	19.5	0.1301	81.7	0.0521	45.7	30.8	47.1	0.3154
24	0.3062	34.2	21.7	0.1302	80.8	0.0521	46.6	32.5	48.0	0.3150

4.3.5   Parametric Study

4.3.5.1   Analysis Parameters

Three parameters are used to evaluate the performance of the regenerator. The first parameter is regeneration rate (RR), which can be expressed as:

$$RR = G_a \left(w_{a,out} - w_{a,in} \right), \tag{4.53}$$

where G_a is mass flow of regeneration air in kg/s and $w_{a,in}$ and $w_{a,out}$ are humidity of regeneration air (in kg/(kg dry air)) entering and leaving the regenerator, respectively.

The RR reflects directly the dehydration rate of desiccant solution in a regeneration process. A higher RR means a larger capacity of dehumidification in a specific time. In a real desiccant system, the minimum RR is required for the need of the actual running. Normally, the RR can be improved through altering the regeneration conditions (e.g., increase the temperature of the

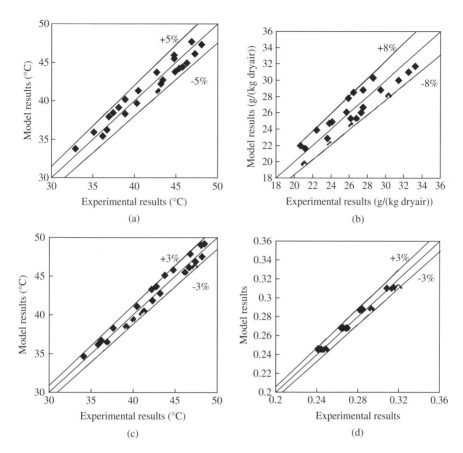

Figure 4.14   Comparisons of model results with the experimental data (LiCl solution). (a) For exit regeneration air temperature (°C), (b) for exit regeneration air humidity (g/(kg dry air)), (c) for exit solution temperature (°C), and (d) for exit solution mass concentration

solution or the regeneration air) and the physical structure of regenerator (e.g., increase the contact area or contact time between the solution and regeneration air).

The second parameter is the regeneration effectiveness (*RE*), which is defined by Equation (4.54).

$$RE = \frac{w^*_{a,in} - w^*_{a,out}}{w^*_{a,in} - w_{a,in}}. \tag{4.54}$$

where $w^*_{a,in}$ and $w^*_{a,out}$ are air humidity (in kg/(kg dry air)) in equilibrium with the desiccant solution entering and leaving the regenerator, respectively; and they are dependent of the temperature and mass concentration of solution.

According to Equation (4.54), the regeneration effectiveness (*RE*) will be affected directly by the inlet solution state (temperature and mass concentration) and the inlet regeneration air conditions (temperature and humidity). Essentially, it is closely related to the structure of the regenerator and the way of regeneration. The *RE* reflects the regeneration degree of the liquid

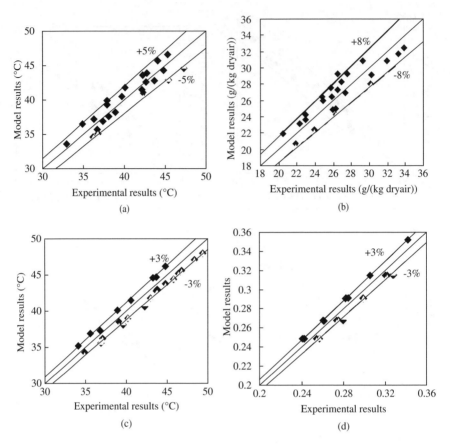

Figure 4.15 Comparisons of model results with the experimental data (LiCl-CaCl$_2$ mixture solution). (a) For exit regeneration air temperature (°C), (b) for exit regeneration air humidity (g/(kg dry air)), (c) for exit solution temperature (°C), and (d) for exit solution mass concentration

desiccant by the regenerator. A higher *RE* means that the heat and mass transfer between the solution and the regeneration air in the regenerator proceeds more fully.

The third parameter is the specific energy consumption (SEC: J/(kg water)), which is defined as:

$$SEC = \frac{E_{\text{reg}}}{m_{\text{reg}}}, \tag{4.55}$$

where E_{reg} is energy consumption (in J) for a specific time period of regeneration, which includes the energy consumption by the solution heater, the ultrasonic atomizer, the fan, and the solution pump and m_{reg} is mass of moisture desorption during the regeneration in kg.

The power of the solution heater (P_{heater}) can be calculated by:

$$P_{\text{heater}} = c_{\text{s}} S_{\text{reg}} G_{\text{s}} \left(t_{\text{s,in,heater}} - t_{\text{s,out,heater}} \right) / \eta_{\text{heater}}, \tag{4.56}$$

where c_{s} is the specific heat of the solution, J/(kg·°C); S_{reg} is the cross-sectional area of the regenerator, m^2; G_{s} is the mass flow rate of the solution passing through the solution heater, kg/(m^2·s); $t_{s,in,heater}$ and $t_{s,out,heater}$ are the inlet and the outlet solution temperature of the solution heater, °C; and η_{heater} is the heating efficiency of the heater.

The power of the fan (P_{fan}) and the solution pump (P_{pump}) can be calculated by Equations (4.57) and (4.58), respectively.

$$P_{fan} = \frac{S_{reg}G_a\Delta p_{packing}}{\eta_{fan}\rho_{a,reg}},$$ (4.57)

$$P_{pump} = S_{reg}G_s\Delta p_{pump}/\left(\eta_{pump}\rho_{s,reg}\right),$$ (4.58)

where, G_a and G_s are mass flow rate of regeneration air and solution passing through the regenerator, respectively, kg/(m^2·s); $\rho_{a,reg}$ and $\rho_{s,reg}$ are density of air and solution in the regenerator, respectively, kg/m^3; and η_{fan} and η_{pump} are efficiency of fan and pump, respectively.

Δp_{pump} is pressure head required for solution spraying and overcoming the fluid flow resistances including frictional drag and local resistance; $\Delta p_{packing}$ is air pressure drop in the packing stuff (Pa), which can be calculated by Equation (4.59).

$$\Delta p_{packing} = \frac{C_f G_a{}^2}{\rho_a}.$$ (4.59)

where the drag coefficient C_f for different packing stuffs can be available in Ref. [47].

4.3.5.2   Simulation Results

The parametric studies on the performance of the solution regenerator with ultrasonic atomizing are made under different conditions. The baseline simulation conditions are listed as follows: 25 °C and 0.01 kg/(kg dry air) in the inlet air temperature and humidity, respectively; 3.1847 and 0.796 kg/(m^2·s) in the regeneration air mass flow rate and the solution mass flow rate, respectively; 60 °C in the inlet solution temperature; 0.28 in the inlet solution mass concentration; 0.1 m in the height of regeneration space. The influence of each parameter on the performance of the regenerator is to be investigated while the other parameters are held as the baseline conditions. Other conditions for the simulation are given as below: the height of the packing stuff in the regenerator (wire mesh demister) is 0.05 m; the drag coefficient C_f is 909; the ultrasonic frequency is 30 kHz; the efficiency of fan and pump is 0.7 and 0.85, respectively; and the LiCl solution is used for this study.

Influence of Inlet Solution Temperature

Figure 4.16 shows the influence of the inlet solution temperature on the outlet parameters of the regenerator with ultrasonic atomizing. It can be seen from Figure 4.16 that all the outlet variables, including the exit regeneration air temperature and humidity, and the exit solution temperature and mass concentration, increase linearly with the inlet solution temperature. The increase of the inlet solution temperature causes the vapor pressure on the solution to rise and brings about an increase in the vapor pressure difference between the regeneration air and the desiccant solution. As a result, the exit regeneration air humidity and the exit solution concentration all increase. Meanwhile, the increase of the inlet solution temperature will result in an increased heat transfer rate from the solution to the air, and hence, the exit air humidity increases. Although the increased water desorption rate will take away more heat from the solution, the increased heat loss of solution is still smaller than the increased heat gain due to

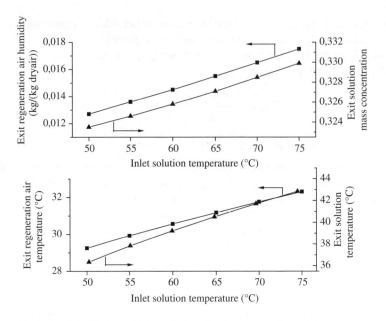

Figure 4.16   Influence of inlet solution temperature on the outlet parameters

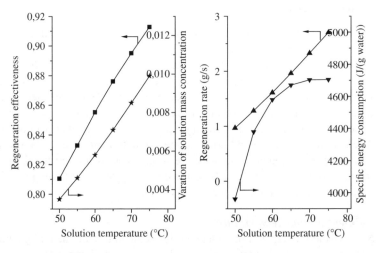

Figure 4.17   Influence of inlet solution temperature on the performance of regenerator with ultrasonic atomizing

the increase of inlet solution temperature. As a result, the exit solution temperature increases with the inlet temperature rising.

The inlet solution temperature will produce great influence on the performances of the regenerator with ultrasonic atomizing. As shown in Figure 4.17, the RR, the regeneration effectiveness (RE), and the variations of the solution mass concentration all increase linearly with the inlet solution temperature rising. In this case, the RE increases from about 0.81 to 0.91 and the

RR increases from about 1.0 to 3.0 g/s as the inlet solution temperature rises from 50 to 80 °C. However, increasing the inlet solution temperature needs more energy consumption. As seen from the curve of the specific energy consumption (SEC) in Figure 4.17, the SEC has a significant increase when the inlet solution temperature increases from 50 to 65 °C. Afterwards, the change rate of the SEC against the inlet solution temperature becomes increasingly smaller.

Influence of Inlet Solution Mass Concentration

Figure 4.18 shows the influence of inlet solution concentration on the outlet parameters of the regenerator with ultrasonic atomizing. From Figure 4.18, we can see that the exit regeneration air humidity decreases while the other three exit parameters increase linearly with the inlet solution concentration increasing. This is because the increase in concentration of inlet solution will reduce the vapor pressure of the solution and lead to a decrease in the moisture transfer rate from the desiccant to the air.

The influence of the inlet solution concentration on the performances of the regenerator is presented in Figure 4.19. As seen from Figure 4.19, the RR and the RE as well as the variations of the solution mass concentration all decrease with the increase of the inlet solution concentration. In this case, the RR, the RE and the variation of solution mass concentration decreases by about 13, 0.5, and 6.4%, respectively, as the inlet solution mass concentration increases from 0.30 to 0.32. By contrast, the SEC increases as the inlet solution concentration increases. This indicates that a lower inlet solution concentration will be conductive to improving the energy efficiency of the regenerator with ultrasonic atomizing.

Influence of Inlet Air Temperature

The results in Figure 4.20 show that all the outlet parameters of the regenerator increase with the inlet regeneration air temperature increasing. The increase of the air temperature results in the decrease of water vapor pressure in the air, and this will increase the moisture transfer rate

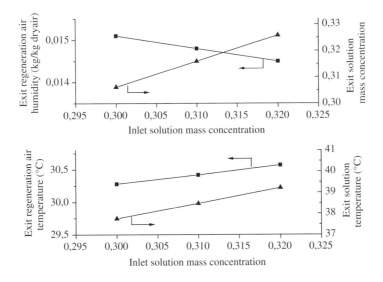

Figure 4.18   Influence of inlet solution mass concentration on the outlet parameters

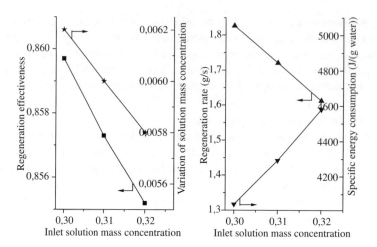

Figure 4.19 Influence of inlet solution mass concentration on the performance of regenerator with ultrasonic atomizing

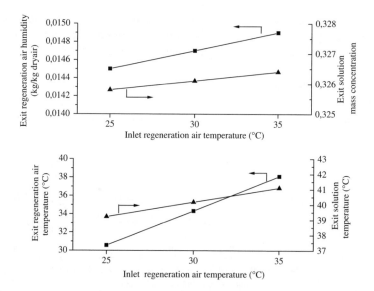

Figure 4.20 Influence of inlet regeneration air temperature on the outlet parameters

from the desiccant solution to the regeneration air. So, it is reasonable to see that the exit air humidity and the exit solution concentration have a slight increase as the inlet regeneration air temperature rises. On the other hand, the increase of the air temperature will decrease the heat transfer rate from the solution to the air. As a result, the exit temperatures of the air and the solution increase with the increase of the inlet air temperature.

The influence of the inlet regeneration air temperature on the performances of the regenerator with ultrasonic atomizing is shown in Figure 4.21. Although the RR and the variation of the solution concentration increase as the inlet air temperature increases, the RE decreases as a

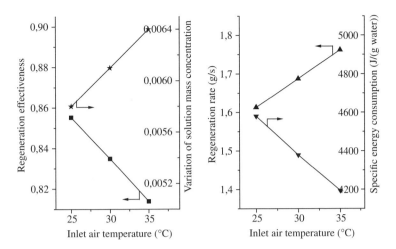

Figure 4.21　Influence of inlet regeneration air temperature on the performance of regenerator with ultrasonic atomizing

result of offsetting effects between the numerator and denominator of the definition of RE (Equation (4.54)). The increase of the inlet air temperature means more thermal energy of solution can be utilized for the moisture desorption, which leads to a decrease in the SEC during the desiccant regeneration.

Influence of Inlet Air Humidity

From Figure 4.22, the exit solution concentration decreases while the other outlet parameters increase as the inlet regeneration air humidity increases. The higher humidity of air means the higher partial water vapor pressure in the air, which will decrease the moisture transfer rate from the solution to the air. So, the exit solution concentration decreases. The decreased moisture transfer rate results in the decreased heat loss of solution due to the water evaporation, and hence, the exit solution temperature increases. And the increased solution temperature will make the air gain more sensible heat. As a result, the exit air temperature increases with the increase of the inlet regeneration air humidity.

The influence of the inlet air humidity on the performances of the regenerator is presented in Figure 4.23. The RE and the SEC are shown to increase with the inlet air humidity increasing, while it is the opposite for the RR and the variations of solution concentration. Although the higher inlet air humidity can bring about the higher RE, it is not preferable for the energy efficiency of regeneration. In the actual situations, we expect the SEC to be as low as possible.

Influence of Regenerator Height

The outlet parameters of the regenerator will be affected as well by the height of regenerator. As shown in Figure 4.24, the exit regeneration air temperature and humidity and the exit solution mass concentration all increase, while the exit solution temperature decreases as the height of regenerator increases. This is because increasing the regenerator height means prolonging the contact time between the solution and the regeneration air, hence increasing the heat and mass transfer capacity from the solution to the regeneration air. As can be seen as well from Figure 4.24, the change in the exit regeneration air humidity and the exit solution concentration

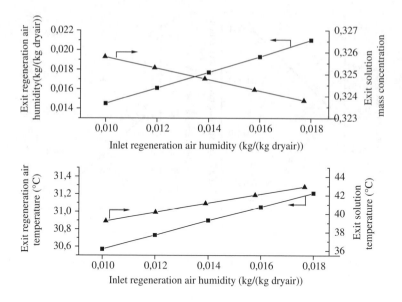

Figure 4.22   Influence of inlet regeneration air humidity on the outlet parameters

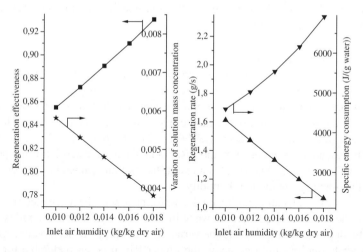

Figure 4.23   Influence of inlet regeneration air humidity on the performance of regenerator with ultrasonic atomizing

becomes increasingly smaller after the regenerator height attains one specific value. For this case, the exit regeneration air humidity and the exit solution concentration hardly increases after 0.125 m in the regenerator height.

From Figure 4.25, the RE and the RR all increase with the height of regenerator increasing, but their increase rates tend to become small. It is also reasonable to see that the SEC of regeneration decreases with the increase of regenerator height and the decrease rate becomes increasingly smaller. Although increasing the regenerator height can improve the performances

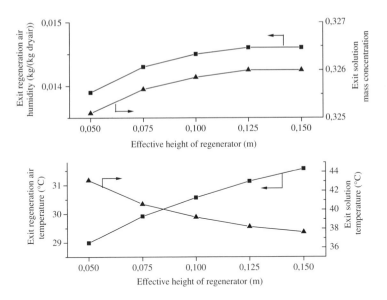

Figure 4.24   Influence of regenerator height on the outlet parameters

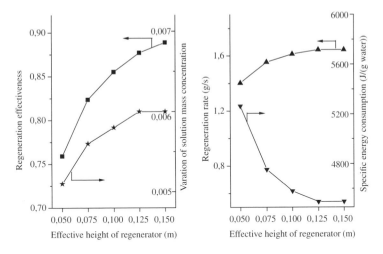

Figure 4.25   Influence of regenerator height on the performance of regenerator with ultrasonic atomizing

of the regenerator, the initial cost will increase at the same time. So, there is a balance between them in the design of the regenerator height. For this case, the regenerator height may be favorably as 0.125 m.

Influence of Gas–Liquid Mass Flow Ratio
The gas–liquid ratio is another key operation parameter affecting the regenerator's performance. Since the gas–liquid ratio is related to the mass flow rate of the regeneration air and

the solution in the regenerator, it varies based on the following two situations: one is that the air mass flow rate is held constant (3.1847 kg/(m²·s)), and the other is that the solution mass flow rate is held constant (0.796 kg/(m²·s)). Figures 4.26 and 4.27 show the influence of the gas–liquid ratio (ranging from 1.0 to 4.0) on the outlet parameters and performances of the regenerator in terms of the condition that the air mass flow rate is kept as 3.1847 kg/(m²·s). As shown in Figure 4.26, the exit regeneration air temperature and humidity and the exit solution temperature decreases with the increase in the gas–liquid ratio, and the exit solution concentration increases with the gas–liquid ratio increasing. This is because the increase in the

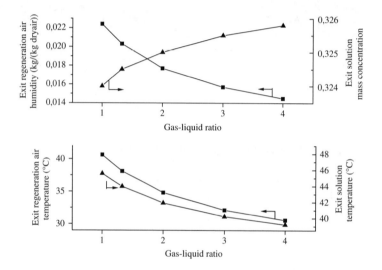

Figure 4.26   Influence of gas–liquid ratio on the outlet parameters (air flow rate: 3.1847 kg/(m²·s))

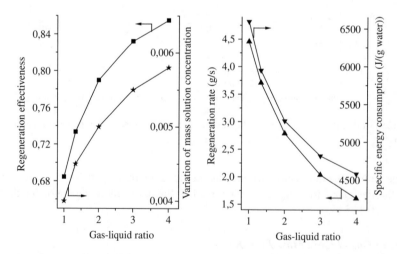

Figure 4.27   Influence of gas–liquid ratio on the performance of regenerator with ultrasonic atomizing (air flow rate: 3.1847 kg/(m²·s))

gas–liquid ratio means the decrease of the solution mass flow rate in the regenerator as the air mass flow rate is unchanged.

Figures 4.28 and 4.29 show the influence of the gas–liquid ratio (ranging from 3.0 to 7.0) on the outlet parameters and performances of the regenerator in terms of the condition that the solution mass flow rate is kept as 0.796 kg/(m²·s). Under such circumstance, the exit regeneration air temperature and humidity decreases with the increase of the gas–liquid ratio; the exit solution mass concentration first increases and then decreases with the gas–liquid ratio increasing, and it is the opposite for the exit solution temperature.

Figure 4.29 shows the influence of the gas–liquid ratio on the performances of the regenerator with ultrasonic atomizing. It can be inferred from Figure 4.29 that there should be a best

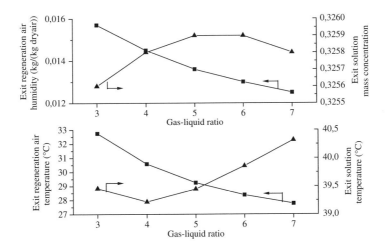

Figure 4.28 Influence of gas–liquid ratio on the outlet parameters (solution flow rate: 0.796 kg/(m²·s))

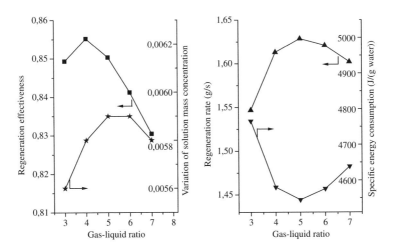

Figure 4.29 Influence of gas–liquid ratio on the performance of regenerator with ultrasonic atomizing (solution flow rate: 0.796 kg/(m²·s))

gas–liquid ratio under which the RR and the RE will be the highest, and the SEC will be the lowest. According to the results in Figure 4.29, the gas–liquid ratio for this case is preferably about 5.0.

Influence of Ultrasonic Frequency

Known from Equation (4.22), the acoustic frequency is one of the most important parameters that affect the atomized droplet size, and then impact the outlet parameters and performances of the regenerator. As indicated from Figure 4.30, the influence of ultrasonic frequency on the exit regeneration air humidity and the exit solution mass concentration is very small. The change percentage is only within 2 and 0.3%, respectively, for the exit regeneration air humidity and the exit solution mass concentration as the ultrasonic frequency changes from 20 to 50 kHz. As also seen from Figure 4.30, the exit regeneration air humidity and the exit solution mass concentration increases firstly and then decreases with the ultrasonic frequency increasing, This phenomenon can be explained by the following reason: the moisture mass transfer rate between the droplet and the regeneration air increases greatly with the decrease of the droplet size as a result of an increase in the ultrasonic frequency, and the solution droplet may be equilibrium with the regeneration air in the mass transfer before leaving the regenerator. But the heat transfer from the solution droplet to the regeneration air still goes on, and the solution droplet temperature continues decreasing and the regeneration air continues increasing until the partial water vapor pressure on the solution droplet is lower than that in the regeneration air. Under such circumstance, the desiccant solution will absorb moisture from the regeneration air before leaving the regenerator, and this causes the solution mass concentration to drop. The influence of the ultrasonic frequency on the performances of the regenerator is shown in Figure 4.31, which indicates that the acoustic frequency is another important parameter that should be seriously considered in the design of the regenerator. According to the results

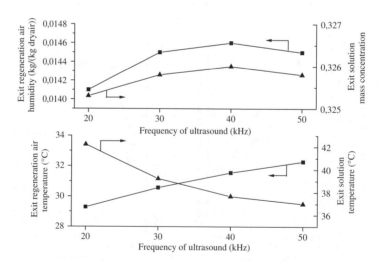

Figure 4.30   Influence of ultrasonic frequency on the outlet parameters

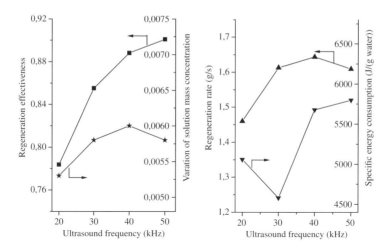

Figure 4.31 Influence of ultrasonic frequency on the performance of regenerator with ultrasonic atomizing

in Figure 4.31, the acoustic frequency used for the regenerator is preferably about 30 kHz in this case.

LiCl Solution versus LiCl-CaCl₂ Mixture Solution

LiCl Solution versus LiCl-CaCl$_2$ Mixture Solution

Figure 4.32 compares the performances of the ultrasonic atomizing regenerator with LiCl solution to that with LiCl-CaCl$_2$ mixture solution. The basic calculation conditions are as below: regeneration air mass flow rate: 3.1847 kg/(m²·s); solution mass flow rate: 0.796 kg/(m²·s); inlet solution temperature: 60 °C; inlet solution mass concentration (total salt mass concentration): 0.24; height of regeneration space: 0.15 m; height of the packing stuff in the regenerator (wire mesh demister): 0.05 m; the drag coefficient C_f: 909; ultrasonic frequency: 30 kHz. It can be seen from Figure 4.32 that the regeneration performances of the two types of liquid desiccants share the same regular pattern of variation with the inlet regeneration air conditions (i.e., temperature and humidity), and the LiCl-CaCl$_2$ mixture solution has a better regenerability than the LiCl solution under the same regeneration conditions.

4.4 Performance Analysis of Liquid-Desiccant Dehumidification System with Ultrasound-Atomizing Regeneration

4.4.1 The Ultrasound-Atomizing Regenerator versus the Packed One

Different from the conventional packed regenerator, the ultrasound-atomizing regenerator has no packing stuff in the solution regeneration zone. The model for calculating the ultrasound-atomizing regenerator can be applied as well to the packed regenerator except for some key model parameters, including the specific contact area (S_V) and the coefficients of heat and mass transfer (i.e., H_m and K_m), between the regeneration air and the desiccant solution in the regenerator.

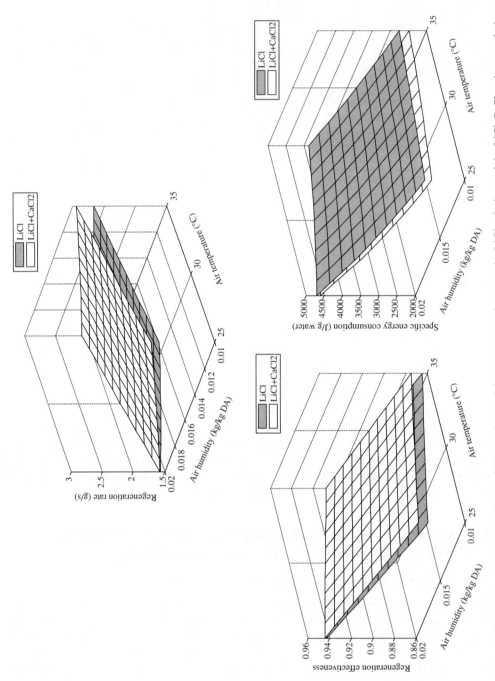

Figure 4.32 Performance comparisons of ultrasonic atomizing regenerator between the LiCl solution and the LiCl-CaCl$_2$ mixture solution

For the packed regenerator, the specific contact area (S_V) is gotten by Equation (4.60) [47], and the heat and mass transfer coefficients (i.e., H_m and K_m) are expressed by Equations (4.61) and (4.62), respectively [48].

$$S_V = n_1 \left(808 G_s / \rho_a^{0.5}\right)^{n_2} \left(G_s\right)^{n_3}, \tag{4.60}$$

$$H_m = 38.34 \left(G_s\right)^{0.23} \left(G_a\right)^{0.64} e^{0.0004 t_a} / S_V, \tag{4.61}$$

$$K_m = 1.458 \left(G_s\right)^{0.23} \left(G_a\right)^{0.64} e^{0.00034 t_a} / S_V. \tag{4.62}$$

In Equations (4.60)–(4.62), G_s and G_a are mass flow rate (kg/(m^2 s)) of the liquid desiccant and the regeneration air, respectively; t_a is the regeneration air temperature in °C; the coefficients n_1, n_2, and n_3 in Equation (4.60) are given as below: for $0.68 < G_s < 2.0$ kg/(m^2·s): $n_1 = 28.01$, $n_2 = 0.2323 G_s - 0.3$, $n_3 = -1.04$; and for $2.0 < G_s < 6.1$ kg/(m^2·s): $n_1 = 14.69$, $n_2 = 0.01114 G_s + 0.148$, $n_3 = -0.111$. In the following comparison analysis, ceramic is used as the packing stuff in the packed regenerator, and the corresponding height and drag coefficient (C_f) are 0.15 and 580 m, respectively. For the ultrasound-atomizing regenerator (ultrasonic frequency: 30 kHz), the packing stuff is actually the wire mesh demister whose height and drag coefficient (C_f) are considered as 0.05 and 909 m, respectively; and the height of regeneration zone is also 0.15 m. The LiCl solution and the LiCl-CaCl$_2$ solution are employed, respectively, for the comparison study, and the inlet solution temperature and mass concentration are identically taken as 60 °C and 0.30, respectively.

Figure 4.33 shows the performance comparisons between the ultrasound-atomizing regenerator and the packed one for different gas–liquid ratios (fixed solution mass flow rate of

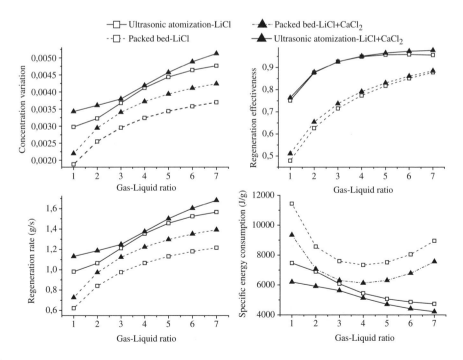

Figure 4.33 Performance comparisons between the ultrasound-atomizing regenerator and the packed one under different gas–liquid ratios (fixed solution mass flow rate; typical spring conditions)

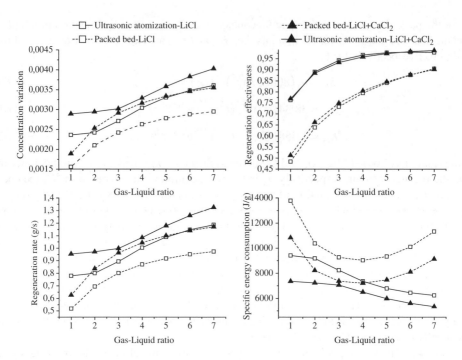

Figure 4.34 Performance comparisons between the ultrasound-atomizing regenerator and the packed one under different gas–liquid ratios (fixed solution mass flow rate; typical summer conditions)

0.796 kg/(m²·s)) under the typical spring conditions (24 °C and 85%, respectively, in the ambient air temperature and relative humidity), and Figure 4.34 shows that under the typical summer conditions (32 °C and 80%, respectively, in the ambient air temperature and relative humidity).

Basically, with the gas–liquid ratio increasing from 1.0 to 7.0, the RR, the RE, and the variations of solution mass concentration of both regenerators all increase; the SEC for the ultrasound-atomizing regenerator decreases all along, and that for the packed regenerator decreases firstly and then increases (the lowest SEC of the packed regenerator occurs at the gas–liquid ratio of around 4.0). For the ultrasound-atomizing regenerator, the change rate of RE becomes inappreciable after 4.0 in the gas–liquid ratio. Although the lower SEC of the ultrasound-atomizing regenerator may be achieved under the higher gas–liquid ratio, the air flow rate should not be too large in order to prevent the solution droplets from being carried over by the regeneration air.

Comparing the results in Figure 4.33 with that in Figure 4.34, we can see that the RR of the regenerators under the typical spring conditions is higher than that under the typical summer ones, and the RE and the SEC under the former conditions are lower than that under the latter ones. This is because the ambient air conditions will produce great influence on the performances of the regenerator. As illustrated in Figure 4.35, the higher humidity of air (in kg/(kg dry air)) will result in the lower RR and the higher RE or SEC. In the actual operations, we always expect the RR of regenerator to be as high as possible and the SEC to be as low as possible.

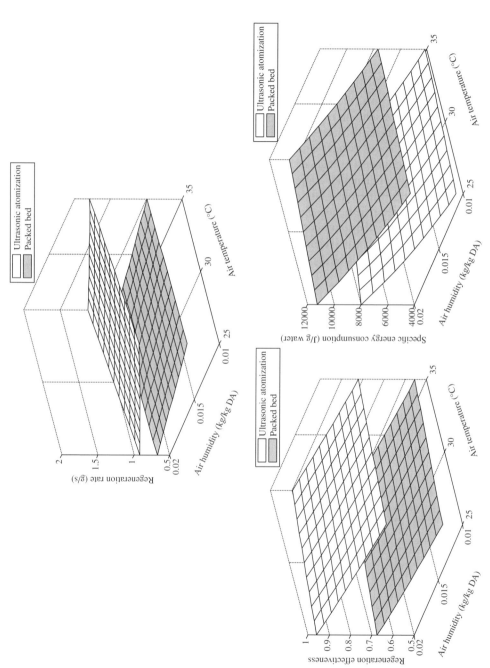

Figure 4.35 Performance comparisons between the ultrasound-atomizing regenerator and the packed one under different ambient air temperatures

From the results in Figures 4.33–4.35, we can safely conclude that the ultrasound-atomizing regenerator will have a considerably better performance than the packed one under different gas–liquid ratios (fixed solution mass flow rate, typical spring conditions).

4.4.2 Performance of Liquid Desiccant System with Different Regenerators

4.4.2.1 Methods

Figure 4.36 shows the schematic diagram for a liquid desiccant system. The system mainly consists of a solution regenerator, an air dehumidifier, a solution heater, a solution-solution heat exchanger, a water-to-solution heat exchanger and two solution pumps. The solution-solution heat exchanger is used for the heat recovery from the regenerated solution. The water-to-solution heat exchanger is used for cooling the strong solution to further gain a better performance of dehumidification.

The models for the ultrasound-atomizing and the packed regenerator/dehumidifier have been illustrated in the earlier sections, and the effectiveness-NTU model is employed for the heat exchangers (including the solution-to-solution and the water-to-solution one). The effectiveness ($\varepsilon_{\text{heat exchanger}}$) of a heat exchanger is defined as the ratio of the actual heat transfer rate (Q_{act}) to the maximum possible heat transfer rate (Q_{max}):

$$\varepsilon_{\text{heat exchanger}} = Q_{\text{act}}/Q_{\text{max}}. \tag{4.63}$$

The maximum possible heat transfer rate (Q_{max}) can be written as:

$$Q_{\text{max}} = C_{\text{min}}\left(t_{h,i} - t_{c,i}\right). \tag{4.64}$$

The actual heat transfer rate (Q_{max}) can be expressed by Equation (4.65) in terms of the hot fluid or Equation (4.66) in terms of the cool fluid.

$$Q_{\text{act}} = C_h\left(t_{h,i} - t_{h,o}\right), \tag{4.65}$$

$$Q_{\text{act}} = C_c\left(t_{c,i} - t_{c,o}\right). \tag{4.66}$$

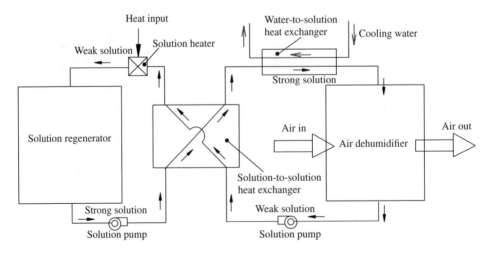

Figure 4.36   Schematic diagram for a liquid desiccant system

Normally, the counter flow heat exchanger is preferred because of its relatively high efficiency of heat exchange, and its effectiveness ($\varepsilon_{heat\,exchanger}$) can be obtained by Equation (4.67a) [49].

$$\varepsilon_{heat\,exchanger} = \frac{1 - \exp\left[-NTU\left(1 - C_r\right)\right]}{1 - C_r \exp\left[-NTU\left(1 - C_r\right)\right]}, \tag{4.67}$$

$$C_r = C_{min}/C_{max}, \tag{4.67a}$$

$$NTU = \frac{U_{hx}S_{hx}}{C_{min}}. \tag{4.67b}$$

In Equations (4.65)–(4.67a), C_{min} and C_{max} stand for the smaller and the larger fluid heat capacity rate, respectively. C_{min}/C_{max} is equal to C_c/C_h or C_h/C_c, depending on the relative magnitudes of the hot (denoted by the subscript "h") and cold (denoted by the subscript "c") fluid heat capacity rates. U_{hx} is the overall heat transfer coefficient of the heat exchanger in W/(m²·°C). A is the total surface area of the heat exchanger.

For a specific heat exchanger, the exit temperature of the hot fluid and the cool fluid can be counted by the effectiveness-NTU model illustrated above with respect to the inlet temperatures and mass flow rate of the two fluids of the heat exchanger.

The COP (coefficient of performance) is used to evaluate the energy efficiency of the desiccant system, which is defined as:

$$COP = \frac{S_{a,deh}\,G_{a,deh}\left(h_{a,in,deh} - h_{a,out,deh}\right)}{P_{heater} + P_{fan} + P_{pump}}, \tag{4.68}$$

where $G_{a,deh}$ is mass flow rate of air passing through the air dehumidifier, kg/(m² s); $S_{a,deh}$ is cross-sectional area of the air dehumidifier, m²; $h_{a,in,deh}$ and $h_{a,out,deh}$ are the inlet and the outlet air enthalpy (J/kg) of the air dehumidifier, respectively; and P_{heater}, P_{fan}, and P_{pump} are power (W) of the solution heater, the fans, and the solution pumps in the system, respectively. The models for P_{heater}, P_{fan}, and P_{pump} have been given by Equations (4.56)–(4.58), respectively.

The calculation procedure for the desiccant system simulation is illustrated in Figure 4.37.

4.4.2.2 Case Study

The performance of the liquid desiccant (LiCl-CaCl$_2$ mixture with 1 : 1 by mass fraction) system with the ultrasound-atomizing regenerator is compared that with the packed regenerator. The two systems share the same air dehumidifier with the height of 0.5 m and the cross-sectional area of 0.1256 m². The packed regenerator has the height of 0.15 m and the cross-sectional area of 0.1256 m², and the ultrasound-atomizing one has the height of 0.05 m and the cross-sectional diameter of 0.4 m. The Raschig ring (whose drag coefficient C_f is considered as 909) is employed as the packing material for the packed regenerator or dehumidifier. The acoustic frequency of 30 kHz is employed for the ultrasound-atomizing regenerator. The solution-to-solution and the water-to-solution heat exchanger adopts the stainless steel plate type, and their total heat transfer coefficients are about 1800 and 2400 W/(m²·°C), respectively. The circulation flow rate of the desiccant solution for the two systems are taken identically as 0.3 kg/s, and the air flow rate passing through the dehumidifier is identically as 0.2 kg/s. The

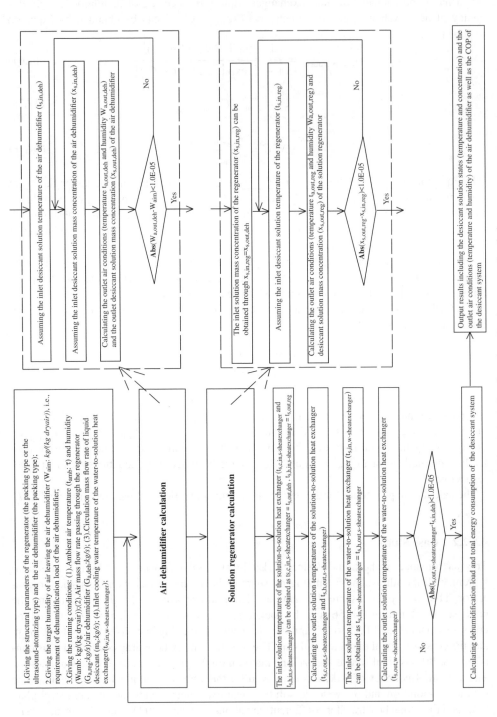

Figure 4.37 Calculation flowchart for the system simulation

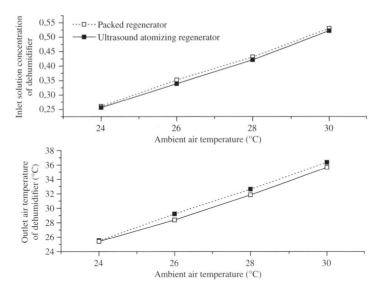

Figure 4.38 Comparisons of some key variables between the two systems with different regenerators under different ambient air temperatures

coolant water flow rate and the inlet water temperature of the water-to-solution heat exchanger are 0.5 kg/s and 25 °C, respectively.

The systems are simulated under different ambient air conditions (temperature or humidity) and gas–liquid ratios of regenerator. The outlet air of the dehumidifier meets the requirement of humidity of 0.014 kg/(kg dry air) for all the simulation conditions. So, the solution temperature required for the regeneration should be adjusted with the change of the ambient air conditions.

Figure 4.38 gives the comparisons of the inlet solution mass concentration and outlet air temperature of dehumidifier between the two systems with different regenerators under different ambient air temperatures. It shows that the two variables increase with the ambient air temperature increasing. As illustrated earlier, the higher ambient air temperature will be conducive to improving the desiccant RR, and hence, the higher mass concentration of the regenerated solution (i.e., the inlet solution concentration of the air dehumidifier) can be achieved. The influence of the inlet air temperature on the outlet air temperature of the air dehumidifier is decisive. So, the outlet air temperature will increase with the increase of the inlet air temperature. It can be seen as well from Figure 4.38 that the inlet solution concentration of dehumidifier with the ultrasound-atomizing regenerator is slightly lower than that with the packed one. This is because the outlet solution temperature of the ultrasound-atomizing regenerator (or the inlet solution temperature of dehumidifier) is lower than that of the packed regenerator. As we know, the lower desiccant solution temperature will have a better performance of dehumidification. So, the desiccant solution concentration required for the same dehumidification load can be reduced if the solution temperature decreases. In addition, the lower inlet solution temperature of the dehumidifier will result in the lower outlet air temperature of the dehumidifier. Therefore, the air dehumidifier with the ultrasound-atomizing regenerator has the lower outlet air temperature than that with the packed regenerator.

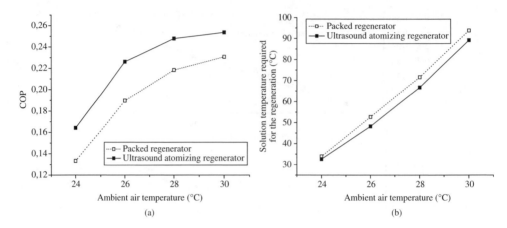

Figure 4.39 Comparisons of COP and desiccant regeneration temperature between the two systems with different regenerators under different ambient air temperatures. (a) COP against ambient air temperature and (b) required regeneration temperature against ambient air temperature. Ambient air relative humidity: 80%; gas–liquid ratio of regenerator: 2.5

From Figure 4.39, we can see that the desiccant system with the ultrasound-atomizing regenerator has a higher COP and a lower regeneration temperature compared with the packed regenerator. This manifests that the ultrasound-atomizing regeneration, on the one hand, will improve the energy performance of the desiccant system directly, and on the other hand, will reduce the requirement of regeneration temperature of the desiccant and promote the utilization of the low grade energy for the dehumidification applications. In addition, the COPs of the two desiccant systems are found to increase with the increase of ambient air temperature, but the increase rate tends to be small.

The comparison analysis between the two systems with different regenerators is also made under different ambient air humidity ratios, as shown in Figures 4.40 and 4.41. Likewise, the inlet solution mass concentration, the outlet air temperature of the dehumidifier, the required solution regeneration temperature (i.e., the inlet solution temperature of regenerator) and the COP of the system all increase with the ambient air relative humidity rising. Meanwhile, the system with the ultrasound atomizing regenerator is shown to have an obviously higher COP than that with the packed regenerator under different ambient air humidity ratios, especially under the lower relative humidity of ambient air.

Figure 4.42 shows the comparisons of the inlet solution mass concentration and the outlet air temperature of the dehumidifier between the two systems with different regenerators under different gas–liquid ratios of regenerator. It can be seen that the two variables (i.e., the inlet solution mass concentration and outlet air temperature of dehumidifier) of the systems both decrease with the gas–liquid ratio of regenerator increasing, and the influences of the gas–liquid ratio of regenerator on these variables are much more significant for the system with the packed regenerator than that for the system with the ultrasound-atomizing regenerator.

Figure 4.43 shows the influence of the gas–liquid ratio of regenerator on the COP and required desiccant regeneration temperature of the two systems. It manifests that the gas–liquid ratio of regenerator produces little influence on the COP of the system with the

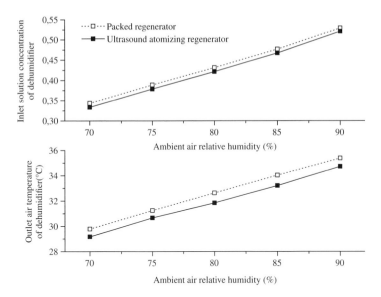

Figure 4.40 Comparisons of some key variables between the two systems with different regenerators under different ambient air humidity ratios. Ambient air temperature: 28 °C; gas–liquid ratio of regenerator: 2.5

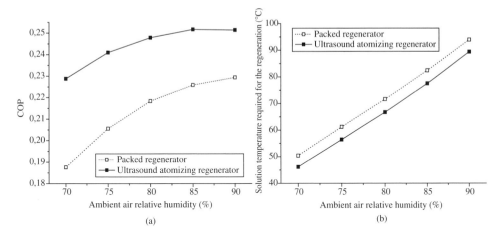

Figure 4.41 Comparisons of COP and desiccant regeneration temperature between the two systems with different regenerators under different ambient air humidity ratios (ambient air temperature is held as 28 °C). (a) COP against ambient air humidity and (b) required regeneration temperature against ambient air humidity. Ambient air temperature: 28 °C; gas–liquid ratio of regenerator: 2.5

ultrasound-atomizing regenerator, while it affects greatly the COP of the system with the packed regenerator. As shown in Figure 4.43a, for the packed regenerator system, the COP decreases by about 37% as the gas–liquid ratio of regenerator increases from 2.0 to 5.0. And from Figure 4.43b, we can see that the higher gas–liquid ratio of regenerator will be favorable for reducing the desiccant regeneration temperature required for the system running.

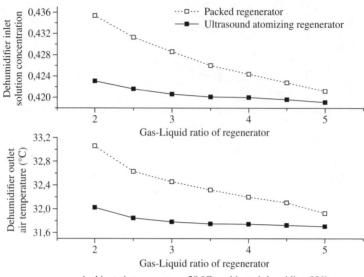

Figure 4.42 Comparisons of some key variables between the two systems with different regenerators under different gas–liquid ratios of regenerator. Ambient air temperature: 28 °C; ambient air humidity: 80%

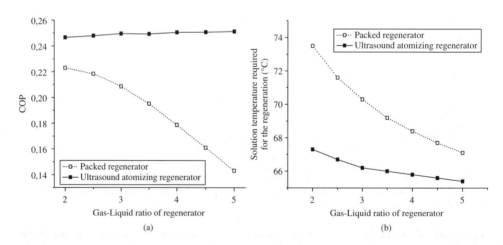

Figure 4.43 Comparisons of COP and desiccant regeneration temperature between the two systems with different regenerators under different gas–liquid ratios. (a) COP against ambient gas–liquid ratio of regenerator and (b) required regeneration temperature against gas–liquid ratio of regenerator. Ambient air temperature: 28 °C; ambient air humidity: 80%

References

[1] Lof, G.O.G. (1955) Cooling with solar energy. World Symposium on Applied Solar Energy, Phoenix, AZ, 1955, pp. 171–189.

[2] Uemura, T. and Hasaba, S. (1964) Studies on the lithium bromide-water absorption refrigerating machines. *Technological Report of Kansai University*, **6**, 31–55.

[3] Mcneely, L.A. (1979) Thermodynamic properties of aqueous solutions of lithium bromide. *ASHRAE Transactions*, **85** (Part 2), 413–434.

[4] Patil, K.R., Tripathi, A.D. and Pathak, G. (1990) Thermodynamic properties of aqueous electrolyte solutions. 1. Vapor pressure of aqueous solutions of LiCl, LiBr and LiI. *Journal of Chemical and Engineering Data*, **35** (2), 166–168.

[5] Chua, H.T., Toh, H.K., Malek, A. *et al.* (2000) Improved thermodynamic property fields of LiBr-H2O solution. *International Journal of Refrigeration*, **23** (6), 412–429.

[6] Yuan, Z. and Herold, K.E. (2005) Thermodynamic properties of aqueous lithium bromide using a multi-property free energy correlation. *International Journal of HVAC&R Research*, **11** (3), 377–393.

[7] Conde, M.R. (2004) Properties of aqueous solutions of lithium and calcium chlorides: formulations for use in air conditioning equipment design. *International Journal of Thermal Sciences*, **43** (4), 367–382.

[8] American Society of Heating, Refrigerating and Air-Conditioning Engineers, Inc. (2003) *ASHRAE Handbook of Fundamentals*, ASHRAE, Atlanta.

[9] Ertas, A., Anderson, E.E. and Kiris, I. (1992) Properties of a new liquid desiccant solution lithium chloride and calthium chloride mixture. *Solar Energy*, **49** (3), 205–212.

[10] (1976) The Editorial Team of Ship Auxiliary Mechanical and Electrical Equipment. Domestic Lithium Bromide Aqueous Solution Property Graph Set (In Chinese).

[11] Patek, J. and Klomfar, J. (2006) Solid-liquid phase equilibrium in the systems of LiBr-H2O and LiCl-H2O. *Fluid Phase Equilibria*, **250** (1-2), 138–149.

[12] Lower, H. (1960) Thermodynamische und physlkalische eigenschaften der wassrigen lithium bromide-losung. PhD thesis. Technischen Hochschule Karlsruhe.

[13] Feuerecker, G., Scharfe, J., Greiter I. et al. (1994) Measurement of thermo physical properties of aqueous LiBr-solutions at high temperature and concentrations. Proceedings of the International Absorption Heat Pump Conference, New Orleans, LA, AES, 1994, vol. **31**, pp. 493–499.

[14] Wimby, J.M. and Berntsson, T.S. (1994) Viscosity and density of aqueous solutions of LiBr, LiCl, ZnBr2, CaCl2, and LiNO3. 1. Single salt solutions. *Journal of Chemical and Engineering Data*, **39** (1), 68–72.

[15] Zhao, Y. and Shi, H.M. (2001) The choice of the desiccant solutions in the liquid desiccant air conditioning system. *Journal of Engineering Thermophysics*, **22** (Supplementary Issue), 165–168.

[16] Studak, J.W. and Perterson, J.L. (1988) A preliminary evaluation of alternative liquid desiccants for a hybrid desiccant air conditioner. Proceedings of the Fifth Annual Symposium on Improving Building Energy Efficiency in Hot and Humid Climates, Houston, TX, 1988, vols. 13–14, pp. 155–159.

[17] Yi, X.Q. (2009) Study on the characteristics of conventional liquid desiccants. Master dissertation. Tsinghua University.

[18] Ahmed, S.Y. and Gandhidasan, P. (1998) Thermodynamic analysis of liquid desiccant. *Solar Energy*, **62** (1), 11–18.

[19] Jain, S., Dhar, P.L. and Kaushik, S.C. (1999) Experimental studies on the dehumidifier and regenerator of a liquid desiccant cooling system. *Applied Thermal Engineering*, **20** (3), 253–267.

[20] Scalabrin, G. and Scaltri, G. (1990) A liquid sorption-desorption system for air conditioning with heat at lower temperature. *ASME, Journal of Solar Energy Engineering*, **112** (1), 70–75.

[21] Chung, T.W., Lai, C. and Wu, H. (1999) Analysis of mass transfer performance in an air stripping tower. *Separation Science and Technology*, **34** (14), 2837–2851.

[22] Factor, H.M. and Grossman, G. (1980) A packed bed dehumidifier/regenerator for solar air conditioning with liquid desiccants. *Solar Energy*, **24** (5), 541–550.

[23] Etras, A., Gandhidsan, P., Kiris, I. and Anderson, E.E. (1994) Experimental study on the performance of a regeneration tower for various climatic conditions. *Solar Energy*, **53** (1), 125–130.

[24] Potniz, S.V. and Lenz, T.G. (1996) Dimensionless mass transfer correlations for packed-bed liquid-desiccant contactors. *Industrial and Engineering Chemistry Research*, **35** (11), 4185–4193.

[25] Martin, V. and Goswami, D.Y. (1999) Heat and mass transfer in packed bed liquid desiccant regenerators e an experimental investigation. *Transactions of the ASME-Journal of Solar Energy Engineering*, **121** (3), 162–170.

[26] Oberg, V.M. and Goswami, D.Y. (2000) Effectiveness of heat and mass transfer processes in packed bed liquid desiccant dehumidifier/regenerator. *HVAC & Research*, **6** (1), 22–39.

[27] Fumo, N. and Goswami, D.Y. (2002) Study of an aqueous lithium chloride desiccant system: air dehumidification and desiccant regeneration. *Solar Energy*, **72** (4), 351–361.

[28] Gandhidsan, P. (2005) Quick performance predictions of liquid desiccant regeneration in a packed bed. *Solar Energy*, **79** (1), 47–55.

[29] Elsarrag, E. (2006) Performance study on a structured packed liquid desiccant regenerator. *Solar Energy*, **80** (12), 1624–1631.

[30] Longo, G.A. and Gasparella, A. (2009) Experimental analysis on desiccant regeneration in a packed column with structured and random packing. *Solar Energy*, **83** (4), 511–521.

[31] Yin, Y. and Zhang, X. (2010) Comparative study on internally heated and adiabatic regenerators in liquid desiccant air conditioning system. *Building and Environment*, **45** (8), 1799–1807.

[32] Raja, S. and Valsaraj, K.T. (2004) Uptake of aromatic hydrocarbon vapors (Benzene and Phenanthrene) at the air-water interface of micron-size water droplets. *Journal of the Air and Waste Management Association*, **54** (12), 1550–1559.

[33] Chen, X.D. (2005) Lower bound estimates of the mass transfer coefficient from an evaporating liquid droplet – the effect of high interfacial vapor velocity. *Drying Technology*, **23** (1), 59–69.

[34] Davis, E.J. and Schweiger, G. (2002) *The Airborne Microparticle*, Springer, New York.

[35] Newbold, F.R. and Amundson, N.R. (1973) A model for evaporation of a multicomponent droplet. *AIChE Journal*, **19** (1), 22–30.

[36] Jennings, S. (1988) The mean free path in air. *Journal of Aerosol Science*, **19** (2), 159–166.

[37] Seinfeld, J.H. and Pandis, S.N. (1998) *Atmospheric Chemistry and Physics: From Air Pollution to Climate Change*, John Wiley & Sons, Inc., New York.

[38] Hopkins, R.J. and Reid, J.P. (2006) A comparative study of the mass and heat transfer dynamics of evaporating ethanol/water, methanol/water, and 1-propanol/water aerosol droplets. *Journal of Physical Chemistry B*, **110** (7), 3239–3249.

[39] Yasuda, K., Bando, Y., Yamaguchi, S. *et al.* (2005) Analysis of concentration characteristics in ultrasonic atomization by droplet diameter distribution. *Ultrasonics Sonochemistry*, **12** (1), 37–41.

[40] Su, F. (2002) Experiment and theory research of ultrasonic atomization. Doctor dissertation, Tianjin. Hebei University of Technology.

[41] Chen, C. and Evans, L.B. (1986) A local composition model for the excess Gibbs energy of aqueous electrolyte systems. *AIChE Journal*, **32** (3), 444–454.

[42] Hsu, H.L., Wu, Y.C. and Lee, L.S. (2003) Vapor pressure of aqueous solution with mixed salts of NaCl + KBr and NaBr + KCl. *Journal of Chemical and Engineering Data*, **48** (3), 514–518.

[43] Neilson, L.E. (1978) *Predicting the Properties of Mixtures Predicting the Properties of Mixtures, Mixing Rules in Science and Engineering*, Marcel Dekker, Inc., New York.

[44] Pitzer, K.S. (1991) *Activity Coefficient in Electrolyte Solution*, 2nd edn, CRC Press, Boca Raton, FL.

[45] Chen, C.C., Britt, H.I., Boston, J.F. and Evans, L.B. (1982) Local composition model for excess Gibbs energy of electrolyte systems. Part 1: single solvent, single completely dissociated electrolyte systems. *AIChE Journal*, **28** (4), 588–596.

[46] Seinfeld, J.H. (1986) *Atmospheric Chemistry and Physics of Air Pollution*, John Wiley & Sons, Inc., New York.

[47] Treybal, R. (1980) *Mass Transfer Operations*, McGraw-Hill, New York.

[48] Gandhidasan, P. (1986) Calculation of heat and mass transfer coefficients in a packed tower operating with a desiccant-air contact system. *Journal of Solar Energy Engineering*, **108** (2), 123–128.

[49] Incropera, F.P., Dewitt, D.P., Bergman, T.L. and Lavine, A.S. (2007) *Fundamentals of Heat and Mass Transfer*, 6th edn, John Willey & Sons, Inc., Hoboken, NJ.

5

Ultrasonic Transducers

5.1 Longitudinal Vibration of Sandwich Piezoelectric Ultrasonic Transducer

5.1.1 Overview

Piezoelectric transducers have been widely applied in the field of power ultrasound. Being different from ultrasonic inspection and measurement, power ultrasonic transducers typically operate at relatively low ultrasonic frequency range (often 15–50 kHz). They require a high power and vibration displacement. Other parameters such as sensitivity, directivity, and resolution are not very demanding for the power ultrasonic transducer.

Based on piezoelectric ceramic vibrator, vibration mode of the piezoelectric ceramic transducer discs or rings and thickness longitudinal vibration mode, electromechanical coupling coefficient ratio is high. Therefore, in order to obtain relatively high electro-acoustic conversion efficiency in the field of power ultrasound, pressure in the electric transducer with the transducing element is essentially axially polarized within piezoelectric ceramic discs or rings. However, for pure piezoelectric ceramic element, a transducer should get a resonance frequency lower than 50 kHz, and the direction of polarization along its thickness should be more than 4 cm. For a transducer with this kind of thickness, the internal impedance is too high and sintering and polarization processes are difficult. To overcome this difficulty, the piezoelectric ceramic wafer is often sandwiched between two metal blocks and fastened by a fastening bolt. In such a way, a sandwich piezoelectric ceramic transducer can be made, as shown in Figure 5.1. Since this kind of structure was proposed by the French physicist Langevin (P. Langevin), it is also known as the Langevin transducer. For the Langevin transducer, the piezoelectric ceramic wafer is polarized along the direction of the transducer's thickness, that is, the piezoelectric ceramic disc or ring is tightly connected with a high-strength plastic or metal block by a stress bolt. The thickness of the transducer is equal to a half wavelength of the fundamental wave. The Langevin transducers have the characteristic that they take advantage of the longitudinal effect of the piezoelectric ceramic vibrator to obtain a lower resonant frequency. Since the piezoelectric ceramic has a poor tensile strength and can be prone to rupture during the work, it is always prestressed by a bolt to avoid the breakdown of the piezoelectric ceramic.

Ultrasonic Technology for Desiccant Regeneration, First Edition. Ye Yao and Shiqing Liu.
© 2014 Shanghai Jiao Tong University Press. All rights reserved. Published 2014 by John Wiley & Sons Singapore Pte Ltd.

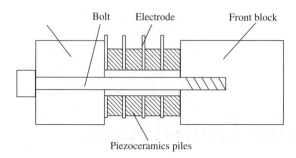

Figure 5.1   Sketch of sandwich piezoelectric transducer

Since the piezoelectric ceramic is an insulating material, it is characterized by poor thermal conductivity. If the heat dissipation during the electro-mechanical conversion cannot be taken away in time, the temperature of the piezoelectric ceramic will become very high, and this will result in poor working efficiency of the piezoelectric ceramic. In the sandwich-type piezoelectric ceramic transducers, the use of a metallic front cover greatly improves the thermal conductivity of the transducer. As long as the metal material, the thickness of the piezoelectric ceramic material, and the lateral dimensions are appropriately selected, the temperature coefficient of the elastic constant of the piezoelectric ceramic material can be compensated by that of the metal material. Consequently, the temperature coefficient of the sandwich piezoelectric ceramic transducer can be very small, and the temperature stability can be improved. Furthermore, through altering the thickness of the piezoelectric ceramic element, the geometry around the metal cover and the shape of the transducer, different operating frequencies can be acquired to meet the need of applications. Since the sandwich piezoelectric transducer structure is simple and its electro-acoustic efficiency is high, it has been widely used in the field of power ultrasound technology.

Basically, the power sandwich piezoelectric ceramic ultrasonic transducer consists of piezoelectric ceramic in the center, a front metal cover, a rear metal cover, a prestressed bolt insulated with bushings and metal electrodes. The crystal stacks (or heaps) in the central part of the transducer are composed of several ceramic piezoelectric ceramic rings. The piezoelectric ceramic crystal stacks (heaps) are connected in form of machinery series and circuit parallel. The direction of polarization of the two adjacent wafers is opposite to each other so that the longitudinal vibration of the two wafers can be superimposed with the phase, and a harmonic vibration of the piezoelectric crystal stacks can be obtained. The number of wafers is generally even so that the front and the rear cover of the transducer can be connected with the electrodes of the same polarity, otherwise an insulating strip is needed between the covers and the wafers of the transducer. For the sake of safety, the front and the rear cover of the transducer are often connected to the negative terminal of the power supply.

To acquire a high power and a high energy conversion efficiency of transducer, the ceramic piezoelectric should use materials with a low mechanical and dielectric loss and a high piezoelectric constant and a high electromechanical conversion factor. The shape, diameter, and number of the piezoelectric ceramic wafer are designed according to the actual requirements including the acoustic frequency, power, and the working mode of the transducer.

For the high-intensity piezoelectric ceramic ultrasonic transducers, the impact of the position of piezoelectric ceramic on their vibration performance is relatively large, and the degree of influence differs for different applications. For the light load operating state in the piezoelectric ceramic composite transducer, such as ultrasonic machining, ultrasonic drilling, and ultrasonic welding of metals and plastics technology, the light load transducer has a very large vibration displacement under low voltage conditions. Under such circumstance, the mechanical loss of the transducer can be reduced as the piezoelectric ceramic stack deviates from the node of displacement, while for the heavy load operating state, such as ultrasonic cleaning, the vibration displacement of the transducer is small and requires a relatively higher driving power. Under such circumstance, the mechanical loss of the transducer is small, while the dielectric loss is relatively large. In order to decrease the dielectric loss of the transducer and improve the effective electromechanical coupling coefficient of transducer, the piezoelectric crystal stack should be near to the displacement node of the transducer.

The design size of the piezoelectric ceramic element mainly refers to the vibration displacement and the total volume of the entire stack of piezoelectric ceramic crystals. In practice, a piezoelectric ceramic element is expected to have a larger power and a smaller size and weight, that is, the ratio of power capacity to the weight is expected to be as high as possible. For a specific size and weight, the power capacity of the transducer will have a maximum value, which is known as "power limit."

The transducer power limit depends on the maximum mechanical strain the transducer can bear (known as "mechanical limit"), the maximum temperature the transducer can withstand (known as "thermal limit") and the maximum electric field that the transducer can accommodate (known as "electric limit").

When the power output exceeds the mechanical limits of the transducer, there will appear fracture and fragmentation in the piezoelectric ceramic. The transducer faults often occur at the place where the internal stress is the biggest. Since the mechanical strength of the metal material is greater than that of the piezoelectric ceramic material, the displacement node of the transducer may be better designed on the metal. However, under such a circumstance, the electromechanical conversion capability, and the load adaptability performance of the transducer will be affected. For the heavy load conditions of the high-power ultrasonic transducer, the transducer vibration displacement is small, and this will reduce the probability of rupture phenomenon. The displacement node of the transducer can be designed on the piezoelectric ceramic in order to achieve the highest electromechanical conversion efficiency.

The mechanical power limit ($P_{m,\max}$) of the composite sandwich longitudinal piezoelectric transducer can be determined by:

$$P_{m,\max} = \frac{\omega}{C_m}\left(\frac{P_{pc,\max}^2}{2\rho_{pc}c_{o,pc}^2}V_{pc}\right), \tag{5.1}$$

where ω is resonance angular frequency of the transducer in rad/s; C_m is mechanical quality factor of the transducer; $P_{pc,\max}$ is the maximum dynamic tensile stress the piezoelectric ceramic can withstand in Pa; ρ_{pc} is density of the piezoelectric ceramic in kg/m^3; $c_{o,pc}$ is sound speed in the piezoelectric ceramic, m/s; and V_p is volume of the piezoelectric ceramic transducer in m^3.

Since the piezoelectric ceramic is made by the ferroelectric material whose ferroelectric hysteresis loss and dielectric loss is very large under the high-power conditions, this will generate a lot of heat dissipation and lead to an increase in the temperature of the piezoelectric ceramic. If no measures are taken to take the heat away, the temperature of the piezoelectric ceramic will continue to rise. When the temperature reaches a half of the Curie temperature of the piezoelectric ceramic, the transducer will be in a serious state of instability and can even lead to a failure of the transducer. To avoid such kinds of failure due to the thermal limit of the transducer, some measures can be taken including the selection of the ceramic materials with a low electromechanical conversion loss and a small temperature coefficient, and the use of the forced cooling method.

The electrical limit of the transducer can be analyzed statically and dynamically. Under the static conditions, the electrical limit of transducer ceramic material is mainly determined by its dielectric loss due to the relaxation loss and the ceramic leakage loss which are mainly related to the composition of the ceramic material, the manufacturing process, and so on. In order to improve the electrical limit of the transducer, the ceramic material should have a good performance of voltage withstanding. The electrical limit of piezoelectric ceramic is relatively more complex, which is not only related to the piezoelectric ceramic material, but also to the operating frequency and the matching conditions of the transducer. When the transducer is operating in a state of electrical mismatch, the reactive power of the transducer increases, resulting in the rapid heating of the transducer, and this may damage the transducer. For different piezoelectric ceramic materials, the voltage required for the vibration of the transducer is not the same. Ideally, the piezoelectric ceramic field excitation voltage should be in the range of 4–8 kV/cm.

The influential factors on the performance of the transducer interact with each other, and they must be considered together. For instance, when the radiation efficiency and the power of the transducer decreases as a result of the presence and the enhancement of the cavitation, the sound power output doesn't increase with the increase of the input electrical power of the transducer. This is because the mechanical loss becomes large when the transducer efficiency decreases, which causes the heat dissipation to rise and results in a temperature rise in the transducer, and this will further bring in the decrease of the transducer performance. In order to increase the radiation power, the input electric power of the transducer must be increased; this increases the electrical power of transducer while at the same time affecting the electrical performance. If no measures are taken, it will eventually affect the normal operation of the transducer.

The transverse dimension of the piezoelectric ceramic element, that is, the diameter of the piezoelectric ceramic element, normally should be smaller than a quarter of the acoustic wavelength in the ceramic material. For the piezoelectric ceramic element with a large diameter, apart from the longitudinal vibration mode, there are some other vibration modes, such as radial vibration mode. In order to avoid the coupling of the resonance frequency of the transducer and the radial piezoelectric ceramic or other vibration mode which may lead to a decrease in the transducer efficiency, the resonance frequency, and the diameter of the piezoelectric ceramic chip of the transducer should be properly designed. Generally, the longitudinal resonance frequency of the transducer is much lower than that of the piezoelectric ceramic chip and the other parts, such as the fundamental resonance vibration frequency of the front and the rear cover to prevent mutual coupling of the running frequency of the transducer and the radial frequency of the electric ceramic chips and the other components.

The number and total volume of the piezoelectric ceramic chips often depends on the power capacity of piezoelectric ceramic materials. According to some researches, the emission power capacity of the Lead Zirconate Titanate material is about 6 W/(cm^3·kHz) [1]. It is therefore clear that, the volume of piezoelectric ceramic can be made very small for a high-frequency transducer. But on the other hand, when the frequency increases, the transducer's internal mechanical and dielectric loss will have a corresponding increase. Under the existing technology, the transducer power capacity is generally taken to be $2-3$ W/(cm^3·kHz).

The thickness and number of the piezoelectric ceramic element used for the transducer will affect greatly the electrical impedance, the mechanical quality factor, and the electromechanical coupling coefficient of the transducer. The thickness of the wafer should not be too thick, otherwise it may be difficult to be motivated. At the same time, the wafer should not be too thin because the use of the thin wafers will increase the number of the wafers and affect the sound propagation. Generally, the thickness of a single piezoelectric ceramic is taken as $5-10$ mm or one-third of the total length of the transducer.

The front cover of the sandwich piezoelectric ceramic transducer is mainly used to emit the ultrasonic radiation. At the same time, the front cover actually acts as an impedance converter. It can adjust the load impedance to achieve an anticipated impedance of the piezoelectric ceramic element, and hence, improve the emission efficiency of the transducer and guarantee a certain frequency bandwidth. These can be realized mainly through selecting proper material for the front cover and optimizing its geometric size and shape. In the field of high-intensity ultrasound applications, the materials of the front cover are basically light metals such as aluminum, magnesium alloys, and titanium alloys.

Concerning the shape of the front cover of the transducer, there can be many choices. The most common shapes of the front cover are cylindrical, conical, exponential, and catenary line. For the convenience of machining, the front covers of the transducer are basically conical in shape for most situations.

The rear cover is designed to make sure that the majority of the ultrasonic radiation is emitted from the front surface instead of the rear surface of the transducer. In order to achieve this, the rear cover of the transducer is often made of some heavy metal materials, such as steel, copper, and so on. Meanwhile, the shape of the rear cover is made to be as simple as possible, normal in the cylindrical or conical shape.

The material selection for the front and rear cover of the sandwich piezoelectric ceramic transducer should follow some basic principles as below: first, the internal mechanical losses of the material should be as small as possible for the operating frequency range of the transducer; secondly, the mechanical fatigue strength must be high, while the acoustic impedance ratio should be relatively small, that is, the product of density and sound velocity of the material is smaller; thirdly, the material should be inexpensive and easy to machine; Lastly, the material should have a good ability of corrosion resistance.

The most suitable materials for the above requirements are mainly aluminum alloy, titanium alloy, copper-nickel alloy, bronze, and so on. Although aluminum alloy is superior to those of the other metals in many performances, it is more difficult to process and its price is relatively higher. Aluminum is easy to process and its price is low, but its anti-cavitations corrosion is poor. For the steel, the price is cheap, but its mechanical loss is large. In the existing sandwich transducers, aluminum is normally used for the front cover of the transducer; and the rear cover of the transducer is mainly made of steel plate.

The tensile strength of the piezoelectric ceramic material is low (only about 5×10^7 Pa) but its compressive strength is high (about 10 times of the tensile strength). In such a case, the piezoelectric ceramic is susceptible to be damaged under the high-power condition. To overcome this shortcoming, the piezoelectric ceramic crystal stack is often preloaded by a prestressed bolt, that is, the piezoelectric ceramic crystal stack is always in a state of compression. At the same time, the tensile stress generated by the vibration is always lower than the critical tensile strength of the material and doesn't hinder the longitudinal vibration to avoid the reduction of the electromechanical coupling coefficient of the stack of piezoelectric ceramic crystals.

The prestressed bolt is required to have a large prestressing force and good flexibility. It is often made of the high-strength metals like chrome steel, tool steel, and titanium alloy. The prestressing force will make an influence on the resonant frequency of the transducer, but it doesn't affect the electro-acoustic efficiency of the transducer. The maximum electric power the transducer can withstand can be increased by several times. There ought to be an appropriate range in the prestressing force which must be larger than the maximum tensile stress of the transducer required for the actual applications. For the sandwich piezoelectric transducer, the prestress of the prestressed bolt should be large enough to avoid the rupture of the piezoelectric ceramic. At the same time, the prestress of the prestressed bolt should not be too large because too much prestress will affect the vibration of the piezoelectric ceramic and might cause the breakage of the piezoelectric ceramic. According to the present applications, the prestress of the prestressed bolt is normally 20–25 MPa.

The production process will have a great influence on the performance of the sandwich piezoelectric transducer. In the process of machining, the following points must be considered: first, the contact surface between the components of the transducer should be as smooth as possible; second, the metal electrode between the adjacent wafers and that between the front or the rear cover and the wafer should have good electrical and thermal conductivity the thickness of the metal electrode is normally about 0.2 mm; third, the epoxy resin adhesives are used to eliminate the presence of an air layer between the contact surfaces of the components of the transducer (i.e., the wafer, the metal electrode, and the covers) to guarantee sound transmission in the transducer. In addition, the prestressed bolts must be perpendicular to the surface of the cross section of each component. Otherwise, the transducer can not work normally or may cause a breakage of the piezoelectric ceramic wafer.

Currently, the operating frequency of the sandwich piezoelectric ceramic transducers is probably in the range of tens to hundreds of kilohertz and the maximum power up to 20 kW or so. Depending on its electro-acoustic conversion efficiency, different states have different working values. For a well-designed sandwich transducer, the vibration mode is relatively pure, and the vibration distribution on the radiating surface is relatively uniform. In general, under the high-power mode of operation, such transducers mainly work near the fundamental frequency vibration mode under which the middle part of the transducer has the largest vibration displacement and the boundary has the smallest vibration displacement, just like a piston radiator. To make a comprehensive analysis on the performance of the sandwich piezoelectric transducer, a theoretical model for this transducer must be employed.

5.1.2 Theoretical Analysis

A sandwich piezoelectric transducer mainly consists of three parts: metal front cover, piezoelectric ceramic crystal stack, and metal rear cover. Hence, the models for the electromechanical vibration characteristics of these components are to be developed for the study of the entire transducer.

For the convenience of modeling, the following assumptions are necessarily made on the vibration mode and geometry of the transducer:

1. The diameter of the transducer is much smaller than the wavelength of the transducer or the longitudinal length of the transducer so that the sandwich piezoelectric ceramic transducer can be approximated as the vibration of an elongated longitudinal vibration of a composite rod.
2. The piezoelectric ceramic stack (heap) is composed of many thin piezoelectric ceramic wafers with a small hole at the center. The influence of the hole on the vibration characteristics of these piezoelectric ceramic wafers can be neglected if the size of the hole is small enough.
3. The displacement and vibration are continuous at the joint interfaces between the components of the transducer.

Thus, the sandwich piezoelectric ceramic transducer can be considered as a slender composite rod vibrator, and the one-dimensional model is employed for the analysis.

5.1.2.1 One-Dimensional Longitudinal Vibration Equation of Tapered Rod and Its Solution

The schematic diagram for a variable cross-section rod is shown in Figure 5.2. Its transverse size is much smaller than the sound wavelength, and the transducer vibrates along the direction of x axis.

For the representative element of dx, the kinetic equation for the tapered rod can be written by Equation (5.2) based on Newton's laws.

$$\frac{\partial \left(S_{rod}\sigma_{rod} \right)}{\partial x} dx = S_{rod}\rho_{rod} dx \frac{\partial^2 \xi_{rod}}{\partial x^2}, \tag{5.2}$$

where $S_{rod} = S(x)$ is the function of the cross-sectional area of the rod along the x axis in m^2; $\xi_{rod} = \xi(x)$ is the displacement function of a specific particle in the rod in m; $\sigma_{rod} = \sigma(x) = E_{rod}\partial\xi_{rod}/\partial x$ is the stress function of rod in Pa, E_{rod} is the Young's modulus of the rod in Pa; and ρ_{rod} is density of the rod in kg/m^3.

In the case of harmonic vibration, Equation (5.2) can be rewritten as:

$$\frac{\partial^2 \xi_{rod}}{\partial x^2} + \frac{1}{S_{rod}} \frac{\partial S_{rod}}{\partial x} \frac{\partial \xi_{rod}}{\partial x} + k_{rod}^2 \xi_{rod} = 0. \tag{5.3}$$

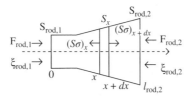

Figure 5.2   Schematic for a longitudinal vibration of thin rods of variable cross-section

Equation (5.3) is a wave equation of one-dimensional longitudinal vibration of a tapered thin rod, in which $k_{rod} = \omega_{rod}/c_{o,rod}$, k_{rod} is the sound wave number in the rod in 1/m, ω_{rod} is the angular frequency of vibration of the rod in rad/s, and $c_{o,rod} = \sqrt{E_{rod}/\rho_{rod}}$ is the sound speed in the rod in m/s.

Let $K_{rod}{}^2 = k_{rod}{}^2 - \dfrac{1}{\sqrt{S_{rod}}} \cdot \dfrac{\partial^2(\sqrt{S_{rod}})}{\partial x^2}$ and $\xi_{rod} = S_{rod}{}^{-\frac{1}{2}}X$. Equation (5.3) can be transformed into the form as below:

$$\frac{\partial^2 X}{\partial x^2} + K_{rod}{}^2 X = 0. \tag{5.4}$$

When the $K_{rod}{}^2$ is positive, Equation (5.4) has a harmonic solution as follows:

$$\xi_{rod} = \frac{1}{\sqrt{S_{rod}}} \left[A_{const} \sin\left(K_{rod}x\right) + B_{const} \cos\left(K_{rod}x\right) \right], \tag{5.5}$$

where A_{const} and B_{const} are undetermined coefficients that are related to the initial conditions.

Let $\beta = \dfrac{1}{\sqrt{S_{rod}}} \cdot \dfrac{\partial^2(\sqrt{S_{rod}})}{\partial x^2}$, then $\beta \le k_{rod}{}^2$ is the condition for a positive value of $K_{rod}{}^2$.

If $\beta < 0$, then, $\sqrt{S_{rod}} = C_{const} \sin(\sqrt{-\beta}x) + D_{const} \cos(\sqrt{-\beta}x)$, where, C_{const} and D_{const} are undetermined coefficients. In this case, the shape of the rod is trigonometric.

If $\beta = 0$, then $\sqrt{S_{rod}} = C_{const}x + D_{const}$. In this case, the cross-section of the rod is conical. As a special case, when the constant $C_{const} = 0$, it is actually a thin rod with uniform section.

If $\beta > 0$, then $\sqrt{S_{rod}} = C_{const} \sinh(\sqrt{\beta}x) + D_{const} \cosh(\sqrt{\beta}x)$. When $C_{const} = 0$, it stands for the catenary line rod. When $C_{const} = D_{const}$ or $C_{const} = -D_{const}$, it stands for the exponential variable cross-section rod.

According to Figure 5.2, The constant, A_{const} and B_{const}, in Equation (5.5) can be obtained as below:

$$A_{const} = -\frac{\dot{\xi}_{rod,2}\sqrt{S_{rod,2}} + \dot{\xi}_{rod,1}\sqrt{S_{rod,1}}\cos\left(K_{rod}l_{rod}\right)}{i\omega_{rod}\sin\left(K_{rod}l_{rod}\right)}, B_{const} = \frac{\sqrt{S_{rod,1}}}{i\omega_{rod}}\dot{\xi}_{rod,1}, \tag{5.6}$$

where $\dot{\xi}_{rod,1}$ and $\dot{\xi}_{rod,2}$ denote the vibration velocity of the left and the right end of the tapered rod (m/s), respectively; $S_{rod,1}$ and $S_{rod,2}$ are cross-sectional area of the left and the right end of the tapered rod (m^2), respectively; i is imaginary number; l_{rod} is the length of the rod (m). According to the equilibrium condition of forces at both ends of the rod, that is, $F_{rod,1} = -(F_{rod})_{x=0}$, $F_{rod,2} = -(F_{rod})_{x=l_{rod}}$, the following two formulas can be obtained:

$$F_{rod,1} = \frac{\rho_{rod}c_{o,rod}}{2ik_{rod}}\left(\frac{\partial S_{rod}}{\partial x}\right)_{x=0}\dot{\xi}_{rod,1} + \frac{\rho_{rod}c_{o,rod}K_{rod}S_{rod,1}}{ik_{rod}}\cot\left(K_{rod}l_{rod}\right)\cdot\dot{\xi}_{rod,1}$$

$$+\frac{\rho_{rod}c_{o,rod}K_{rod}\sqrt{S_{rod,1}S_{rod,2}}}{ik_{rod}\sin\left(K_{rod}l_{rod}\right)}\dot{\xi}_{rod,2}. \tag{5.7}$$

$$F_{rod,2} = -\frac{\rho_{rod}c_{o,rod}}{2ik_{rod}}\left(\frac{\partial S_{rod}}{\partial x}\right)_{x=l}\dot{\xi}_{rod,2} + \frac{\rho_{rod}c_{o,rod}K_{rod}S_{rod,2}}{ik_{rod}}\cot\left(K_{rod}l_{rod}\right)\cdot\dot{\xi}_{rod,2}$$

$$+\frac{\rho_{rod}c_{o,rod}K_{rod}\sqrt{S_{rod,1}S_{rod,2}}}{ik_{rod}\sin\left(K_{rod}l_{rod}\right)}\dot{\xi}_{rod,1}. \tag{5.8}$$

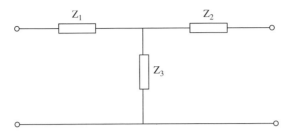

Figure 5.3 Longitudinal vibration electro-mechanical equivalent circuit of a variable cross-section rod

The longitudinal vibration of the tapered rod can be represented by an equivalent circuit as shown in Figure 5.3. Z_1, Z_2, and Z_3 represent parallel (or series) impedances in the equivalent circuit, and they can be expressed as below:

$$Z_1 = \frac{\rho_{rod}c_{o,rod}}{2ik_{rod}}\left(\frac{\partial S_{rod}}{\partial x}\right)_{x=0} + \frac{\rho_{rod}c_{o,rod}K_{rod}S_{rod,1}}{ik_{rod}}\cot(K_{rod}l_{rod}) - \frac{\rho_{rod}c_{o,rod}K_{rod}\sqrt{S_{rod,1}S_{rod,2}}}{ik_{rod}\sin(K_{rod}l_{rod})}.$$

$$(5.9)$$

$$Z_2 = -\frac{\rho_{rod}c_{o,rod}}{2ik_{rod}}\left(\frac{\partial S_{rod}}{\partial x}\right)_{x=l_{rod}} + \frac{\rho_{rod}c_{o,rod}K_{rod}S_{rod,2}}{ik_{rod}}\cot(K_{rod}l_{rod}) - \frac{\rho_{rod}c_{o,rod}K_{rod}\sqrt{S_{rod,1}S_{rod,2}}}{ik_{rod}\sin(K_{rod}l_{rod})}.$$

$$(5.10)$$

$$Z_3 = \frac{\rho_{rod}c_{o,rod}K_{rod}\sqrt{S_{rod,1}S_{rod,2}}}{ik_{rod}\sin(K_{rod}l_{rod})}.$$

$$(5.11)$$

Although the equivalent impedances in the equivalent circuit are different for different shapes of the rod, they share the same form. For any shapes of the rod, the equivalent circuit of one-dimensional vibration can be expressed with a T-network as shown in Figure 5.3. In the following part, the equivalent impedances for different shapes of rod, such as the conical, the exponential, and the catenary line, are presented.

The Cone Type

For the cone-type rod with the cross-section shape function of $S = S_{rod,1}(1 - \varpi x)^2$, the equivalent impedances can be expressed, respectively, as

$$Z_1 = -i\frac{\rho_{rod}c_{o,rod}S_{rod,1}}{k_{rod}l_{rod}}\left(\sqrt{\frac{S_{rod,2}}{S_{rod,1}}} - 1\right) - i\rho_{rod}c_{o,rod}S_{rod,1}\cot\left(k_{rod}l_{rod}\right)$$

$$+ i\frac{\rho_{rod}c_{o,rod}\sqrt{S_{rod,1}S_{rod,2}}}{\sin\left(k_{rod}l_{rod}\right)}.$$

$$(5.12)$$

$$Z_2 = -i\frac{\rho_{rod}c_{o,rod}S_{rod,2}}{k_{rod}l_{rod}}\left(\sqrt{\frac{S_{rod,1}}{S_{rod,2}}} - 1\right) - i\rho_{rod}c_{o,rod}S_{rod,2}\cot\left(k_{rod}l_{rod}\right)$$

$$+ i\frac{\rho_{rod}c_{o,rod}\sqrt{S_{rod,1}S_{rod,2}}}{\sin\left(k_{rod}l_{rod}\right)}.$$

$$(5.13)$$

$$Z_3 = \frac{\rho_{rod} c_{rod,o} \sqrt{S_{rod,1} S_{rod,2}}}{i \sin\left(k_{rod} l_{rod}\right)}. \tag{5.14}$$

In Equations (5.12)–(5.14),

$$\varpi = \frac{d_{rod,1} - d_{rod,2}}{d_{rod,1} l_{rod}} = \frac{X_d - 1}{X_d l_{rod}}, X_d = \frac{d_{rod,1}}{d_{rod,2}} = \frac{\sqrt{S_{rod,1}}}{\sqrt{S_{rod,2}}}, S_{rod,2} = S_{rod,1}\left(1 - \varpi l_{rod}\right)^2.$$

For a conical rod, we usually use a concept called extension factor ϑ, defined as:

$$\vartheta = \frac{d_{rod,1}}{d_{rod,2} - d_{rod,1}}, \tag{5.15}$$

where $d_{rod,1}$ and $d_{rod,2}$ are the diameter of the left end and right end of the conical rod, m; Thus, the equivalent impedances of a conical rod can also be expressed by the following equations:

$$Z_1 = -i \frac{\rho_{rod} c_{o,rod} S_{rod,1}}{k_{rod} l_{rod} \vartheta} - i\rho_{rod} c_{rod,o} S_{rod,1} \cot\left(k_{rod} l_{rod}\right) + i\frac{\vartheta}{\vartheta + 1} \cdot \frac{\rho_{rod} c_{rod,o} S_{rod,2}}{\sin\left(k_{rod} l_{rod}\right)}, \tag{5.16}$$

$$Z_2 = -i \frac{\rho_{rod} c_{rod,o} S_{rod,2}}{k_{rod} l_{rod} (\vartheta + 1)} - i\rho_{rod} c_{rod,o} S_{rod,2} \cot\left(k_{rod} l_{rod}\right) + i\frac{\vartheta}{\vartheta + 1} \cdot \frac{\rho_{rod} c_{rod,o} S_{rod,2}}{\sin\left(k_{rod} l_{rod}\right)}, \tag{5.17}$$

$$Z_3 = \frac{\vartheta}{\vartheta + 1} \cdot \frac{\rho_{rod} c_{rod,o} S_{rod,2}}{i \sin\left(k_{rod} l_{rod}\right)}. \tag{5.18}$$

The Exponential Type

For the exponential-type rod with the cross-section shape function of $S = S_{rod,1} e^{-2\beta x}$, the equivalent impedances are given, respectively, by:

$$Z_1 = i\frac{z_{rod,1}}{k_{rod}}\beta - iz_{rod,1}\frac{k'_{rod}}{k_{rod}} \cot\left(k'_{rod} l_{rod}\right) + i\frac{k'_{rod}}{k_{rod}} \cdot \frac{\sqrt{z_{rod,1} z_{rod,2}}}{\sin\left(k'_{rod} l_{rod}\right)}, \tag{5.19}$$

$$Z_2 = -i\frac{z_{rod,2}}{k_{rod}}\beta - iz_{rod,2}\frac{k'_{rod}}{k_{rod}} \cot\left(k'_{rod} l_{rod}\right) + i\frac{k'_{rod}}{k_{rod}} \cdot \frac{\sqrt{z_{rod,1} z_{rod,2}}}{\sin\left(k'_{rod} l_{rod}\right)}, \tag{5.20}$$

$$Z_3 = \frac{k'_{rod}}{ik_{rod}} \cdot \frac{\sqrt{z_{rod,1} z_{rod,2}}}{\sin\left(k'_{rod} l_{rod}\right)}, \tag{5.21}$$

where $k'_{rod} = \sqrt{k_{rod}^2 - \beta^2}$, $z_{rod,1} = \rho_{rod} c_{o,rod} S_{rod,1}$, $z_{rod,2} = \rho_{rod} c_{o,rod} S_{rod,2}$, $k_{rod} = \omega_{rod}/c_{o,rod}$, $S_{rod,2} = S_{rod,1} e^{-2\beta l_{rod}}$. Equations (5.19)–(5.21) are only for the case when $k_{rod} > \beta$. And when $k_{rod} < \beta$, the impedances for an exponential rod are expressed by:

$$Z_1 = i\frac{z_{rod,1}}{k_{rod}}\beta - iz_{rod,1}\frac{\beta'}{k_{rod}}\frac{1}{\tanh\left(\beta' l_{rod}\right)} + i\frac{\beta'}{k_{rod}} \cdot \frac{\sqrt{z_{rod,1} z_{rod,2}}}{\sinh\left(\beta' l_{rod}\right)}, \tag{5.22}$$

$$Z_2 = -i\frac{z_{rod,2}}{k_{rod}}\beta - iz_{rod,2}\frac{\beta'}{k_{rod}}\frac{1}{\tanh\left(\beta' l_{rod}\right)} + i\frac{\beta'}{k_{rod}} \cdot \frac{\sqrt{z_{rod,1} z_{rod,2}}}{\sinh\left(\beta' l_{rod}\right)}, \tag{5.23}$$

$$Z_3 = \frac{\beta'}{ik_{\text{rod}}} \cdot \frac{\sqrt{z_{\text{rod},1} z_{\text{rod},2}}}{\sinh\left(\beta' l_{\text{rod}}\right)}. \tag{5.24}$$

In Equations (5.22)–(5.24), $\beta' = \sqrt{\beta^2 - k_{\text{rod}}{}^2}$.

The Catenary Line Type

For the catenary line rod with the cross-section shape function of $S = S_{\text{rod},2}\cosh^2[\gamma(l_{\text{rod}} - x)]$, The equivalent impedances are given, respectively, as:

$$Z_1 = i\frac{z_{\text{rod},1}}{k_{\text{rod}}}\gamma \tanh\left(\gamma l_{\text{rod}}\right) - iz_{\text{rod},1}\frac{k'_{\text{rod}}}{k_{\text{rod}}}\cot\left(k'_{\text{rod}} l_{\text{rod}}\right) + i\frac{k'_{\text{rod}}}{k_{\text{rod}}} \cdot \frac{\sqrt{z_{\text{rod},1} z_{\text{rod},2}}}{\sin\left(k'_{\text{rod}} l_{\text{rod}}\right)}, \tag{5.25}$$

$$Z_2 = -iz_{\text{rod},2}\frac{k'_{\text{rod}}}{k_{\text{rod}}}\cot\left(k'_{\text{rod}} l_{\text{rod}}\right) + i\frac{k'_{\text{rod}}}{k_{\text{rod}}} \cdot \frac{\sqrt{z_{\text{rod},1} z_{\text{rod},2}}}{\sin\left(k'_{\text{rod}} l_{\text{rod}}\right)}, \tag{5.26}$$

$$Z_3 = \frac{k'_{\text{rod}}}{ik_{\text{rod}}} \cdot \frac{\sqrt{z_{\text{rod},1} z_{\text{rod},2}}}{\sin\left(k'_{\text{rod}} l_{\text{rod}}\right)}, \tag{5.27}$$

where $k'_{\text{rod}} = \sqrt{k^2{}_{\text{rod}} - \gamma^2}$, $z_{\text{rod},1} = \rho_{\text{rod}} c_{\text{o,rod}} S_{\text{rod},1}$, and $z_2 = \rho_{\text{rod}} c_{\text{o,rod}} S_2$. The above formulas are only for $k_{\text{rod}} > \gamma$. In situations when $k_{\text{rod}} < \gamma$, the equivalent impedances of the catenary line rod are given, respectively, as:

$$Z_1 = i\frac{z_{\text{rod},1}}{k_{\text{rod}}}\gamma \tanh\left(\gamma l_{\text{rod}}\right) - iz_{\text{rod},1}\frac{\gamma'}{k_{\text{rod}}}\frac{1}{\tanh\left(\gamma' l_{\text{rod}}\right)} + i\frac{\gamma'}{k_{\text{rod}}} \cdot \frac{\sqrt{z_{\text{rod},1} z_{\text{rod},2}}}{\sinh\left(\gamma' l_{\text{rod}}\right)}, \tag{5.28}$$

$$Z_2 = -iz_{\text{rod},2}\frac{\gamma'}{k_{\text{rod}}} \cdot \frac{1}{\tanh\left(\lambda' l_{\text{rod}}\right)} + i\frac{\gamma'}{k_{\text{rod}}} \cdot \frac{\sqrt{z_{\text{rod},1} z_{\text{rod},2}}}{\sinh\left(\gamma' l_{\text{rod}}\right)}, \tag{5.29}$$

$$Z_3 = \frac{\gamma'}{ik_{\text{rod}}} \cdot \frac{\sqrt{z_{\text{rod},1} z_{\text{rod},2}}}{\sinh\left(\gamma' l_{\text{rod}}\right)}, \tag{5.30}$$

where $\gamma' = \sqrt{\gamma^2 - k_{\text{rod}}{}^2}$, $S_{\text{rod},1} = S_{\text{rod},2}\cosh^2(\gamma l_{\text{rod}})$. For a straight rod with uniform cross section, for example, $S_{\text{rod},1} = S_{\text{rod},2} = S_{\text{rod}}$, the equivalent impedances are given, respectively, as:

$$Z_1 = Z_2 = i\rho_{\text{rod}} c_{\text{o,rod}} S_{\text{rod}} \cdot \tan\left(\frac{k_{\text{rod}} l_{\text{rod}}}{2}\right), \tag{5.31}$$

$$Z_3 = \frac{\rho_{\text{rod}} c_{\text{o,rod}} S_{\text{rod}}}{i \sin\left(k_{\text{rod}} l_{\text{rod}}\right)}. \tag{5.32}$$

5.1.2.2 Electromechanical Piezoelectric Crystal Stack State Equation and Its Equivalent Circuit

Known from the above analysis and assumptions, for the sandwich of piezoelectric ceramic composite transducer, they are only stress waves along the axial direction in the piezoelectric ceramic crystal stack and the other stress components are zero, that is, $\sigma_{pc,1} = \sigma_{pc,2} = 0, \sigma_{pc,3} \neq 0$ ($\sigma_{pc,1}$ and $\sigma_{pc,2}$ stand for stress along the transverse direction, respectively, and $\sigma_{pc,3}$ for stress along the axial direction). The same is the case for the electric field, that is, $E_1 = E_2 = 0$, $E_3 \neq 0$. Meanwhile, assuming the ceramics to be an insulating medium and no space free charges, the electric displacement vector can be considered to be uniform, that is, $\partial\phi_3/\partial z = 0$.

Thus, the following piezoelectric equations can be used:

$$F_3 = s_{33}^D \sigma_{pc,3} + g_{33}\phi_3, \tag{5.33}$$

$$E_3 = \beta_{33}^T \phi_3 - g_{33}\sigma_{pc,3}, \tag{5.34}$$

where, F_3 is axial strain, N and s_{33}^D, g_{33}, β_{33}^T are the elastic flexibility coefficient under the condition of constant electric displacement, the voltage constant, and the clamped impermeability of the piezoelectric ceramic, respectively.

According to Newton's law, the equation of motion for the piezoelectric ceramic can be expressed as:

$$\frac{\partial^2 \xi_{pc}}{\partial \tau^2} = c_{o,pc}^2 \frac{\partial^2 \xi_{pc}}{\partial x^2}, \tag{5.35}$$

where ξ_{pc} is vibration displacement of the piezoelectric ceramic in m; $c_{o,pc} = \sqrt{1/(\rho_{pc} s_{33}^D)}$ is sound speed in the piezoelectric ceramic, m/s; ρ_{pc} is density of the piezoelectric ceramic, kg/m^3; and τ is time, s.

For the simple harmonic vibration, the solution to the above motion equation, that is, Equation (5.35), can be obtained as:

$$\xi_{pc} = \left[A_{const} \sin\left(k_{pc}x\right) + B_{const} \cos\left(k_{pc}x\right)\right] e^{i\omega_{pc}\tau}, \tag{5.36}$$

where A_{const} and B_{const} are constants which are determined by boundary conditions; k_{pc} is sound wave number in the piezoelectric ceramic in 1/m, $k_{pc} = \omega_{pc}/c_{o,pc}$, ω_{pc} is angular frequency of vibration of the piezoelectric ceramic in rad/s.

Thus, the displacement distribution in the piezoelectric ceramic can be obtained as:

$$\xi_{pc} = \frac{\xi_{pc,1} \sin\left[k_{pc}\left(l_{pc} - x\right)\right] - \xi_{pc,2} \sin\left(k_{pc}x\right)}{\sin\left(k_{pc}l_{pc}\right)}, \tag{5.37}$$

where $\xi_{pc,1}$ and $\xi_{pc,2}$ are vibration displacement of the left end and the right end of the piezoelectric ceramic, respectively. l_{pc} is length of the piezoelectric ceramic, m.

According to the mechanical boundary conditions, that is, $F_{pc,1} = -S_{pc}\sigma_{pc,3}|_{x=0}, F_{pc,2} = -S_{pc}\sigma_{pc,3}|_{x=l_{pc}}$, the mechanical forces on both ends of the piezoelectric ceramic can be obtained as below:

$$F_{pc,1} = \left(\frac{\rho_{pc}c_{o,pc}S_{pc}}{i\sin\left(k_{pc}l_{pc}\right)} - \frac{n^2}{i\omega_{pc}C_0}\right)\left(\dot{\xi}_{pc,1} + \dot{\xi}_{pc,2}\right) + i\rho_{pc}c_{o,pc}S_{pc} \tan\left(\frac{k_{pc}l_{pc}}{2}\right)\dot{\xi}_{pc,1} + nV_{pc}, \tag{5.38}$$

$$F_{pc,2} = \left(\frac{\rho_{pc}c_{o,pc}S_{pc}}{i\sin\left(k_{pc}l_{pc}\right)} - \frac{n^2}{i\omega_{pc}C_0}\right)\left(\dot{\xi}_{pc,1} + \dot{\xi}_{pc,2}\right) + i\rho_{pc}c_{o,pc}S_{pc} \tan\left(\frac{k_{pc}l_{pc}}{2}\right)\dot{\xi}_{pc,2} + nV_{pc}, \tag{5.39}$$

where S_{pc} is cross-section area of the piezoelectric ceramic, m^2; $n = \frac{g_{33}S_{pc}}{l_{pc}s_{33}^D\overline{\beta}_{33}}$ is the electrome-

chanical conversion factor, $\overline{\beta}_{33} = \beta_{33}^T(1 + \frac{g_{33}^2}{s_{33}^D\beta_{33}^T})$; and V is the electrical voltage across the

ceramic sheet, which is given by the following formula:

$$V = \int_0^{l_{pc}} E_3 dx = I_{pc}/\left(i\omega_{pc}C_0\right) - n\left(\dot{\xi}_{pc,1} + \dot{\xi}_{pc,2}\right), \tag{5.40}$$

where I_{pc} is electric current passing through the piezoelectric ceramic, A and $C_0 = S_{pc}/(l_{pc}\overline{\beta}_{33})$ is the one-dimensional cutoff capacitance of the piezoelectric ceramic, F.

Likewise, the electromechanical equivalent circuit for a single piezoelectric ceramic can be obtained as presented in Figure 5.4a in which $Z_1 = i\rho_{pc}c_{o,pc}S_{pc}\tan(k_{pc}l_{pc}/2)$, $Z_2 = \rho_{pc}c_{o,pc}S_{pc}/[i\sin(k_{pc}l_{pc})] - n^2/\left(i\omega_{pc}C_0\right)$.

For the piezoelectric ceramic crystal heap with multiple wafers (the number is N), these wafers are mechanically connected in series, while they are connected in parallel in circuit. According to the cascade theory of circuit, we can obtain the electromechanical equivalent circuit of the piezoelectric ceramic stack as shown in Figure 5.4b in which $Z_{1p} = Z_{2p} = i\rho_{pc}c_{e,pc}S_{pc}\tan(Nk_{e,cp}l_{pc}/2)$, $Z_{3p} = \rho_{pc}c_{e,pc}S_{pc}/[i\sin(Nk_{e,cp}l_{pc})]$, and $k_{e,cp} = \omega_{pc}/c_{e,cp}$. $c_{e,cp}$ is equivalent longitudinal vibration velocity of piezoelectric ceramic crystal heap, which can be calculated by

$$c_{e,cp} = c_{o,cp}\sqrt{1 - k_{33}^2\tan\left(k_{cp}l_{cp}/2\right)/\left(k_{cp}l_{cp}/2\right)}. \tag{5.41a}$$

In the case of $kl/2 \ll \pi$, that is, the thickness of each wafer in the piezoelectric ceramic crystal heap is much smaller than the wave length of sound. The equivalent vibration velocity of piezoelectric ceramic crystal heap, $c_{e,cp}$, can be simplified as:

$$c_{e,cp} = c_{o,cp}\sqrt{1 - k_{33}^2} = \frac{1}{\sqrt{\rho_{cp}s_{33}^E}}, \tag{5.41b}$$

where k_{33} is longitudinal electro-mechanical coupling coefficient and s_{33}^E is elastic flexibility coefficient under constant electric field.

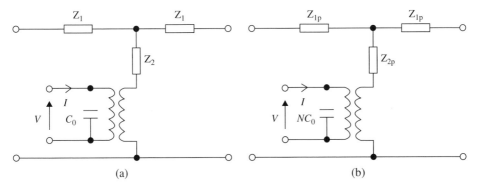

Figure 5.4 Electromechanical equivalent circuit of piezoelectric ceramic piece. (a) Single piezoelectric ceramic wafer and (b) piezoelectric ceramic crystal heap

It can be seen that the sound velocity in the piezoelectric crystal stack is equal to that in the electric field with constant intensity. This is reasonable since the electric field inside the ceramic can be considered to be uniform when the piezoelectric ceramic plate is thin.

5.1.3   State Equations of Sandwich Piezoelectric Electromechanical Transducer

As mentioned earlier, the sandwich piezoelectric ceramic transducer mainly consists of a piezoelectric ceramic crystal stack sandwiched by two tapered rods (i.e., the front cover and the rear cover). The equivalent circuit for the entire sandwich piezoelectric ceramic transducer can be obtained based on the equivalent circuit of the three components of the transducer, which is shown in Figure 5.5. The Z_{11}, Z_{12}, and Z_{13} represent parallel (or series) impedances of the rear cover (refer to Figure 5.3). The Z_{21}, Z_{22}, and Z_{23} represent parallel (or series) impedances of the front cover (refer to Figure 5.3). The Z_{1p} and Z_{2p} represent parallel (or series) impedances of the piezoelectric ceramic stack (refer to Figure 5.4).

The block, Z_{fl} and Z_{bl}, is load impedances of the front and the rear cover of the transducer, respectively. Under normal circumstances, the rear cover of the transducer can be considered as no-load state, that is, $Z_{bl} = 0$, while the front cover of the transducer is connected to the load, and hence, Z_{fl} has different values corresponding to different loads. However, the transducer load is difficult to determine in the actual situations. Thus, the load impedance on the front cover of the transducer is often neglected during the transducer design.

5.1.3.1   Frequency Equation

The power ultrasonic transducers mostly work in a resonant state where the transducer can radiate the maximum sound energy. The frequency equation of the transducer is often used to analyze the relationships among the shape, the geometrical dimensions, and the working frequency of the transducer and obtain the resonance frequency of the transducer. In principle, the frequency equation can be derived from the equivalent circuit of the transducer. The basic steps are as follows: first, by using the circuit theory, we can obtain the transducer's equivalent input impedance after a series of transformations; next, let the reactant part of the input impedance

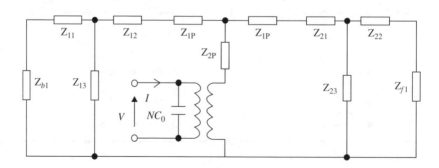

Figure 5.5   Electromechanical equivalent circuit of sandwich piezoelectric ceramic transducer

of the transducer be zero, and the frequency equation can thus be obtained; finally, the resonant frequency of the transducer can be acquired through solving the frequency equation according to the specific shape and dimensions of the transducer. However, it is found that the frequency equation of the transducer obtained by this method is a very complex transcendental equation which is very difficult to solve.

To simplify the analysis, another method is employed to derive the frequency equation of the transducer. The sandwich piezoelectric ceramic transducers are basically half-wave vibrators. For the half-wave vibrator, there exists a cross section in the transducer whose vibration displacement is zero during the vibration, and this cross section is called "nodal plane." The position of the nodal plane in the transducer is affected by many factors such as the vibration frequency, the shape, the geometrical dimensions, and the materials of the covers and the piezoelectric ceramic. In the design of the composite sandwich piezoelectric ceramic transducer, the nodal plane can be taken as an interface, and the entire transducer is considered to be composed of two quarter-wavelength vibrators. Thus, the frequency equation of the entire transducer can be obtained based on that of the two vibrators. Taking a composite transducer shown in Figure 5.6a, for example, AB represents the displacement nodal plane of the transducer, and it divides the entire transducer into two quarter-wavelength vibrators. The equivalent circuit of the vibrator on the right quarter-wavelength side of the displacement nodal plane can be depicted by Figure 5.6b.

In Figure 5.6, l_{c1} and l_{c2} are length (in m) of the piezoelectric ceramic crystal stack in the left and the right quarter-wavelength vibrator, respectively; Z_{m2} is the input impedance of the front cover of the transducer, that is, the load impedance of piezoelectric ceramic crystal stack; $Z_{1p} = i\rho_{pc}C_{o,pc}S_{pc}\tan(k_{e,cp}l_{c2}/2)$; $Z_{2p} = \rho_{pc}C_{e,cp}S_{pc}/[i\sin(k_{e,cp}l_{c2})]$. Since the displacement vibration velocity at the nodal plane is zero, the equivalent circuit of a quarter-wavelength vibrator on the left side can be regarded as an open circuit (see Figure 5.6b). Thus, the frequency equation can be derived on the condition that the total reactance in the circuit be zero.

Let the input impedance of the front cover of the transducer (Z_{m2}) be given as: $Z_{m2} = R_{m2} + X_{m2}i$ (where, R_{m2} is the resistance and X_{m2} is the reactance), we can have the following equation:

$$Z_{1p} + Z_{2p} + X_{m2}i = 0. \tag{5.42}$$

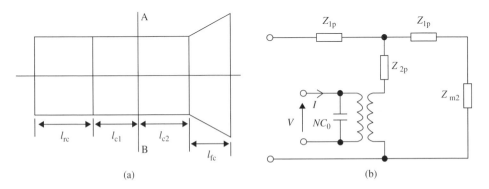

(a)   (b)

Figure 5.6 Composite transducer and its equivalent circuit. (a) Composite transducer and (b) equivalent circuit

Thus, the frequency equation of the quarter-wavelength vibrator on the right side of the nodal plane can be obtained as:

$$\tan\left(k_{e,cp}l_{c2}\right) = \frac{\rho_{pc}c_{e,pc}S_{pc}}{X_{m2}}. \tag{5.43}$$

Likewise, let the input impedance of the rear cover of the transducer (Z_{m1}) be given as: $Z_{m1} = R_{m1} + X_{m1}i$ (where R_{m1} is resistance and X_{m1} is reactance), the frequency equation of the quarter-wavelength vibrator on the left side of the nodal plane can be obtained as:

$$\tan\left(k_{e,cp}l_{c1}\right) = \frac{\rho_{pc}c_{e,cp}S_{pc}}{X_{m1}}. \tag{5.44}$$

In a situation where the covers of the transducer are cylindrical with the uniform cross section, the input impedance of the front and the rear cover can be simplified as:

$$Z_{m1} = iX_{m1} = i\rho_{rc}c_{o,rc}S_{rc}\tan\left(k_{rc}l_{rc}\right), \tag{5.45}$$

$$Z_{m2} = iX_{m2} = i\rho_{fc}c_{o,fc}S_{fc}\tan\left(k_{fc}l_{fc}\right), \tag{5.46}$$

where the subscript "rc" and "fc" stand for the rear and the front cover of the transducer, respectively.

Combining Equations (5.43)–(5.46), the frequency equation of the sandwich piezoelectric transducer with uniform cross-section cylindrical covers can be derived as:

$$\tan\left(k_{e,cp}l_{c1}\right) = \frac{\rho_{cp}c_{e,cp}S_{cp}}{X_{m1}} = \frac{\rho_{cp}c_{e,cp}S_{cp}}{\rho_{rc}c_{o,rc}S_{rc}}\cot\left(k_{rc}l_{rc}\right), \tag{5.47}$$

$$\tan\left(k_{e,pc}l_{c2}\right) = \frac{\rho_{pc}c_{e,pc}S_{pc}}{X_{m2}} = \frac{\rho_{pc}c_{e}S}{\rho_{fc}c_{o,fc}S_{fc}}\cot\left(k_{fc}l_{fc}\right). \tag{5.48}$$

Equations (5.47) and (5.48) can be used for the design of the sandwich piezoelectric ceramic transducer including the shape, the size, and the resonant frequency. In general, the frequency equation of the transducer can be applied to the transducer design in two ways: one is that the size and the shape of the transducer needs to be designed for a specific resonant frequency; the other is that the resonant frequency of the transducer needs to be designed for a specific material and physical dimension of the transducer.

Known from the above analysis, the input mechanical impedance of the transducer's front and rear cover must be obtained in advance for acquiring the resonant frequency of the entire transducer. For the cylindrical covers, their input impedances have been illustrated earlier. Using the same method, the mechanical input impedances of the other shapes of cover can be expressed as below:

For the conical cover:

$$Z_m = iX_m = i\rho_{rod}c_{o,rod}S_{rod,1} \cdot \frac{X_d k_{rod}l_{rod}\tan\left(k_{rod}l_{rod}\right) + \left(X_d - 1\right)^2\left(\dfrac{\tan\left(k_{rod}l_{rod}\right)}{k_{rod}l_{rod}} - 1\right)}{X_d k_{rod}l_{rod} + X_d\left(X_d - 1\right)\tan\left(k_{rod}l_{rod}\right)}. \tag{5.49}$$

For the exponential cover:

$$Z_m = iX_m = i\rho_{rod}c_{o,rod}S_{rod,1} \cdot \frac{k'_{rod}}{k_{rod}}\left[1+\left(\frac{\beta}{k'_{rod}}\right)^2\right] \cdot \frac{\tan\left(k'_{rod}l_{rod}\right)}{1+\frac{\beta}{k'_{rod}}\tan\left(k'_{rod}l_{rod}\right)} \left(k_{rod} > \beta\right).$$

$$(5.50)$$

$$Z_m = iX_m = i\rho_{rod}c_{o,rod}S_{rod,1} \cdot \frac{\beta'}{k_{rod}}\left[\left(\frac{\beta}{\beta'}\right)^2 - 1\right] \cdot \frac{\tanh\left(\beta'l_{rod}\right)}{1+\frac{\beta}{\beta'}\tanh\left(\beta'l_{rod}\right)} \left(k_{rod} < \beta\right).$$

$$(5.51)$$

For the catenary line cover:

$$Z_m = iX_m = i\rho_{rod}c_{o,rod}S_{rod,1} \cdot \frac{k'_{rod}}{k_{rod}}\left(\frac{\gamma}{k'_{rod}} \cdot \frac{\sqrt{X_d^2 - 1}}{X_d} + \tan\left(k'_{rod}l_{rod}\right)\right)\left(k_{rod} > \gamma\right). \quad (5.52)$$

$$Z_m = iX_m = i\rho_{rod}c_{o,rod}S_{rod,1} \cdot \frac{\gamma'}{k_{rod}}\left(\frac{\gamma}{\gamma'} \cdot \frac{\sqrt{X_d^2 - 1}}{X_d} - \tanh\left(\gamma'l_{rod}\right)\right)\left(k_{rod} < \gamma\right). \quad (5.53)$$

Using the input impedance of covers given above, the frequency equation of the transducer whose displacement node is located in the sandwich piezoelectric ceramic can be acquired.

For the transducer whose displacement node is located in the front or the rear cover, the same method can be applied to derive its frequency equation. Figure 5.7 presents a sandwich-type transducer whose displacement node is located in its front cover. In Figure 5.7, the dotted line denotes the nodal plane, l_2 represents the length of the piezoelectric ceramic, and the rest are the front and the rear metal cover. From Figure 5.7, the part on the left of the displacement node can be taken as a quarter-wavelength sandwich transducer, and that on the right of the displacement node can be taken as a quarter-wavelength stepped amplitude rod.

The frequency equation of the quarter-wavelength sandwich transducer is written as:

$$\frac{z_3}{z_2} \tan\left(k_2l_2\right) \tan\left(k_3l_3\right) + \frac{z_3}{z_1} \tan\left(k_1l_1\right) \tan\left(k_3l_3\right) + \frac{z_2}{z_1} \tan\left(k_2l_2\right) \tan\left(k_1l_1\right) = 1, \quad (5.54)$$

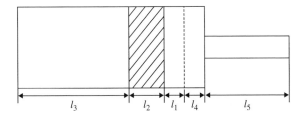

Figure 5.7 The sandwiched transducer with its displacement node located in the front cover

where, z, k, and l are impedance, sound wave number, and length of the component in the quarter-wavelength sandwich transducer. The subscript "1," "2," and "3" correspond to the front cover, the piezoelectric ceramic, and the rear cover of the quarter-wavelength sandwich transducer (refer to Figure 5.7).

For the quarter-wavelength stepped amplitude rod, the frequency equation is given as:

$$\tan\left(k_4 l_4\right) \tan\left(k_5 l_5\right) = \frac{z_4}{z_5}, \tag{5.55}$$

where the subscript "4" and "5" correspond to the rear end and the front end of the quarter-wavelength stepped amplitude rod (refer to Figure 5.7).

In the above analysis, the load influence has been ignored. However, in actual situations, all transducers have load, and the load may produce significant influence on the frequency of the transducer. Although the transducer load is very complicated and difficult to obtain, it is necessarily taken into account during the theoretical analysis of a transducer. In the following part, the frequency equation of transducer under load is to be analyzed. For the convenience of analysis, it assumes that the rear cover of the transducer be cylindrical and be of no load. This is consistent with the real situations.

According to Figure 5.5, the input impedance of the front cover of the transducer can be written as:

$$Z_{m2} = R_f + iX_f = Z_{21} + \frac{Z_{23}\left(Z_{22} + Z_{fl}\right)}{Z_{22} + Z_{23} + Z_{fl}}. \tag{5.56}$$

Using the Mobius transform, Equation (5.56) can be rewritten in the following form:

$$Z_{m2} = R_f + iX_f = \frac{A_{mobius}Z_{fl} - iB_{mobius}}{iC_{mobius}Z_{fl} - D_{mobius}}, \tag{5.57}$$

where Z_{fl} is the transducer load impedance of the front cover; A_{mobius}, B_{mobius}, C_{mobius}, D_{mobius} are the coefficients of Mobius transform, which are determined by the front cover material, the shape, and the dimensions of the transducer. For the rear cover with no load, the input impedance can be written as:

$$Z_{m1} = iz_{rc} \tan\left(k_{rc}l_{rc}\right). \tag{5.58}$$

From Figure 5.5, we can derive as well the mechanical impedance of the transducer Z_m as:

$$Z_m = R_m + iX_m = Z_{2p} + \frac{\left(Z_{1p} + Z_{m2}\right)\left(Z_{1p} + Z_{m1}\right)}{2Z_{1p} + Z_{m2} + Z_{m1}}. \tag{5.59}$$

Substituting Equations (5.52)–(5.58) into Equation (5.59), we can obtain the specific expression of mechanical impedance of the transducer as:

$$R_m = \frac{R_f\left[z_{pc} \tan\left(k_{e,pc}l_{pc}/2\right) + z_{rc} \tan k_{rc}l_{rc}\right]^2}{R_f^2 + \left[X_f + 2z_{pc} \tan\left(k_{e,pc}l_{pc}/2\right) + z_{rc} \tan\left(k_{rc}l_{rc}\right)\right]^2}, \tag{5.60}$$

$$X_m = -\frac{z_{pc}}{\sin\left(k_{e,pc}l_{pc}\right)} + X_{m1}, \tag{5.61}$$

where

$$
X_{\mathrm{ml}} = \cfrac{
\begin{bmatrix}
R_{\mathrm{f}}^2 + X_{\mathrm{f}}^2 + X_{\mathrm{f}} \left(3z_{\mathrm{pc}} \tan \left(\frac{k_{\mathrm{e,pc}}l_{\mathrm{pc}}}{2} \right) + z_{\mathrm{rc}} \tan \left(k_{\mathrm{rc}}l_{\mathrm{rc}} \right) \right) \\[2mm]
+ z_{\mathrm{pc}} \tan \left(\frac{k_{\mathrm{e,pc}}l_{\mathrm{pc}}}{2} \right) \left(2z_{\mathrm{pc}} \tan \left(\frac{k_{\mathrm{e,pc}}l_{\mathrm{pc}}}{2} \right) + z_{\mathrm{rc}} \tan \left(k_{\mathrm{rc}}l_{\mathrm{rc}} \right) \right)
\end{bmatrix} Z_R
}{
R_{\mathrm{f}}^2 + \left[X_{\mathrm{f}} + 2z_{\mathrm{cp}} \tan \left(k_{\mathrm{e,cp}}l_{\mathrm{cp}}/2 \right) + z_{\mathrm{rc}} \tan \left(k_{\mathrm{rc}}l_{\mathrm{rc}} \right) \right]^2
};
$$

$$
Z_R = z_{\mathrm{pc}} \tan \left(k_{\mathrm{e,pc}}l_{\mathrm{pc}}/2 \right) + z_{\mathrm{rc}} \tan \left(k_{\mathrm{rc}}l_{\mathrm{rc}} \right).
$$

When the imaginary part of the mechanical impedance (i.e., the reactance) of the transducer is equal to zero, that is, $X_m = 0$, the transducer achieves the state of mechanical resonance. After a series of transformations, the resonance frequency equation of the transducer can be expressed as:

$$
H' \cdot M \cdot Y = 0. \tag{5.62}
$$

where $H = \begin{bmatrix} R_f^2 + X_f^2 \\ X_f \\ 1 \end{bmatrix}$, $M = \begin{bmatrix} M_{11} & M_{12} & M_{13} \\ M_{21} & M_{22} & M_{23} \\ M_{31} & M_{32} & M_{33} \end{bmatrix}$, $Y = \begin{bmatrix} Y^2 \\ Y \\ 1 \end{bmatrix}$, H' is the transposed matrix of H.

Each element in the above matrices is expressed, respectively, as below:

$$
M_{11} = 0, M_{12} = - \sin \left(k_{\mathrm{e,pc}}l_{\mathrm{pc}} \right), M_{13} = z_{\mathrm{pc}} \cos(k_{\mathrm{e,pc}}l_{\mathrm{pc}}),
$$

$$
M_{21} = - \sin \left(k_{\mathrm{e,pc}}l_{\mathrm{pc}} \right), M_{22} = 2z_{\mathrm{pc}} \left(\cos \left(k_{\mathrm{e,pc}}l_{\mathrm{pc}} \right) - \sin \left(k_{\mathrm{e,pc}}l_{\mathrm{pc}} \right) \cdot \tan \frac{k_{\mathrm{e,pc}}l_{\mathrm{pc}}}{2} \right),
$$

$$
M_{23} = 2z_{\mathrm{pc}}^2 \cdot \tan \left(\frac{k_{\mathrm{e,pc}}l_{\mathrm{pc}}}{2} \right) \cdot \left(\cos^2 \frac{k_{\mathrm{e,pc}}l_{\mathrm{pc}}}{2} + \cos \left(k_{\mathrm{e,pc}}l_{\mathrm{pc}} \right) \right),
$$

$$
M_{31} = z_{\mathrm{pc}} \cos \left(k_{\mathrm{e,pc}}l_{\mathrm{pc}} \right), M_{32} = 2z_{\mathrm{pc}}^2 \cdot \tan \left(\frac{k_{\mathrm{e,pc}}l_{\mathrm{pc}}}{2} \right) \cdot \left(\cos^2 \frac{k_{\mathrm{e,pc}}l_{\mathrm{pc}}}{2} + \cos \left(k_{\mathrm{e,pc}}l_{\mathrm{pc}} \right) \right),
$$

$$
M_{33} = 4z_{\mathrm{pc}}^2 \left(1 - \cos^2 \frac{k_{\mathrm{e,pc}}l_{\mathrm{pc}}}{2} \right), Y = z_{\mathrm{rc}} \tan \left(k_{\mathrm{rc}}l_{\mathrm{rc}} \right).
$$

It can be seen from the above analysis that, the frequency equation of the sandwich-type piezoelectric ceramic composite transducer consists of three components which represent the three main components of a transducer, that is, the H matrix stands for the front cover of the transducer and the load, M for the piezoelectric ceramic element, and Y matrix for the rear cover of the transducer. When all the parameters of the transducer satisfy the frequency equation, the transducer will achieve a state of resonance.

5.1.3.2 Oscillation Speed Ratio of Both Ends of Crystal Stack in Sandwich Composite Piezoelectric Ceramic Transducer

Known from the above analysis, the material and geometrical dimension of the covers as well as the external load of the transducer will affect the load of the piezoelectric ceramic stack and vibration distribution in the transducer. Let the vibration velocity at both ends of the piezoelectric crystal stack be $\dot{\xi}_{pc,1}$ and $\dot{\xi}_{pc,2}$ respectively and the displacement node of the transducer be located at the piezoelectric ceramic stack. The vibration speed ratio at both ends of the piezoelectric crystal stack can be derived according to the equivalent electromechanical circuit of the transducer, which is given as below:

$$\frac{\dot{\xi}_{pc,2}}{\dot{\xi}_{pc,1}} = \frac{\sin\left(k_{e,pc}l_{c2}\right)}{\sin\left(k_{e,pc}l_{c1}\right)}, \tag{5.63}$$

where, l_{c1} is the distance between the displacement nodal plane and the rear end of the piezo-electric ceramic stack and l_{c2} is the distance between the displacement nodal plane and the rear end of the piezoelectric ceramic stack.

It can be inferred from Equation (5.63) that the vibration speed ratio of the front crystal stack to the rear one will be relatively large when the displacement node of the transducer are in the latter part of the ceramic crystal stack. In addition, if the acoustic impedance of the front cover of the transducer is small and that of the rear cover of the transducer is large, the radiation power from the front surface will be improved greatly.

5.1.3.3 The Vibration Speed Ratio of a Sandwich Piezoelectric Ceramic Composite Transducer

The vibration speed ratio of the front surface to the rear surface of the transducer (R_V) is defined as:

$$R_V = \frac{\dot{\xi}_{fc}}{\dot{\xi}_{rc}}, \tag{5.64}$$

where $\dot{\xi}_{fc}$ and $\dot{\xi}_{rc}$ are vibration velocity of the front and the rear surface of the transducer, respectively.

After some transformation, Equation (5.64) can be rewritten as:

$$R_V = \frac{\dot{\xi}_{fc}}{\dot{\xi}_{pc,2}} \cdot \frac{\dot{\xi}_{pc,2}}{\dot{\xi}_{pc,1}} \cdot \frac{\dot{\xi}_{pc,1}}{\dot{\xi}_{rc}}, \tag{5.65}$$

where $\dot{\xi}_{pc,1}$ and $\dot{\xi}_{pc,2}$ are vibration velocity of the front and the rear surface of the piezoelectric ceramic stack,

$$\frac{\dot{\xi}_f}{\dot{\xi}_2} = \frac{Z_{23}}{Z_{22} + Z_{23} + Z_{fl}}, \tag{5.66}$$

$$\frac{\dot{\xi}_1}{\dot{\xi}_b} = \frac{Z_{11} + Z_{13} + Z_{bl}}{Z_{13}}, \tag{5.67}$$

$$\frac{\dot{\xi}_2}{\dot{\xi}_1} = -\frac{Z_{12} + Z_{1p} + Z_{13} - \dfrac{Z_{13}^2}{Z_{11} + Z_{13} + Z_{bl}}}{Z_{21} + Z_{2p} + Z_{23} - \dfrac{Z_{23}^2}{Z_{22} + Z_{23} + Z_{fl}}}. \tag{5.68}$$

Normally, the rear surface of the transducer is of empty load, that is, $Z_{bl} = 0$. For most applications of the high-intensity ultrasound, such as metal and plastic ultrasonic welding, ultrasonic drilling, and ultrasonic cell crushing, the load of the front surface of the transducer is relatively small and can be neglected, that is, $Z_{fl} = 0$. Under such circumstance, the R_V of the transducer can be expressed as:

$$R_V = -\frac{Z_{23}}{Z_{22} + Z_{23}} \cdot \frac{Z_{11} + Z_{13}}{Z_{13}} \cdot \frac{Z_{12} + Z_{1p} + Z_{13} - \dfrac{Z_{13}^2}{Z_{11} + Z_{13}}}{Z_{21} + Z_{2p} + Z_{23} - \dfrac{Z_{23}^2}{Z_{22} + Z_{23}}}. \tag{5.69}$$

Inferred from Equation (5.69), the R_V of the transducer will be related to the material, the shape and the geometrical dimension of the covers and the piezoelectric ceramic as well as the vibration frequency of the transducer. Since the sandwich ultrasonic transducers mostly work in a resonant state, the effect of frequency on the R_V of the transducer is often neglected. In the case of cylindrical and circular-ring shape in the covers and the piezoelectric ceramic stack, Equation (5.69) can be simplified as below:

$$R_V = \cos\left(k_{rc} l_{rc}\right) \cdot \frac{Z_{23}}{Z_{22} + Z_{23}} \cdot \frac{iz_{rc} \tan\left(k_{rc} l_{rc}\right) + iz_{pc} \tan\left(k_{e,pc} l_{pc}/2\right)}{\dfrac{Z_{23}^2}{Z_{22} + Z_{23}} - Z_{23} - Z_{21} - iz_{pc} \tan\left(k_{e,pc} l_{pc}/2\right)}. \tag{5.70}$$

Using Equation (5.70), the R_V of the transducer for different shapes of the front cover can be obtained as below:

1. For the conical

$$R_V = -\cos\left(k_{rc} l_{rc}\right) \cdot \frac{z_{rc} \tan\left(k_{rc} l_{rc}\right) + z_{pc} \tan\left(k_{e,pc} l_{pc}/2\right)}{\dfrac{\vartheta}{\vartheta+1} \cdot \dfrac{z_{fc}}{\sin(k_{fc} l_{fc})} + \left(\dfrac{\vartheta+1}{\vartheta} \cos(k_{fc} l_{fc}) - \dfrac{\sin(k_{fc} l_{fc})}{\vartheta k_{fc} l_{fc}}\right)}$$
$$\times \left[z_{pc} \tan\frac{k_{e,pc} l_{pc}}{2} - \frac{z_{rc}}{\vartheta k_{fc} l_{fc}} - z_{rc} \cot(k_{fc} l_{fc})\right] \tag{5.71}$$

2. For the exponential (Index)

$$R_V = -\frac{\cos\left(k_{rc} l_{rc}\right)\left[z_{rc} \tan\left(k_{rc} l_{rc}\right) + z_{pc} \tan\left(k_{e,pc} l_{pc}/2\right)\right]}{\sqrt{\dfrac{z_{fc}}{z_{rc}}}\left(\cos\left(k'_{fc} l_{fc}\right) + \dfrac{\beta}{k'_{fc}} \sin\left(k'_{fc} l_{fc}\right)\right)} \quad (k_{fc} > \beta).$$
$$\times \left[z_{pc} \tan\left(k_{e,pc} l_{pc}/2\right) + \frac{z_{rc}\beta}{k_{fc}} - \frac{z_{rc} k'_{fc}}{k_{fc} \tan\left(k'_{fc} l_{fc}\right)}\right] + \frac{k'_{fc}\sqrt{z_{rc} z_{fc}}}{k_{fc} \sin\left(k'_{fc} l_{fc}\right)} \tag{5.72}$$

$$R_V = -\frac{\cos\left(k_{rc} l_{rc}\right)\left[z_{rc} \tan\left(k_{rc} l_{rc}\right) + z_{pc} \tan\left(k_{e,pc} l_{pc}/2\right)\right]}{\sqrt{\dfrac{z_{fc}}{z_{rc}}}\left(\cosh\left(\beta' l_{fc}\right) + \dfrac{\beta}{\beta'} \sinh\left(\beta' l_{fc}\right)\right)} \quad (k_{fc} < \beta).$$
$$\times \left[z_{pc} \tan\left(k_{e,pc} l_{pc}/2\right) + \frac{z_{rc}\beta}{k_{fc}} - \frac{z_{rc}\beta'}{k_{fc} \tanh(\beta' l_{fc})}\right] + \frac{\beta'\sqrt{z_{rc} z_{fc}}}{k_{fc} \sinh(\beta' l_{fc})} \tag{5.73}$$

3. For the Catenary line

$$R_V = \cfrac{-\cos\left(k_{rc}l_{rc}\right)\left[z_{rc}\tan\left(k_{rc}l_{rc}\right) + z_{pc}\tan\left(k_{e,pc}l_{pc}/2\right)\right]}{\sqrt{\frac{z_{fc}}{z_{rc}}}\cos\left(k'_{fc}l_{fc}\right)\left[z_{pc}\tan\left(k_{e,pc}l_{pc}/2\right) + \frac{z_{rc}\gamma\tanh(\gamma l_{fc})}{k_{fc}} - \frac{z_{rc}k'_{fc}}{k_{fc}\tan\left(k'_{fc}l_{fc}\right)}\right] + \frac{k'_{fc}\sqrt{z_{rc}z_{fc}}}{k_{fc}\sin\left(k'_{fc}l_{fc}\right)}}$$

$$\times\left(k_{fc} > \gamma\right). \tag{5.74}$$

$$R_V = \cfrac{-\cos\left(k_{rc}l_{rc}\right)\left[z_{rc}\tan\left(k_{rc}l_{rc}\right) + z_{pc}\tan\left(k_{e,pc}l_{pc}/2\right)\right]}{\sqrt{\frac{z_{fc}}{z_{rc}}}\cos\left(\gamma'l_{fc}\right)\left[z_{pc}\tan\left(k_{e,pc}l_{pc}/2\right) + \frac{z_{rc}\gamma\tanh(\gamma l_{fc})}{k_{fc}} - \frac{z_{rc}\gamma'}{k_{fc}\tanh\left(\gamma'l_{fc}\right)}\right] + \frac{\gamma'\sqrt{z_{rc}z_{fc}}}{k_{fc}\sinh\left(\gamma'l_{fc}\right)}}$$

$$\times\left(k_{fc} < \gamma\right). \tag{5.75}$$

In Equations (5.71)–(5.75), the subscript "rc," "fc," and "pc" denote the rear cover, the front cover, and the piezoelectric ceramic of the transducer, respectively. The meaning of the other letters can be found in the earlier Section 5.1.2.1.

Using the above formula (Equations (5.71)–(5.75)), the relationships between the shape of the front cover and the R_V of transducer have been studied, and several key points can be obtained as below:

1. For any specific shape of the front cover, such as the conical, the exponential, and the catenary line, there exists a best design scheme under which the maximum R_V of the transducer can be acquired.
2. To improve the R_V of the transducer, the front cover should use light metal with a lower density, and the rear cover should use heavy metal with a higher density.
3. Among the three commonly used shapes (i.e., the conical, the exponential, and the catenary line) of the front cover, the catenary line shape will have the largest R_V, and then followed by the exponential and the conical.

5.1.4   Design Case

As an example, the design of a half-wavelength cylindrical sandwich composite transducer is illustrated here. The basic conditions for the design are listed as below: the rear and the front cover of the transducer are made of mild steel and hard aluminum, respectively; the nodal plane is located at the front block; the resonant frequency of the transducer without external load is set as 20 kHz; two pieces of piezoelectric ceramic chips (PZT-4, which is made from lead, zirconate, and titanate) with the thickness of 5 mm and the diameter of 60 mm are used for the transducer design. Under the above conditions, find the length of the front and the rear cover of the transducer as well as the vibration speed ratio of the front surface to the rear surface of the transducer.

Analysis: As shown in Figure 5.8, the transducer is divided into three parts, and the total length is equal to a half of the wavelength. The nodal plane is bounded before and after the quarter-wavelength.

When the front cover is considered as a quarter wavelength rod, we can have Equation (5.76).

$$k_1l_1 = \frac{\pi}{2} \Rightarrow l_1 = \lambda_1/4 \tag{5.76}$$

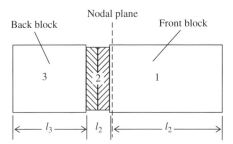

Figure 5.8 Schematic for the case study

where λ_1 is the acoustic wavelength in the front cover whose material is hard aluminum. Density of the hard aluminum: $\rho_1 = 2.79 \times 10^3$ kg/m^3; acoustic velocity in the hard aluminum: $u_1 = 5150$ m/s. If the acoustic frequency is set as 20 kHz, the wavelength of sound in the hard aluminum (λ_1) can be then calculated by Equation (5.77).

$$\lambda_1 = \frac{u_1}{f} = 257.5 \text{ mm.} \tag{5.77}$$

Thus, the length of the front cover l_1 can be obtained as 64.4 mm. The rear cover plate and the piezoelectric ceramic stack (two sheets) forms a longitudinal quarter-wavelength vibrator, and its frequency equation can be written as below according to Equation (5.55).

$$\tan(k_2 l_2) \tan(k_3 l_3) = \frac{z_2}{z_3}, \tag{5.78}$$

where $z_2 = \rho_2 c_2 S_2$ and $z_3 = \rho_3 c_3 S_3$.

Taking into account the influence of the thin electrode and the cementing layer, the equivalent sound velocity in the piezoelectric ceramic stack is estimated as 2950 m/s (i.e., $c_2 = 2950$ m/s), and that in the steel is estimated as 5200 m/s (i.e., $c_3 = 5200$ m/s). Thus, the length of the rear cover can be obtained as about 32.8 mm (i.e., $l_3 \approx 32.8$ mm).

According to Equations (5.63)–(5.70), the vibration velocity ratio of the front surface to the rear surface of the transducer (R_V) can be expressed by:

$$R_V = \frac{\dot{\xi}_f}{\dot{\xi}_b} = \frac{z_2}{z_1} \frac{1}{\sin(k_2 l_2) \sqrt{1 + \left(\frac{z_2}{z_3}\right)^2 \cot^2(k_2 l_2)}}, \tag{5.79}$$

where $z_i = \rho_i c_i S_i$ and $i = 1, 2, 3$.

Using the relevant data, R_V can be obtained as -1.71 (i.e., $R_V = -1.71$). The negative result of R_V indicates that vibration directions of the front and the rear surface of the transducer are opposite.

To further improve the R_V, the front cover can be designed to have a variable cross-section shape, such as the conical or the exponential, which has been illustrated in Ref. [2]. Some other methods to improve the R_V include using the multi-stage cascaded half wavelength horn. However, the number of the cascades cannot be too large since the larger number of the cascades will result in more damping loss of the transducer.

5.2   Radial Vibration Ultrasonic Transducer

5.2.1   Overview

For different applications of the power ultrasound, we often need to design different types of shapes and vibration modes of the ultrasonic transducer. In addition to the longitudinal vibration, the radial vibration is another commonly used mode. The radial vibration composite piezoelectric transducer is featured by a high power density, a large radiation area, and a uniform radiation, and it is very suitable for ultrasonic cleaning, chemical reactions, and other processing technologies.

To obtain a high-power radial vibration energy output, two methods may be employed as below:

- One is using the longitudinal vibration sandwich piezoelectric ultrasonic transducer to drive the circular tube. In such way, the longitudinal vibrations are transformed into radial ones [3, 4]. As shown in Figure 5.9, the circular tube can be hollow, solid, or a stepped rod, and its length should be the integral multiple of half wavelength of corresponding ultrasound. The power of a single-tube ultrasonic transducer can be as high as 2 kW. Since the ultrasound is radiated from the tubular surface, the radiation surface area will be much larger than a single longitudinal sandwich ultrasonic transducer, and more uniform ultrasonic radiation will be acquired. Essentially, such a tubular ultrasonic transducer is of a one-dimensional and longitudinal composite structure and can be designed by using the theory of the longitudinal ultrasonic transducer.

- Another way to achieve a large power of radial vibration is adopting a radial composite transducer [5, 6]. This type of transducer is made of a metal tube or ring and a piezoelectric ceramic tube or plate. The piezoelectric ceramic tube or plate may be polarized in the direction of thickness or radial direction. As shown in Figure 5.10, the radial composite piezoelectric transducer is made of a radial polarized piezoelectric ceramic tube and a fastening metal tube. The role of the metal tube, on one hand, is to offer sufficient radial prestress so as to improve the power density or power capacity of the transducer, and on the other hand, it can prevent the piezoelectric ceramic tube from being ruptured due to the high-power vibration. To further enhance the power density of this tubular transducer, the ternary composite structure can be used, that is, the piezoelectric ceramic tube is externally and internally fastened with a prestressed metal tube. In the following part, the radial composite transducer is to be illustrated.

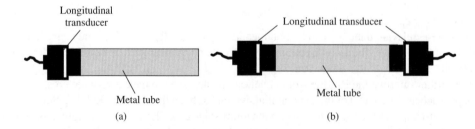

Figure 5.9   Push-pull transducer. (a) Unidirectional vibration and (b) bidirectional vibration

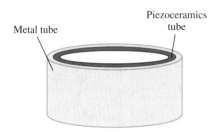

Figure 5.10   Short tubular radial composite piezoelectric ultrasonic transducer

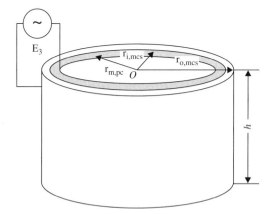

Figure 5.11   Cylindrical radial composite piezoelectric ceramic transducer

5.2.2   *Theoretical Analysis and Design of a Binary Radial Transducer*

Figure 5.11 presents the schematic diagram of a cylindrical radial composite piezoelectric ceramic transducer. The inner part is the piezoelectric ceramic thin-walled circular tube polarized in its radial direction, and the outer part is the metal thin-walled circular cylindrical shell. They are tightly composed in the radial direction and their height or length is assumed to be the same. The wall thickness (δ_{pc}) of the piezoelectric ceramic circular tube is much less than its radius ($r_{m,pc}$). The outer and the inner radius of the metal thin-walled cylindrical shell are $r_{o,mcs}$ and $r_{i,mcs}$, respectively. The height of the cylindrical transducer is assumed to be less than its diameter and denoted by h. E_3 is external excitation electric field. If the driving frequency of the external excitation electric field can make the transducer vibrate in radial resonance, the axial vibration can be ignored. In the following analysis, it is assumed that the radial vibration of the cylindrical radial composite piezoelectric transducer should be axis-symmetric, and the vibration displacement, and velocity be both in the radial direction.

5.2.2.1   Radial Vibration of a Metal Thin-Walled Cylindrical Shell

Figure 5.12a shows a metal cylindrical shell in radial vibration. Its height, wall thickness, outer and inner radius are h, δ_{mcs}, $r_{o,mcs}$, and $r_{i,mcs}$, respectively. $F_{i,mcs}$ and $\dot{\xi}_{i,mcs}$ represent external radial force and radial vibration velocity at the inner surface of the metal cylindrical shell.

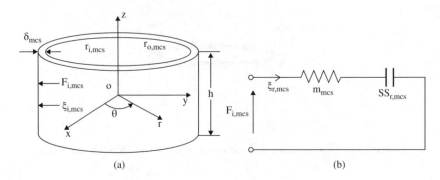

Figure 5.12 Schematic for a thin-walled metal cylindrical shell in radial vibration and its equivalent circuit. (a) Thin-walled metal cylindrical shell and (b) equivalent circuit

In cylindrical coordinates, the wave equations for a metal cylindrical shell can be expressed as follows:

$$\rho_{mcs}\frac{\partial^2 \xi_{r,mcs}}{\partial \tau^2} = \frac{\partial \sigma_{r,mcs}}{\partial r} + \frac{1}{r}\frac{\partial \sigma_{r\theta,mcs}}{\partial \theta} + \frac{\partial \sigma_{rz,mcs}}{\partial z} + \frac{\sigma_{r,mcs} - \sigma_{\theta,mcs}}{r}, \tag{5.80}$$

$$\rho_{mcs}\frac{\partial^2 \xi_{\theta,mcs}}{\partial \tau^2} = \frac{\partial \sigma_{r\theta,mcs}}{\partial r} + \frac{1}{r}\frac{\partial \sigma_{\theta,mcs}}{\partial \theta} + \frac{\partial \sigma_{\theta z,mcs}}{\partial z} + \frac{2\sigma_{r\theta,mcs}}{r}, \tag{5.81}$$

$$\rho_{mcs}\frac{\partial^2 \xi_{z,mcs}}{\partial \tau^2} = \frac{\partial \sigma_{rz,mcs}}{\partial r} + \frac{1}{r}\frac{\partial \sigma_{\theta z,mcs}}{\partial \theta} + \frac{\partial \sigma_{z,mcs}}{\partial z} + \frac{\sigma_{rz,mcs}}{r}, \tag{5.82}$$

where $\xi_{r,mcs}$, $\xi_{\theta,mcs}$, and $\xi_{z,mcs}$ are radial, tangential, and axial displacement in m and $\sigma_{r,mcs}$, $\sigma_{\theta,mcs}$, $\sigma_{z,mcs}$, $\sigma_{r\theta,mcs}$, $\sigma_{rz,mcs}$, and $\sigma_{\theta z,mcs}$ represent different kinds of stresses in Pa.

In cylindrical coordinates, the relationships between the strains and displacements are as follows:

$$s_{r,mcs} = \frac{\partial \xi_{r,mcs}}{\partial r}, \tag{5.83}$$

$$s_{\theta,mcs} = \frac{1}{r}\frac{\partial \xi_{\theta,mcs}}{\partial \theta} + \frac{\xi_{r,mcs}}{r}, \tag{5.84}$$

$$s_{z,mcs} = \frac{\partial \xi_{z,mcs}}{\partial z}, \tag{5.85}$$

$$s_{r\theta,mcs} = \frac{1}{r}\frac{\partial \xi_{r,mcs}}{\partial \theta} + \frac{\partial \xi_{\theta,mcs}}{\partial r} - \frac{\xi_{\theta,mcs}}{r}, \tag{5.86}$$

$$s_{\theta z,mcs} = \frac{1}{r}\frac{\partial \xi_{z,mcs}}{\partial \theta} + \frac{\partial \xi_{\theta,mcs}}{\partial z}, \tag{5.87}$$

$$s_{rz,mcs} = \frac{\partial \xi_{r,mcs}}{\partial z} + \frac{\partial \xi_{z,mcs}}{\partial r} \tag{5.88}$$

Here, $s_{r,mcs}$, $s_{\theta,mcs}$, $s_{z,mcs}$, $s_{r\theta,mcs}$, $s_{\theta z,mcs}$, and $s_{rz,mcs}$ represent different kind of strains.

According to Hooke's law, the relationship between the stresses and the strains can be expressed as

$$S_{r,mcs} = \frac{1}{E_{mcs}}\left[\sigma_{r,mcs} - \nu_{mcs}\left(\sigma_{\theta,mcs} + \sigma_{z,mcs}\right)\right], \tag{5.89}$$

$$S_{\theta,mcs} = \frac{1}{E_{mcs}}\left[\sigma_{\theta,mcs} - \nu_{mcs}\left(\sigma_{r,mcs} + \sigma_{z,mcs}\right)\right], \tag{5.90}$$

$$S_{z,mcs} = \frac{1}{E_{mcs}}\left[\sigma_{z,mcs} - \nu_{mcs}\left(\sigma_{\theta,mcs} + \sigma_{r,mcs}\right)\right], \tag{5.91}$$

$$S_{r\theta,mcs} = \frac{\sigma_{r\theta,mcs}}{\Upsilon_{mcs}}, S_{rz,mcs} = \frac{\sigma_{rz,mcs}}{\Upsilon_{mcs}}, S_{\theta z,mcs} = \frac{\sigma_{\theta z,mcs}}{\Upsilon_{mcs}}, \tag{5.92}$$

where Υ_{mcs} is shear modulus of material, $\Upsilon_{mcs} = E_{mcs}/[2(1 + \nu_{mcs})]$, E_{mcs} and ν_{mcs} are the Young's modulus and Poisson's ratio of material, respectively.

For the symmetrical radial vibration of a short cylindrical shell with very thin wall thickness, it is assumed that the wall thickness (δ_{mcs}) of the metal cylindrical shell is much smaller than its radius ($r_{o,mcs}$ or $r_{i,mcs}$). In this case, $\sigma_{r,mcs}$, $\sigma_{z,mcs}$, $\sigma_{r\theta,mcs}$, $\sigma_{rz,mcs}$, $\sigma_{\theta z,mcs}$, $S_{r,mcs}$, $S_{z,mcs}$, $S_{r\theta,mcs}$, $S_{\theta z,mcs}$, $S_{rz,mcs}$ can be ignored, and $\partial\xi_{\theta,mcs}/\partial\theta = 0$, $\partial\sigma_{\theta,mcs}/\partial\theta = 0$. Using Equations (5.80)–(5.92) and the assumptions, the wave equations of pure radial vibration for the short and thin-walled cylindrical shell can be obtained as:

$$\rho_{mcs}\frac{\partial^2\xi_{r,mcs}}{\partial\tau^2} = -\frac{\sigma_{\theta,mcs}}{r_{m,mcs}}, \tag{5.93}$$

$$\sigma_{\theta,mcs} = E_{mcs}S_{\theta,mcs}, \tag{5.94}$$

$$S_{\theta,mcs} = \frac{\xi_{r,mcs}}{r_{m,mcs}}. \tag{5.95}$$

Here, $r_{m,mcs} = (r_{i,mcs} + r_{o,mcs})/2$.

For harmonic vibration, the radial force on the inside wall of the metal cylindrical shell $F_{i,mcs}$ by the inner piezoelectric tube can be written as:

$$F_{i,mcs} = F_{max}\exp\left(i\omega_{mcs}\tau\right). \tag{5.96}$$

Using Equations (5.93)–(5.96), the wave equation of a thin-walled cylindrical shell in radial vibration can be written as:

$$m_{mcs}\frac{\partial^2\xi_{r,mcs}}{\partial\tau^2} + \frac{1}{ss_{r,mcs}}\xi_{r,mcs} = F_{i,mcs}, \tag{5.97}$$

where, $m_{mcs} = 2\rho_{mcs}\pi r_{m,mcs}h_{mcs}\delta_{mcs}$ and $ss_{r,mcs} = r_{m,mcs}/\left(2\pi h_{mcs}\delta_{mcs}E_{mcs}\right)$ are the equivalent mass and elastic flexibility coefficient of the metal thin-walled cylindrical shell, respectively. For harmonic vibration, we have $\xi_{r,mcs} = \xi_{max}\exp(i\omega_{mcs}\tau)$. Thus the solution to Equation (5.97) is:

$$\xi_{max} = \frac{F_{max}}{-\omega^2_{mcs}m_{mcs} + 1/ss_{r,mcs}}. \tag{5.98}$$

With the expression of $\dot{\xi}_{i,\text{mcs}} = i\omega_{\text{mcs}}\xi_{r,\text{mcs}}$, Equation (5.98) can be transformed into:

$$F_{i,\text{mcs}} = \left(i\omega_{\text{mcs}}m_{\text{mcs}} + \frac{1}{i\omega_{\text{mcs}}ss_{r,\text{mcs}}} \right)\dot{\xi}_{i,\text{mcs}}. \tag{5.99}$$

According to Equation (5.99), the electro-mechanical equivalent circuit for a metal thin-walled cylindrical shell in radial vibration can be obtained as shown in Figure 5.12b.

5.2.2.2   Radial Vibration of a Short Piezoelectric Ceramic Thin-Walled Circular Tube

A radially polarized piezoelectric ceramic thin-walled tube is shown in Figure 5.13a. In Figure 5.13a, δ_{pc}, $r_{m,pc}$, and h_{pc} stand for the wall thickness, the mean radius, and the height of the piezoelectric ceramic tube, respectively. $F_{o,pc}$ and $\dot{\xi}_{o,pc}$ represent the external radial force and vibration velocity on the outer surface of the piezoelectric ceramic tube. For a short thin-walled circular tube, the wall thickness (δ_{pc}) and the height (h_{pc}) of the piezoelectric ceramic is assumed to be much smaller than its radius ($r_{m,pc}$). In such a case, the vibration in the axial direction of the piezoelectric ceramic circular tube are very weak, and it can be considered as an ideal axis-symmetric radial vibration. The radial equation can be written as:

$$\rho_{pc}\frac{\partial^2 \xi_{r,pc}}{\partial \tau^2} = -\frac{\sigma_{\theta,pc}}{r_{m,pc}}. \tag{5.100}$$

Here, ρ_{pc} is density of the piezoelectric ceramic material in kg/m^3.

Correspondingly, the piezoelectric constitutive equations can be expressed as:

$$s_{\theta,pc} = s_{33,pc}^{E}\sigma_{\theta,pc} + s_{c,pc}E_{r,pc}, \tag{5.101}$$

$$\phi_{r,pc} = s_{c,pc}\sigma_{\theta,pc} + \varepsilon_{33,pc}^{T}E_{r,pc}. \tag{5.102}$$

In Equations (5.101) and (5.102), $s_{33,pc}^{E}$ is the elastic flexibility coefficient of the piezoelectric ceramic; $s_{c,pc}$ is the piezoelectric ceramic strain constant; $\varepsilon_{33,pc}^{T}$ is the piezoelectric ceramic

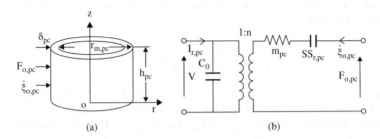

(a)   (b)

Figure 5.13   Schematic for a short thin-walled piezoelectric ceramic tube in radial vibration and its electro-mechanical equivalent circuit. (a) Short thin-walled circular tube and (b) electro-mechanical equivalent circuit

dielectric constant; $E_{r,pc}$ is radial external exciting electric field acting on the piezoelectric ceramic; and $\phi_{r,pc}$ represents radial electric displacement.

By using Equation (5.101) and the expression $s_{\theta,pc} = \xi_{r,pc}/r_{m,pc}$, we can obtain,

$$\sigma_{\theta_{pc}} = \frac{\xi_{r,pc}}{r_{m,pc}s_{33,pc}^{E}} - \frac{s_{c,pc}}{s_{33,pc}^{E}}E_{r,pc}. \tag{5.103}$$

Substituting Equation (5.103) into Equation (5.100), and considering the external radial force at the outer surface of the piezoelectric ceramic tube, yields

$$\rho_{pc}\frac{\partial^2 \xi_{r,pc}}{\partial \tau^2} + \frac{1}{r_{m,pc}^2 s_{33}^{E}}\xi_{r,pc} = \frac{s_{c,pc}}{r_{m,pc}s_{33}^{E}}E_{r,pc} - \frac{F_{o,pc}}{S_{m,pc}\delta_{pc}}. \tag{5.104}$$

where $S_{m,pc}$ is the side face area of the piezoelectric ceramic tube, $S_{m,pc} = 2\pi r_{m,pc}h_{pc}$. When the piezoelectric ceramic tube is in the harmonic radial vibration, we have $\dot{\xi}_{o,pc} = i\omega_{pc}\xi_{r,pc}$ and $\ddot{\xi}_{o,pc} = -\omega_{pc}^2\xi_{r,pc}$. From Equation (5.104), it yields:

$$nV = \left(i\omega_{pc}m_{pc} + \frac{1}{i\omega_{pc}ss_{r,pc}}\right)\dot{\xi}_{o,pc} + F_{o,pc}, \tag{5.105}$$

where V is the voltage applied to the piezoelectric ceramic tube, $V = E_{r,pc}\delta_{pc}$; n is the electro-mechanical conversion coefficient, $n = 2\pi h_{pc}s_{c,pc}/s_{33,pc}^{E}$, $m_{pc} = \rho_{pc}S_{m,pc}\delta_{pc}$, and $ss_{r,pc} = s_{33,pc}^{E}r_{m,pc}/(2\pi\delta_{pc}h_{pc})$ are the equivalent mass and elastic flexibility coefficient of the piezoelectric ceramic tube, respectively. From Equations (5.102) and (5.103), the electric displacement can be obtained as:

$$\phi_{r,pc} = \frac{s_{c,pc}}{s_{33}^{E}r_{m,pc}}\xi_{r,pc} + \varepsilon_{33,pc}^{T}\left(1 - \varepsilon_{m-e}^{2}\right)E_{r,pc}, \tag{5.106}$$

where ε_{m-e} is the transverse electro-mechanical coupling coefficient, $\varepsilon_{m-e}^{2} = s_{c,pc}^{2}/(s_{33,pc}^{E}\varepsilon_{33,pc}^{T})$.

The electric displacement flux flowing out of the electrode surface of the piezoelectric ceramic tube can be expressed by

$$\overline{\phi}_{pc} = 2\pi r_{m,pc}h_{pc}\phi_{r,pc} = \frac{2\pi r_{m,pc}h_{pc}\varepsilon_{33}^{T}\left(1 - \varepsilon_{m-e}^{2}\right)}{\delta_{pc}}V + n\xi_{o,pc}. \tag{5.107}$$

For the harmonic vibration, the electric current entering into the piezoelectric ceramic circular tube $I_{r,pc}$ can be expressed as:

$$I_{r,pc} = d\overline{\phi}_{pc}/d\tau = \omega_{pc}C_0 V + n\dot{\xi}_{o,pc}, \tag{5.108}$$

where C_0 is clamped capacitance of the piezoelectric ceramic circular tube in radial vibration, $C_0 = 2\pi r_{m,pc}h_{pc}\varepsilon_{33,pc}^{T}(1 - \varepsilon_{m-e}^{2})/\delta_{pc}$.

According to Equations (5.105) and (5.108), the electro-mechanical equivalent circuit for the radial vibration of the thin-walled piezoelectric ceramic tube can be obtained as shown in Figure 5.13b.

5.2.2.3 Equivalent Circuit and Resonance Frequency Equation of the Cylindrical Radial Composite Piezoelectric Ceramic Transducer

In the cylindrical radial composite piezoelectric ceramic transducer, the inner piezoelectric ceramic tube and the outer metal cylindrical shell are tightly fastened together in the radial direction. At the interface between the inner piezoelectric ceramic tube and the outer metal cylindrical shell, the radial vibration velocity, and force are continuous. Using the equivalent circuit of the metal thin-walled cylindrical shell in radial vibration as shown in Figure 5.12b and the equivalent circuit of the piezoelectric ceramic thin-walled circular tube in radial vibration as shown in Figure 5.13b, the composite electro-mechanical equivalent circuit for a cylindrical radial composite piezoelectric transducer can be obtained as shown in Figure 5.14 where z_r represents the load impedance on the outer surface of the transducer. For a light-load transducer, $z_r = 0$.

According to Figure 5.14, the input electric impedance z_E of the cylindrical radial composite piezoelectric ceramic transducer in radial vibration can be expressed as

$$z_E = \frac{V}{I_{r,pc}} = \frac{z_{pc} + z_{mcs} + z_r}{n^2 + i\omega_{pc}C_0\left(z_{pc} + z_{mcs} + z_r\right)}. \tag{5.109}$$

Here, $z_{pc} = i\omega_{pc}m_{pc} + 1/(i\omega_{pc}ss_{r,pc})$, $z_{mcs} = i\omega_{mcs}m_{mcs} + 1/(i\omega_{mcs}ss_{r,mcs})$. z_{pc} and z_{mcs} are mechanical impedance of the piezoelectric ceramic circular tube and the short-and-thin-wall metal cylindrical shell in radial vibration, respectively.

From Equation (5.109), we can obtain the resonance frequency and anti-resonance frequency equations of the radial vibration of the cylindrical radial composite piezoelectric ceramic transducer. The resonance frequency equation is $z_E = 0$, and the anti-resonance frequency equation is $z_E \rightarrow \infty$.

For the piezoelectric ceramic ultrasonic transducers, the effective electro-mechanical coupling coefficient ε_e is an important parameter for evaluating the vibration performance of the transducer. It can be expressed as:

$$\varepsilon_e^2 = \frac{f_{parallel}^2 - f_{series}^2}{f_{parallel}^2}. \tag{5.110}$$

Here $f_{parallel}$ and f_{series} are the parallel and series resonance frequencies of the transducer, respectively. They can be approximated by the resonance and anti-resonance frequencies when the transducer has a high mechanical quality factor.

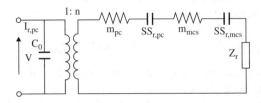

Figure 5.14 Composite electro-mechanical equivalent circuit of a cylindrical radial composite piezoelectric ceramic transducer

According to the frequency equations, the theoretical relationship among the resonance frequency, the anti-resonance frequency, the effective electro-mechanical coupling coefficient and the wall thickness of metal cylinder shell of the cylindrical radial composite piezoelectric transducer can be investigated by using the MATHEMATICA software. The results are shown in Figures 5.15–5.17. In the numerical calculations, the material of the piezoelectric ceramic tube is PZT-4, and the materials of the metal cylinder shell are steel and aluminum. Their basic material parameters are given as below: For the piezoelectric ceramic, $\rho_{pc} = 7500 \, \text{kg/m}^3$, $s_{33}^E = 12.3 \times 10^{-12} \, \text{m}^2/\text{N}$, $s_{c,pc} = -123 \times 10^{-12} \, \text{C/N}$, $\varepsilon_{m-e} = 0.32$, $v_{pc} = 0.33$, $\varepsilon_{33}^T/\varepsilon_0 = 1300$, $\varepsilon_0 = 8.85 \times 10^{-12} \, \text{C}^2/(\text{N·m}^2)$. For the steel, $\rho_{steel} = 7800 \, \text{kg/m}^3$, $E_{steel} = 2.09 \times 10^{11} \, \text{N/m}^2$, and $v_{steel} = 0.28$. For the aluminum, $\rho_{Al} = 2700 \, \text{kg/m}^3$, $E_{Al} = 7.15 \times 10^{10} \, \text{N/m}^2$, and $v_{Al} = 0.34$. The geometrical dimensions of $r_{m,pc}$, h_{mcs}, h_{pc}, and δ_{pc} are 23.5, 30, 30, and 5 mm, respectively.

From Figures 5.15 and 5.16, it can be seen that when the wall thickness of the outer metal cylindrical shell is smaller than a specific value, the radial resonance frequency, and anti-resonance frequency both increase with the increase of wall thickness, and it is reverse when the wall thickness of the metal cylindrical shell is larger than this value. The reason for this may be that the equivalent elastic constant of the composite transducer is the highest at this value of wall thickness that results in the maximum resonance frequency. The value of wall thickness corresponding to the peak point of resonance frequency is related to the materials and geometrical dimensions of the transducer. In this case, it is about 7 mm for the aluminum cylinder shell and about 5 mm for the steel cylinder shell. On the other hand, the difference between the resonance frequency and the anti-resonance frequency decreases with the increase of the wall thickness of the outer metal cylindrical shell of the transducer. It indicates that the effective electro-mechanical coupling coefficient will decrease as the wall thickness of the outer metal cylinder shell increases. It can be found as well from Figure 5.17 that the effective electro-mechanical coupling coefficient of the transducer with aluminum cylindrical shell is obviously larger than that of the transducer with steel cylindrical shell.

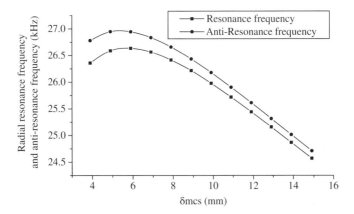

Figure 5.15 Theoretical relationships between the resonance frequency, the anti-resonance frequency, and the wall thickness of steel cylinder shell of a cylindrical radial composite piezoelectric ceramic transducer

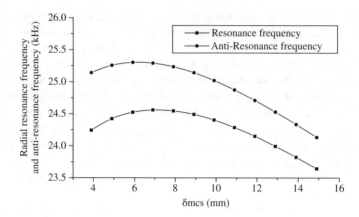

Figure 5.16 Theoretical relationship between the resonance frequency, the anti-resonance frequency, and the wall thickness of aluminum cylinder shell of a cylindrical radial composite piezoelectric ceramic transducer

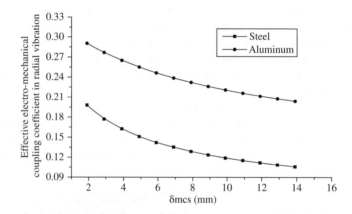

Figure 5.17 Theoretical relationships between the effective electro-mechanical coupling coefficient and the wall thickness of steel and aluminum cylinder shell of a cylindrical radial composite piezoelectric ceramic transducer

5.2.2.4 Experimental Study

In order to verify the above theoretical analysis, some cylindrical radial composite piezoelectric ceramic transducers have been investigated. The materials of the piezoelectric ceramic tube and the metal cylindrical shell are PZT-4 and aluminum, respectively. The heights of the inner piezoelectric ceramic tube and the outer aluminum cylindrical shell of the trial-made transducers are 30 and 40 mm, respectively. Their material parameters and the other geometrical dimensions of the transducer are the same as those used in the above calculations. The radial resonance frequencies and anti-resonance frequencies of the trial-made transducers are measured by an Agilent 4294A Precision Impedance Analyzer. The theoretical and experimental resonance frequencies, anti-resonance frequency, and the effective electro-mechanical coupling coefficient are listed in Table 5.1. From Table 5.1, it can be found that the theoretical

Table 5.1 Theoretical and measured radial resonance frequencies, the anti-resonance frequencies, and the effective electro-mechanical coupling coefficient of a cylindrical radial composite piezoelectric ceramic transducer

No.	γ	δ_{mcs} (mm)	$f_{r,cal}$ (Hz)	$f_{a,cal}$ (Hz)	$\varepsilon_{e,cal}$	$f_{r,exp}$ (Hz)	$f_{a,exp}$ (Hz)	$\varepsilon_{e,exp}$	Δ (%)
1	1.15	3.9	246 99	252 84	0.2139	237 65	244 05	0.2257	3.93
2	1.25	6.5	249 52	254 48	0.1965	245 71	249 43	0.1668	1.55
3	1.35	9.1	247 99	252 15	0.1810	244 53	248 11	0.1641	1.41

Here, γ is the ratio of the outer radius to the inner radius of the metal cylindrical shell, that is, $\gamma = r_{o,mcs}/r_{i,mcs}$. $f_{r,cal}$, $f_{a,cal}$, and $\varepsilon_{e,theory}$ are theoretical resonance frequency, anti-resonance frequency, and effective electro-mechanical coupling coefficient, respectively. $f_{r,exp}$, $f_{a,exp}$, and $\varepsilon_{e,exp}$ are the measured resonance frequency, anti-resonance frequency, and effective electro-mechanical coupling coefficient, respectively. $\Delta = \left| f_{r,exp} - f_{r,cal} \right| / f_{r,exp} \times 100\%$.

radial resonance frequencies and anti-resonance frequencies have a good agreement with the experimental results. However, the theoretical and experimental effective electro-mechanical coupling coefficients of the trial-made cylindrical radial composite piezoelectric ceramic transducers are not in good agreement with each other.

As for the experimental errors, the following factors should be taken into account. Firstly, the standard material parameters used in the numerical calculations are not the same as the real values of the materials. Secondly, the outside diameter of the inner piezoelectric ceramic tube and the inside diameters of the outer hollow cylindrical shell are difficult to control during mechanical processing. Meanwhile, the radial prestresses for these transducers are different from one another during the transducer processing. As a result, different cylindrical radial composite piezoelectric transducers used for the experiments will have different characteristic parameters. Thirdly, in the above theoretical analysis, the mechanical, and dielectric losses in the transducers are not considered. In practical cases, the losses have more influence on the efficiency and effective electro-mechanical coupling coefficient of the transducer, especially at a higher electric power level. These are often ignored in the theoretical analysis for the power piezoelectric transducer. In addition, the mechanical and dielectric losses in a composite piezoelectric ceramic transducer are not constant under different electric levels.

5.2.3 Radial Vibration Sandwich Piezoelectric Transducer

In the previous section, the binary radial composite ultrasonic transducer has been studied. Such kind of transducer is often made by the expansion-and-contraction method. Although the thermal treatment method is simple, there are some challenges in the real applications. Firstly, the thermal treatment temperature must be controlled to be lower than the Curie temperature of the piezoelectric ceramic to avoid depolarization. Secondly, the interference fit between the piezoelectric ceramic tube and the outer metal cylindrical shell must be accurate enough to meet the actual requirements of prestressing force. Thirdly, the radial compressive strength of the piezoelectric ceramic tube is low, and too large prestressing force will easily cause the rupture of the piezoelectric ceramic tube. It indicates that the power density of the binary radial composite ultrasonic transducer will be limited. Furthermore, the binary radial composite ceramic transducer is often radially polarized, and the electromechanical coupling coefficient of the transverse vibration mode is small, which leads to a poor electromechanical conversion efficiency.

In response to the above problems, the authors of this book proposed a ternary radial composite sandwich piezoelectric ceramic transducer. In the following section, the radial vibration characteristics of this new type transducer are to be analyzed and discussed.

5.2.3.1 Basic Structure of a Radial Vibration Sandwich Piezoelectric Transducer

Figure 5.18 is the schematic diagram of a ternary radial composite sandwich piezoelectric ceramic transducer which consists of three parts: a circular expandable inner core, a circular piezoelectric crystal heap, and a metal thin-walled cylindrical shell. To reduce the weight of the transducer, the circular expandable inner core often adopts a metal thin-walled cylindrical shell. The inner and the outer radius of the inner core are r_d and r_a, respectively. The inner and the outer radius of the outer metal cylindrical shell are r_b and r_c, respectively. The cylindrical piezoelectric ceramic crystal stack consists of several cylindrical ceramic sheets (usually even in number). As shown in Figure 5.19, two adjacent piezoelectric sheets are radially polarized in the opposite direction, and there lays a cylindrical metal sheet (which is generally used as the positive electrode) between the two adjacent piezoelectric sheets.

The radial composite sandwich piezoelectric ceramic transducer has several advantages as below: on one hand, the use of radial stacked piezoelectric crystal stack can improve the power capacity of the radial piezoelectric transducer; on the other hand, the electromechanical coupling coefficient will be improved since the radial vibration mode of the piezoelectric ceramic tube can be extended as the axial vibration mode, and hence, the electromechanical conversion capacity and efficiency will be improved as well. In addition, the piezoelectric ceramic elements are radially prestressed on the two sides, and this can help to increase the power capacity of the transducer.

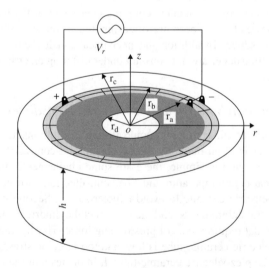

Figure 5.18   Sandwiched radial vibration piezo-transducer

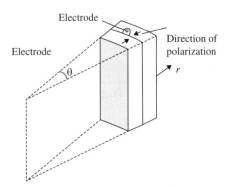

Figure 5.19   Cylindrical piezoceramic piles

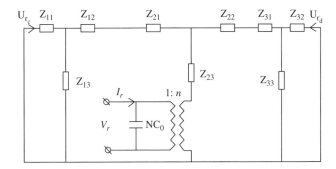

Figure 5.20   The equivalent circuit of a sandwiched radial composite piezotransducer in radial vibration

5.2.3.2   Vibration Analysis of a Radial Vibration Sandwich Piezoelectric Transducer

Based on the equivalent circuit of each component in the transducer, the equivalent circuit of radial free vibration of the sandwich radial composite piezoelectric ultrasonic transducer can be described by Figure 5.20. In Figure 5.20, u_{rc} is that of the external surface of the outer metal cylindrical shell; u_{rd} is the vibration velocity of the internal surface of the inner metal cylindrical shell of the transducer; N is the number of the piezoelectric ceramic circular tube in the transducer; Z_{11}, Z_{12}, and Z_{13} stand for the mechanical impedances of the outer metal cylindrical shell; Z_{21}, Z_{22}, and Z_{23} stand for the mechanical impedances of the piezoelectric ceramic stack; and Z_{31}, Z_{32}, and Z_{33} stand for the mechanical impedances of the inner metal cylindrical shell. These mechanical impedances for different parts of the radial vibration sandwich piezoelectric transducer are listed below as:

$$Z_{11} = i \frac{2z_{rb}}{\pi k_1 r_b \left[J_1(k_1 r_c)Y_1(k_1 r_b) - J_1(k_1 r_b)Y_1(k_1 r_c) \right]} \left[\frac{J_1(k_1 r_c)Y_0(k_1 r_b) - J_0(k_1 r_b)Y_1(k_1 r_c)}{J_1(k_1 r_b)Y_0(k_1 r_b) - J_0(k_1 r_b)Y_1(k_1 r_b)} - 1 \right]$$

$$- i \frac{2z_{rb}(1 - v_1)}{\pi(k_1 r_b)^2 \left[J_1(k_1 r_b)Y_0(k_1 r_b) - J_0(k_1 r_b)Y_1(k_1 r_b) \right]} , \tag{5.111}$$

$$Z_{12} = i \frac{2z_{rb}}{\pi k_1 r_b \left[J_1(k_1 r_c)Y_1(k_1 r_b) - J_1(k_1 r_b)Y_1(k_1 r_c)\right]} \left[\frac{J_1(k_1 r_b)Y_0(k_1 r_c) - J_0(k_1 r_c)Y_1(k_1 r_b)}{J_1(k_1 r_c)Y_0(k_1 r_c) - J_0(k_1 r_c)Y_1(k_1 r_c)} - 1\right]$$

$$+ i \frac{2z_{rc}(1 - v_1)}{\pi(k_1 r_c)^2 \left[J_1(k_1 r_c)Y_0(k_1 r_c) - J_0(k_1 r_c)Y_1(k_1 r_c)\right]}, \qquad (5.112)$$

$$Z_{13} = i \frac{2z_{rb}}{\pi k_1 r_b \left[J_1(k_1 r_c)Y_1(k_1 r_b) - J_1(k_1 r_b)Y_1(k_1 r_c)\right]}, \qquad (5.113)$$

$$Z_{21} = Z_{22} = i\rho_{pc}u_{r,pc}S_{pc} \tan\left(\frac{1}{2}Nk_{pc}h_{pc}\right), \qquad (5.114)$$

$$Z_{23} = \frac{\rho_{pc}u_{r,pc}S_{pc}}{i \sin(Nk_{pc}h_{pc})}, \qquad (5.115)$$

$$Z_{31} = i \frac{2z_{rd}}{\pi k_3 r_d \left[J_1(k_3 r_a)Y_1(k_3 r_d) - J_1(k_3 r_d)Y_1(k_3 r_a)\right]} \left[\frac{J_1(k_3 r_a)Y_0(k_3 r_d) - J_0(k_3 r_d)Y_1(k_3 r_a)}{J_1(k_3 r_d)Y_0(k_3 r_d) - J_0(k_3 r_d)Y_1(k_3 r_d)} - 1\right]$$

$$- i \frac{2z_{rd}(1 - v_3)}{\pi(k_1 r_d)^2 \left[J_1(k_1 r_d)Y_0(k_1 r_d) - J_0(k_1 r_d)Y_1(k_1 r_d)\right]}, \qquad (5.116)$$

$$Z_{32} = i \frac{2z_{rd}}{\pi k_3 r_d \left[J_1(k_3 r_a)Y_1(k_3 r_d) - J_1(k_3 r_d)Y_1(k_3 r_a)\right]} \left[\frac{J_1(k_3 r_d)Y_0(k_3 r_a) - J_0(k_3 r_a)Y_1(k_3 r_d)}{J_1(k_3 r_a)Y_0(k_3 r_a) - J_0(k_3 r_a)Y_1(k_3 r_a)} - 1\right]$$

$$+ i \frac{2z_{ra}(1 - v_3)}{\pi(k_3 r_a)^2 \left[J_1(k_3 r_a)Y_0(k_3 r_a) - J_0(k_3 r_a)Y_1(k_3 r_a)\right]}, \qquad (5.117)$$

$$Z_{33} = i \frac{2z_{rd}}{\pi k_3 r_d \left[J_1(k_3 r_a) Y_1(k_3 r_d) - J_1(k_3 r_d) Y_1(k_3 r_a)\right]}, \qquad (5.118)$$

where, k is the sound wave number in material, 1/m; v is Poisson's ratio of material; r_a, r_b, r_c, and r_d are radius illustrated in Figure 5.20; z_{ra} and z_{rb} are the internal and the external surface characteristic acoustic impedance of the piezoelectric ceramic stack, $z_{ra} = \rho_{pc}c_{o,pc}S_a = 2\rho_{pc}c_{o,pc}\pi r_a h_{pc}$, $z_{rb} = \rho_{pc}c_{o,pc}S_b = 2\rho_{pc}c_{o,pc}\pi r_b h_{pc}$; z_{rc} is the external surface characteristic acoustic impedance of the outer metal cylindrical shell, $z_{rc} = \rho_1 c_{o,1}S_c = 2\rho_1 c_{o,1}\pi r_c h_1$; z_{rd} is the internal surface characteristic acoustic impedance of the inner metal cylindrical shell, $z_{rd} = \rho_3 c_{o,3}S_d = 2\rho_3 c_{o,3}\pi r_d h_3$; and the subscript "1," "2," and "3" stand for the outer metal cylindrical shell, the piezoelectric ceramic stack, and the inner metal cylindrical shell of the transducer, respectively.

$J_0(x)$ and $J_1(x)$ are the zero-order and the first-order Bessel function of the first kind, respectively.

$$J_0(x) = \sum_{k=0}^{\infty} \frac{(-1)^k}{(k!)^2}\left(\frac{x}{2}\right)^{2k}, J_1(x) = \sum_{k=0}^{\infty} \frac{(-1)^k}{k!(k+1)!}\left(\frac{x}{2}\right)^{2k+1}.$$

$Y_0(x)$ and $Y_1(x)$ are the zero-order and the first-order Bessel function of the second kind, respectively.

$$Y_0(x) = \frac{2}{\pi}J_o(x)\left(\ln\frac{x}{2} + c\right) - \frac{2}{\pi}\sum_{k=0}^{\infty}\frac{(-1)^k(x/2)^{2k}}{k!}\sum_{k=0}^{\infty}\frac{1}{k+1},$$

$$Y_1(x) = \frac{2}{\pi} J_1(x) \left(\ln \frac{x}{2} + c \right) - \frac{2}{\pi} \sum_{m=0}^{\infty} \frac{(-m)!(x/2)^{2m-1}}{m!}$$

$$- \frac{1}{\pi} \sum_{m=0}^{\infty} \frac{(-1)^m (x/2)^{2m}}{m!(m+1)!} \left(\sum_{k=0}^{m} \frac{1}{k+1} + \sum_{k=0}^{m-1} \frac{1}{k+1} \right),$$

$$c = \lim \left(1 + \frac{1}{2} + \frac{1}{3} + \dots \frac{1}{n} - \ln n \right).$$

According to the equivalent circuit of the radial vibration sandwich piezoelectric transducer as shown in Figure 5.18, the input electrical impedance of the transducer can be obtained as:

$$Z_i = \frac{V_r}{I_r} = \frac{Z_e}{n^2 + iZ_e \omega N C_0}, \tag{5.119}$$

where

$$Z_e = Z_{23} + \frac{(Z_{1e} + Z_{21})(Z_{22} + Z_{3e})}{Z_{1e} + Z_{21} + Z_{22} + Z_{3e}}. \tag{5.120}$$

In Equation (5.120), Z_{1e} and Z_{3e} are equivalent mechanical impedance of the outer and the inner metal cylindrical shell, respectively.

$$Z_{1e} = \frac{Z_{11}Z_{12} + Z_{11}Z_{13} + Z_{12}Z_{13}}{Z_{11} + Z_{13}}, \tag{5.121}$$

$$Z_{3e} = \frac{Z_{31}Z_{32} + Z_{31}Z_{33} + Z_{32}Z_{33}}{Z_{32} + Z_{33}}. \tag{5.122}$$

Using Equation (5.119), the resonance frequency equation of the free vibration of a radial transducer can be written as:

$$\text{Im}(Z_i) = 0. \tag{5.123}$$

And the anti-resonance frequency equation can be written as:

$$\text{Im}(Z_i) \to \infty. \tag{5.124}$$

The frequency equation reflects the relationships among the resonance frequency, the geometrical dimensions, and the material properties of the transducer.

5.2.3.3 Parametric Study

The frequency equation is used to investigate the relationships among the resonance frequency and the geometrical dimensions of a radial vibration sandwich piezoelectric transducer. The inner core is made of steel, the outer shell is made of aluminum, and the piezoelectric ceramic adopts the material of PZT4. The basic calculation parameters are listed as below: the inner ($r_{i,pc}$) and the outer radius ($r_{o,pc}$) of the piezoelectric ceramic circular tube are 19.5 and 25.5 mm, respectively; the height of the piezoelectric ceramic circular tube (h_{pc}) is 3.0 mm, and that of the inner and the outer shell is 36 mm. The basic material parameters are given as below: For the material of PZT4, $\rho_{pc} = 7500 \text{ kg/m}^3$, $s_{33}^E = 12.3 \times 10^{-12} \text{ m}^2/\text{N}$, $s_{c,pc} = -289 \times 10^{-12} \text{C/N}$, $v_{pc} = 0.38$, $\varepsilon_{33}^T/\varepsilon_0 = 650$, $\varepsilon_0 = 8.85 \times 10^{-12} \text{ C}^2/(\text{N·m}^2)$, $\varepsilon_{m-e} = 0.28$;

for the aluminum, $\rho_{Al} = 2700\,\text{kg/m}^3$, $E_{Al} = 7.15 \times 10^{10}\,\text{N/m}^2$, $v_{Al} = 0.34$; and for the steel, $\rho_{steel} = 7800\,\text{kg/m}^3$, $E_{steel} = 2.09 \times 10^{11}\,\text{N/m}^2$, $v_{steel} = 0.28$.

Figures 5.21 and 5.22 illustrate the influence of the internal radius of the inner prestressing shell (r_d) on the first- and the second-order radial resonance frequency and anti-resonance frequencies of the transducer with the wall thickness of 6 mm in its outer prestressing shell. As seen from Figure 5.21, with the internal radius of the inner prestressing shell increasing, the first-order radial resonance frequency and anti-resonance frequency of the transducer both decreases, and the resonance frequency and anti-resonance frequency tend to be the same. However, the second-order resonance frequency and anti-resonance frequency of the transducer versus the r_d shares different regular patterns. As shown in Figure 5.22, the second-order resonance frequency and anti-resonance frequency of the transducer decreases first and then increases with the increase of the r_d, and the lowest resonance frequency and anti-resonance frequency occur at about 7 mm in the r_d. Figure 5.23 presents the influence of the r_d on the

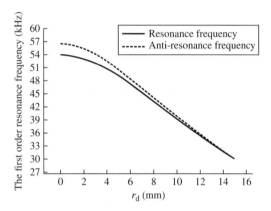

Figure 5.21 The first-order radial vibration resonant frequency and anti-frequency versus the internal radius of the inner prestressing shell of the transducer

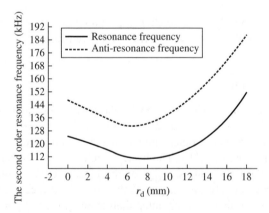

Figure 5.22 The second-order radial vibration resonant frequency and anti-frequency versus the internal radius of the inner prestressing shell of the transducer

effective electromechanical coupling coefficient of the transducer in the case of the first- and the second-order radial vibration. The effective electromechanical coupling coefficient of the transducer (ε_e) in the second-order radial vibration is shown to be much larger than that in the first-order radial vibration. Meanwhile, the ε_e of the transducer in the first-order vibration drops with the increase of the r_d. The ε_e of the transducer in the second-order vibration changes little before 6 mm in the r_d, and it increases gradually as the r_d rises from 6 to 14 mm, and afterwards it decreases with the r_d increasing.

The influences of the wall thickness of the outer prestressing shell of the transducer ($\delta_{o,mcs}$) on the resonant frequency, and the anti-resonant frequency as well as the effective electromechanical coupling coefficient (ε_e) of the first- and the second-order radial vibration, are presented in Figures 5.24–5.26, respectively. The results in Figures 5.24 and 5.25 manifest that the

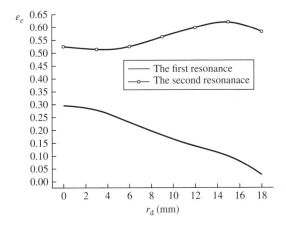

Figure 5.23   Effective electromechanical coupling coefficient versus the internal radius of the prestressing shell of the transducer in the first- and the second-order radial vibration

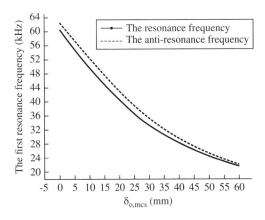

Figure 5.24   The first-order radial vibration resonant frequency and anti-frequency versus the wall thickness of the outer prestressing shell of the transducer

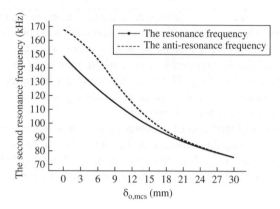

Figure 5.25 The second-order radial vibration resonant frequency and anti-frequency versus the wall thickness of the outer prestressing shell of the transducer

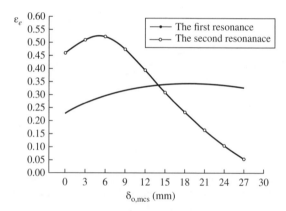

Figure 5.26 The effective electromechanical coupling coefficient versus the wall thickness of the outer prestressing shell of the transducer in first- and the second-order radial vibration

resonant frequency and the anti-resonant frequency of the transducer both decrease with the increase of the wall thickness of the outer prestressing shell. This is because an increase in the wall thickness of the outer prestressing shell will cause the diameter of the radial vibration transducer to rise, and this leads to the decrease of the resonant frequency and the anti-resonant frequency of the transducer. As shown in Figure 5.26, there exists a best value of $\delta_{o,mcs}$ under which the highest ε_e of the transducer can be acquired. The $\delta_{o,mcs}$ corresponding to highest ε_e for the first- and the second-order vibration of the transducer is about 16 and 5 mm, respectively. Moreover, the second-order vibration mode can acquire a much larger value of ε_e than the first-order vibration mode. All the results above indicate that the second-order vibration mode will help to develop a high-performance radial vibration sandwich piezoelectric transducer with a higher acoustic frequency and power as well as a smaller size.

5.2.4   Summary

1. The resonance frequency equations have been derived for the tubular radial composite ultrasonic transducer and the sandwich-type radial composite piezoelectric ultrasonic transducer. Meanwhile, the method of the equivalent electromechanical circuit has been illustrated for the transducer design.
2. The half-wavelength-cascade theory cannot be applied to the radial composite vibration system, which is different from the longitudinal composite vibration systems. Hence, the radial composite ultrasonic vibration system must be designed as a whole.
3. The use of the sandwich composite structure can further improve the power capacity and electromechanical conversion efficiency of the radial vibration transducer.
4. The use of the sandwich radial composite piezoelectric ultrasonic transducer with the second-order radial vibration mode can help to develop a high-frequency-and-high-power radial vibration transducer with a smaller size.

5.3   Ultrasonic Atomization Transducer

5.3.1   Basic Principle of Ultrasonic Atomization

Ultrasonic atomization is one of the important applications of power ultrasound. Under the effect of the high-intensity ultrasound, the fluid is dispersed into the gas phase, and there form numerous fine droplets on the liquid surface, which is known as ultrasonic atomization. The special effect of cavitation induced by the high-intensity ultrasound is the main reason of ultrasonic atomizing. Cavitation is the formation of vapor cavities in a liquid – that is, small liquid-free zones ("bubbles" or "voids") – that are the consequence of forces acting upon the liquid. It usually occurs when a liquid is subjected to rapid changes of pressure that cause the formation of cavities where the pressure is relatively low. When subjected to higher pressure, the voids implode and can generate an intense shockwave. The process of ultrasonic atomization is illustrated in Figure 5.27. To begin with, the ultrasonic waves emitted by the sound source are transmitted to the liquid surface. The big oscillations take place as the ultrasonic power attains a specific value. And then, the liquid drops will appear if the ultrasonic power continues rising. Afterwards, the liquid droplets split and form liquid atomization. Atomizing rate is closely related to the ultrasonic power applied. The other influential factors mainly include surface tension, density, viscosity, vapor pressure, and temperature of the liquid being atomized.

5.3.2   Basic Structure of Ultrasonic Atomizers

There are many ways of classifying ultrasonic atomizers. They are usually divided into two categories. One is a hydrodynamic-type ultrasonic generator whose structure is shown in Figure 5.28. For this atomizer, as sound waves are generated directly from the high pressure fluid, and its frequency is primarily determined by the geometry of the resonant cavity. Since the working efficiency of this kind of atomizer decreases rapidly with the increase of the sound frequency, it generally runs at the lower ultrasonic frequencies. The hydrodynamic-type

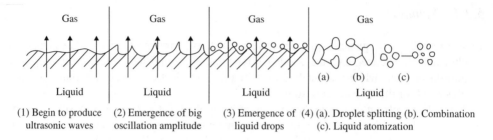

(1) Begin to produce   (2) Emergence of big   (3) Emergence of   (4) (a). Droplet splitting (b). Combination
    ultrasonic waves   oscillation amplitude   liquid drops   (c). Liquid atomization

Figure 5.27   Liquid atomization process by power ultrasound

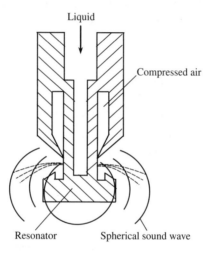

Figure 5.28   Hydrodynamic ultrasonic atomization device

atomizer can be designed with respect to the requirements of the atomized droplet size ranging from several to hundreds micrometers. The atomized size will be affected by the fluid flow rate (velocity), the pressure, the size of the nozzle, and the position of the resonant cavity. The advantage of the hydrodynamic atomizer is its high atomization capacity. So, it is particularly suitable for the industrial sewage treatment. Its shortcomings are low efficiency and high energy consumption.

Another category of ultrasonic atomizers is the use of electric power-driven ultrasonic transducer. It can be either magnetostrictive type or piezoelectric-ceramic-driven type. The frequency of the magnetostrictive transducer generally ranges from 20 to 40 kHz. In contrast, the range of the operating frequency of the piezoelectric transducer is much wider, usually from 20 kHz to 5 MHz. Although the electric power-driven atomizers have a relatively high efficiency, its atomizing capacity is small, and this restricts its applications to some extent.

Ultrasonic atomizers can be of high or low frequency, depending on the type of device and actual requirements of application. The low-frequency ultrasonic atomizing devices are mainly used in the fuel oil combustion or in the preparation of metal powders. To improve the power

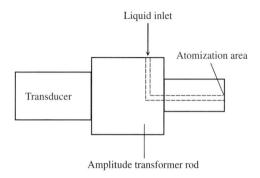

Figure 5.29   Low frequency ultrasonic atomization device

of this kind of ultrasonic atomizer, an amplitude rod is often used to enlarge the displacement amplitude for an easier atomization as shown in Figure 5.29. The high-frequency ultrasonic atomizers normally operate at the frequency ranging from 0.8 to 5 MHz, and the resulting droplet diameter is about 1–5 µm. It is commonly used in household humidifiers and medical aspects. The following sections mainly discuss several piezoelectric transducer driven ultrasonic atomizers.

5.3.3   Research Status and Applications

5.3.3.1   Hydrodynamic Ultrasonic Atomization Device

In early 1980s, the Chinese Academy of Sciences Institute of Acoustics has developed several types of hydrodynamic ultrasonic atomizers. The most representative is the rod-and-whistle structure which mainly consists of three parts, namely the center rod, nozzle, and the cavity. The high-intensity sound is produced by the high-speed air flow which excites high-frequency vibrations of a cavity. In the several structures developed, there is an ultrasonic nozzle whose sound pressure level can be up to 158.5 dB. It has been applied to some industrial applications, like the oil burner and furnace [7]. The sound pressure and operating frequency of the rod-and-whistle atomizer have been studied, and the results are presented as below.

- **Sound Pressure.**   Sound pressure is related to the size of the cavity and the gas pressure. The experimental study manifests that the sound pressure increases with the increase of the cavity volume and the gas pressure. The maximum sound pressure can be acquired at the location right ahead of the nozzle.

    The recommended relationship among dimensions of the rod-and-whistle atomizer can be summarized by Equation (5.125) based on numerous experimental data.

$$d_{\text{cavity}} = 3.4 + d_{\text{nozzle}} = 6 + d_{\text{centralrod}}, \tag{5.125}$$

where d_{cavity}, d_{nozzle}, and $d_{\text{centralrod}}$ are diameter of the resonant cavity, the nozzle, and the central rod, respectively, in mm. The rod-and-whistle atomizer under the dimension relationship of Equation (5.125) can work in a wide range of pressure and acquire a higher sound pressure.

- **Frequency.** The following empirical formula for the frequency calculation has been derived based on a large amount of experiments:

$$f_{\text{rod-whistle}} = c_o / \left[L_{\text{cavity}} + 0.3 \left(d_{\text{cavity}} - d_{\text{centralrod}} \right) \right], \tag{5.126}$$

where c_o is sound speed in air at room temperature, m/s and L_{cavity} is depth of the resonant cavity, mm.

Known from Equation (5.126), the frequency of the rod-and-whistle atomizer will decrease with the increase of the depth and the diameter of the resonant cavity. In real applications, the depth of the resonant cavity should have an appropriate value. A value of L_{cavity} that is too large will make it more difficult to vibration, while a value of L_{cavity} that is too small will limit the sound energy output. According to the experimental study, the design frequency of the rod-and-whistle atomizer is preferably recommended as about 10 kHz.

5.3.3.2 Electrodynamic-Type Ultrasonic Atomization Device

The key technologies for the electrodynamic-type ultrasonic atomizer are the piezoelectric transducer design and the nozzle dimension design. The design theory of the piezoelectric transducer illustrated in the previous sections can be applied to the electrodynamic-type ultrasonic atomizer design. Zhu [8] has studied the operating characteristics of the electrodynamic-type ultrasonic atomizer including the frequency, the amplitude of the excitation voltage, the nozzle dimensions, and the jet velocity, and the following conclusions can be drawn from his study:

1. The higher the excitation voltage, the more likely it is to break the jet, and the performance of the atomizer can be improved through increasing the excitation voltage.
2. The larger piezoelectric strain constant is preferred for the electrodynamic-type ultrasonic atomizer design to improve the injection pressure.
3. Under the lower jet pressure, the influences of the piezoelectric strain constant and the nozzle dimensions on the atomization become smaller. So, the lower jet pressure is conducive to improve the stability of the atomization.
4. It is easier to atomize under the smaller diameter of the nozzle. However, the smaller nozzle easily causes blocking. For the nozzles with very small diameter, a filter is required to remove the solid impurities in the liquid before it is atomized.
5. When selecting operating frequency, the structure match between the nozzle and the ultrasonic transducer should be carefully considered.

5.3.3.3 Applications

Figure 5.30 is a type of medical inhaler with high-frequency ultrasonic atomizer. When the ultrasonic generator and the transducer produce self-excited oscillation, the transducer induces a strong ultrasonic radiation into the water. Then there is formation of numerous tiny droplets at the water–air interface. As air flows through the blower, there will be the formation of droplets derived from the atomization cup, thereby forming a rising mist. The ultrasonic atomizing machine can as well be used in hospitals as an aerosol machine to treat respiratory system illness such as bronchial and lung diseases or isotopic angiography.

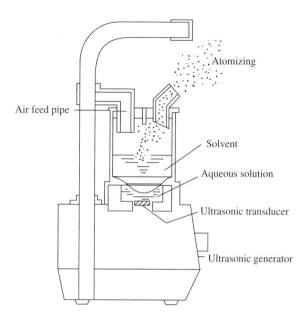

Figure 5.30   High-frequency ultrasonic atomizer

Metal powders can be prepared by using ultrasonic atomization. The molten metal is atomized by ultrasonic vibration, and the droplet-shaped metal particles fall in the air after cooling and solidification, and the metallic powder is formed.

Figure 5.31 is a noncontact ultrasonic apparatus producing metal powders. It uses a parabolic reflector plate from the vibration generated in the air ultrasound energy focused on the molten metal bath surface thereby producing droplets. The vibration plate is driven by an ultrasonic transducer with an amplitude transformer.

Another noncontact method is the use of the high-power ultrasound with its intensity above 170 dB. As shown in Figure 5.32, a titanium horn connected with a sandwich-type piezoelectric ultrasonic transducer has a disc-shaped radiator of which the vibration velocity is about 3–4 m/s. If a reflector is placed at the position $n\lambda/2$ (λ is sound wavelength) away from the disc-shaped radiator, there will form a strong standing wave fields which can be used for the liquid atomization.

Figure 5.33 presents a contact-type low-frequency ultrasonic atomizer whose atomization rate and atomized size can be controllable. A valve is installed at the end of the amplitude horn. During the operation conditions of the transducer, the valve, and the end of the amplitude horn are in a periodical contact. During the separation time, the space between the valve and the amplitude horn will be filled with liquid, and in the following contact time, the liquid in the space will be pressed and atomized by the amplitude horn. The atomized droplets size can be adjusted by altering the pressure in the space between the valve and the amplitude horn.

Figure 5.34 is an improved low-frequency ultrasonic atomizer. The atomization head is in the shape of horn. The main purpose of using the horn is to increase the head of the radiation area of the ultrasonic energy, thereby increasing the amount of atomization. The horn and the vibration transmission rod constitute a half-wavelength of the vibration system which is driven by a

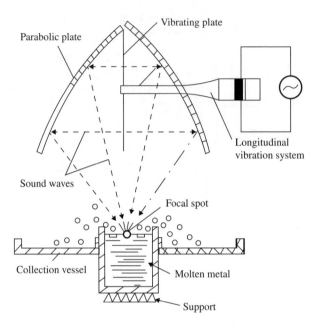

Figure 5.31   Ultrasound device for preparing metal powders

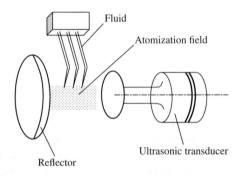

Figure 5.32   Noncontact ultrasonic atomization with standing wave field

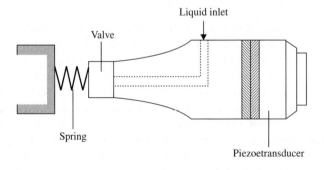

Figure 5.33   Contact-type ultrasonic atomization device

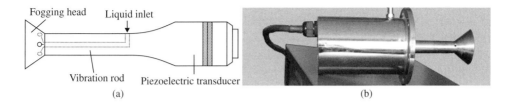

Figure 5.34　(a,b) Low frequency ultrasonic atomization device

sandwich-type piezoelectric transducer. The system consists of numerous half-wave vibrators which can be designed with the one-dimensional longitudinal vibration model developed in the previous section. Despite its high energy conversion efficiency, the power capacity of the piezoelectric transducer is limited, and this will restrict the atomizing rate and capacity. Two measures can be taken to improve the atomizing capacity: one is to develop the ultrasonic transducers with a larger power; the other is to increase the atomization area by using the radial vibration transducer which has been illustrated in the earlier section.

References

[1] Manbachi, A. and Cobbold, R.S.C. (2011) Development and application of piezoelectric materials for ultrasound generation and detection. *Ultrasound*, **19** (4), 187–196.

[2] Lin, S.Y. (2003) *Composite Ultrasonic Transducers*, Shanxi Normal University Press, Xi'an.

[3] Zhou, G.P., Liang, Z.F. and Li, Z.Z. (2007) A tubular focused sonochemistry reator. *Chinese Science Bulletin*, **52** (14), 1902–1905.

[4] Liang, Z.F. (2006) Vibration analysis and sound field characteristics of a tubular ultrasonic radiator. *Ultrasonics*, **45** (1–4), 146–151.

[5] Xu, L., Lin, S.Y. and Hu, W.X. (2011) Optimization design of high power ultrasonic circular ring radiator in coupled vibration. *Ultrasonics*, **51** (7), 815–823.

[6] Lin, S.Y. (2008) The radial composite piezoelectric ceramic transducer. *Sensors and Actuators A*, **141** (2), 136–143.

[7] Chen, Q.Y. and Yan, S.B. (1990) Experimental study on a burner with the rod-whistle ultrasonic atomizer. *Applied Acoustics*, **9** (5), 37–41.

[8] Zhu, Y.G. and Wang, X.F. (1967) An experimental study on ultrasonic injection nozzle. *Journal of Propulsion Technology*, **18** (3), 68–72.

6

Desiccant System with Ultrasonic-Assisted Regeneration

6.1   For Solid-Desiccant System

6.1.1   Based on the Longitudinal Vibration Ultrasonic Transducer

Figure 6.1 presents a temperature-and-humidity independent control air-conditioning system based on the ultrasound-assisted regeneration with the longitudinal ultrasonic transducer. As shown in Figure 6.1, the system mainly consists of a compressor, a condenser, an evaporator, an expansion valve, a dehumidification/regeneration chamber, ultrasonic generators, ultrasonic transducers, air ducts, and dampers. Two adsorption beds packed with silica gel particles are included in the dehumidification/regeneration chamber, and each bed is combined with one longitudinal ultrasonic transducer which is driven by an ultrasonic generator. The compressor, the condenser, the evaporator, and the expansion valve constitute a refrigeration system for the air cooling. The two beds work alternately for the air dehumidification through controlling the dampers and the ultrasonic generators, that is, when one bed is for the air dehumidification, the other one is in the regeneration state. For example, when desiccant in the adsorption bed (A) requires regeneration, the dampers (B), (C), (F), and (G) are closed, while the dampers (A), (D), (E), and (H) are opened. At the same time, the ultrasonic generator (A) starts working, while ultrasonic generator (B) is stopped. The regeneration air, which is firstly heated by the condenser, passes through the adsorption bed (A) and takes away the desiccant moisture desorbed by the ultrasound.

The system in Figure 6.1 is expected to have a high energy efficiency for the following two reasons: (i) the heat rejection by the refrigeration system can be utilized for the desiccant regeneration and (ii) the evaporating temperature of the refrigeration system can be higher than the dew-point temperature of the processed air, and this will improve the energy performance of the refrigeration system.

Figure 6.2 shows an evaporative cooling air conditioning system based on the ultrasound-assisted regeneration with the longitudinal ultrasonic transducer. The system mainly includes a dehumidification/regeneration chamber, an evaporative cooling chamber, and fans. The dehumidification/regeneration chamber is divided into two separate channels by a baffle plate,

Ultrasonic Technology for Desiccant Regeneration, First Edition. Ye Yao and Shiqing Liu.

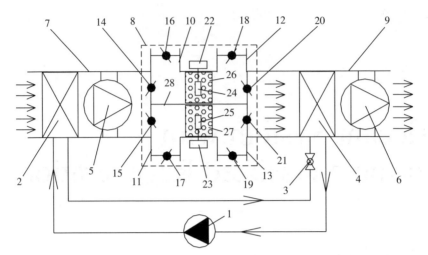

1. Compressor 2. Condenser 3. Expansion valve 4. Vaporator 5. Fan (A) 6. Fan (B) 7. Hot air duct
8. Dehumidification/regeneration chamber 9. Air-delivery duct 10. Indoor return air duct (A)
11. Indoor return air duct (B) 12. Exhaust air duct (A) 13. Exhaust air duct (B) 14. Damper (A)
15. Damper (B) 16. Damper (C) 17. Damper (D) 18. Damper (E) 19. Damper (F) 20. Damper (G)
21. Damper (H) 22. Ultrasound generator (A) 23. Ultrasound generator (B) 24. Ultrasonic transducer (A)
25. Ultrasonic transducer (B) 26. Adsorption bed (A) 27. Adsorption bed (B) 28. Baffle plate

Figure 6.1 Schematic diagram for a temperature-and-humidity independent control air-conditioning system based on the ultrasound-assisted regeneration with the longitudinal ultrasonic transducer

and in each channel there is an ultrasonic dehumidifier whose specific structure is given in Figure 6.2b. The ultrasonic transducer used for the desiccant regeneration is designed as the plate form, and it is tightly connected with the desiccant packed bed. Similarly, the two ultrasonic dehumidifiers alternate air dehumidification work and regeneration treatment of the desiccant. For example, when the ultrasonic dehumidifier (A) is performing the air dehumidification (work), the desiccant in the ultrasonic dehumidifier (B) is in the regeneration state. At this moment, the damper (B) and (C) are opened, and the damper (A) and (D) are closed. The humid air is first dehumidified in the dehumidification/regeneration chamber and then enters into the evaporative cooling chamber where the air is cooled through the evaporation of water with an ultrasonic atomizer.

6.1.2 Based on the Radial Vibration Ultrasonic Transducer

Compared with the two-bed dehumidification system, the rotary dehumidifier has many advantages such as the smaller size, the easier operation, and the more stable running. According to the structure of rotary dehumidifier, we have developed a desiccant wheel with the radial vibration ultrasonic transducer. As shown in Figure 6.3, the desiccant wheel comprises four desiccant packed beds each of which is matched with one ultrasonic transducer. The ultrasonic transducer is in the arc-tube shape, and its radiation cover is in direct contact with the desiccants in the bed.

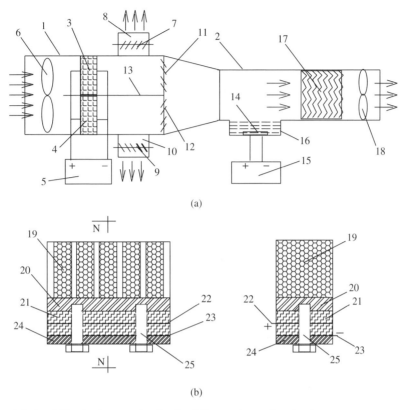

(a)

(b)

1. Dehumidification/regeneration chamber 2. Evaporative cooling chamber
3. Ultrasonic dehumidifier (A) 4. Ultrasonic dehumidifier (B) 5. Ultrasonic generator (A)
6. Fan (A) 7. Damper (A) 8. Exhaust air duct (A) 9. Damper (B) 10. Exhaust air duct (B)
11. Damper (C) 12. Damper (D) 13. Baffle plate 14. Ultrasonic atomizer
15. Ultrasonic generator (B) 16. Atomizer chamber 17.Weir plate 18.Fan (B)
19. Desiccant packed bed 20. Ultrasonic radiation plate 21. Ultrasonic vibration wafer
22. Thin copper sheet 23. Thin copper anode 24. Rear cover plate 25. Connecting bolt

Figure 6.2   Schematic diagram for an evaporative cooling air conditioning system based on the ultrasound-assisted regeneration with the longitudinal ultrasonic transducer. (a) Schematic for system and (b) solid dehumidifier with ultrasonic regeneration

In the real system, the wheel rotates slowly through the shaft driven by the motor. When one of the four ultrasonic transducer enters into the regeneration zone, the positive and negative electrode conductive brush are in contact with the corresponding electrode of ultrasonic transducer, and the ultrasonic transducer starts to work for regenerating the desiccant. The other three ultrasonic transducers stop working, and the desiccant in the corresponding regions of solid desiccant packed bed performs the work of air dehumidification.

An example of the applications of the desiccant wheel with the ultrasonic regeneration in the desiccant air-conditioning system is given in Figure 6.4. This system mainly consists of a

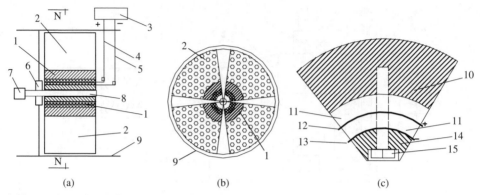

(a)   (b)   (c)

1. Ultrasound transducer 2. Solid desiccant packed bed 3. Ultrasonic generator 4. Positive electrode conductive brush
5. Negative electrode conductive brush 6. Bearing 7. Motor 8. Driven shaft 9. Air duct 10. Ultrasonic radiation cover
11. Vibration wafer 12. Positive copper electrode 13. Negative copper electrode 14. Rear cover 15. Connecting bolt

Figure 6.3   Schematic diagram for a desiccant wheel with the radial vibration ultrasonic transducer.
(a) Elevation view, (b) N-N sectional view, and (c) Structure of ultrasonic transducer

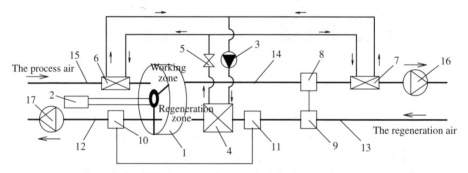

1. Rotary desiccant wheel with ultrasonic regeneration 2. Ultrasonic generator 3. Compressor
4. Condenser 5. Expansion valve 6. Evaporator (A) 7. Evaporator (B) 8. Evaporation end of heat pipe (A)
9. Condensation end of heat pipe (A) 10. Evaporation end of heat pipe (B) 11. Condensation end of heat pipe (B)
12. Exhaust air duct 13. Inlet regeneration air duct 14. Outlet duct of the processed air
15. Inlet duct of the processed air 16. Fan (A) 17. Fan (B)

Figure 6.4   Schematic diagram for a temperature-and-humidity independent control air-conditioning
system based on the rotary desiccant wheel with ultrasound-assisted regeneration

rotary desiccant wheel with ultrasonic regeneration, an ultrasonic generator, a compressor, an
expansion valve, a condenser, two evaporators, two heat pipes, and two fans. The compressor,
the expansion valve, the condenser, and the evaporators constitute a refrigeration system for
the temperature control of the processed air. The two evaporators are located at the inlet and
outlet duct of the processed air, respectively. The purpose of the evaporator (A) is to make
the processed air saturated and improve the air dehumidification efficiency of the desiccant
wheel, and the evaporator (B) is used to cool the processed air after being dehumidified. The

heat pipes here are used for the heat recovery existing in the system, and this can improve the energy performance of the system.

6.2 For Liquid-Desiccant System

Figure 6.5 presents a liquid desiccant regenerator with ultrasonic atomizing. As shown in Figure 6.5, the new regenerator mainly comprises of an inner wall, an outer wall, an ultrasonic atomizer, an ultrasonic generator, a desiccant solution storage tank, a demister, a weir plate, a heat pipe, solution heater, and exhaust fan. The inner wall and the outer wall form an inlet air channel. The heat pipe is used for the heat recovery from the regenerator's exhaust air whose temperature is often higher than the ambient one. The regeneration of the liquid desiccant is carried out mainly through the ultrasonic atomizing. To rinse the desiccant salt crystals retained on the demister, we arrange solution sprinklers to spray the weak desiccant solution on the demister.

A multi-energy-complementary liquid dehumidification system has been developed based on the ultrasonic-atomizing regenerator, as shown in Figure 6.6. The present system mainly comprises of a set of osmotic membrane for air dehumidification, an ultrasound-atomizing desiccant solution regenerator, a wind turbine, a storage cells, a solar collector, a heat exchanger, an energy storage tank, an indoor air circulation fan, two solution pumps, and numerous solution valves. The osmotic membrane is employed here for the air dehumidification, and it can solve the problem of the air entrained with desiccant droplets after the dehumidification with desiccant solution. However, the use of the osmotic membrane in the

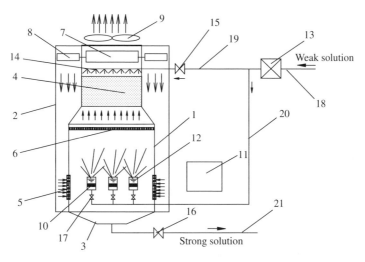

1. Inner wall 2. Outer wall 3. Desiccant solution storage tank 4. Demister 5. Air intake 6. Weir plate
7. Evaporation end of heat pipe 8. Condensation end of heat pipe 9. Exhaust fan 10. Ultrasonic atomizer
11. Ultrasonic generator 12. Atomization tanks 13. Desiccant solution heater 14. Solution sprinkler
15. Solution valve (A) 16. Solution valve (B) 17. Liquid level control valve 18. Inlet liquid pipe
19. Branch liquid pipe (A) 20. Branch liquid pipe (B) 21. Outlet liquid pipe

Figure 6.5   Schematic diagram for a liquid desiccant regenerator with ultrasonic atomizing

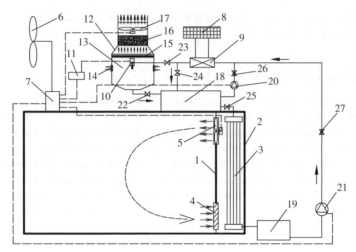

1. Inner partition wall 2. Building external wall of building 3. Osmotic membrane for air dehumidification
4. Indoor return air inlet 5. Indoor circulation fan 6. Wind turbine 7. Storage cells 8. Solar collector 9. Heat exchanger
10. Ultrasonic atomizer 11. Ultrasonic generator 12. Desiccant solution regenerator 13. Atomization chamber
14. Regeneration air inlet 15. Weir plate 16. Demister 17. Exhaust fan 18. Energy storage tank
19. Desiccant solution storage tank 20. Solution circulating pump 21. Solution return pump 22. Valve (A)
23. Valve (B) 24. Valve (C) 25. Valve (D) 26. Valve (E) 27. Valve (F)

Figure 6.6 Schematic diagram for a multi-energy-complementary liquid dehumidification system based on the liquid desiccant regenerator with ultrasonic atomization

air dehumidifier will increase the moisture transfer resistance between the liquid desiccant and the air, and much larger dehumidification area is required for the air dehumidifying. So, we design the osmotic-membrane air dehumidifier combined with the building envelopes. As shown in Figure 6.6, the osmotic membrane is placed between the outer wall and the inner wall, and a new dehumidification wall is formed.

In spring and summer seasons when humidity is high, the valve (A), (B), and (D) are opened while the valve (C) is closed. The indoor circulating fan, the exhaust fan, the ultrasonic generator, the solution circulating pump, and the solution return pump operate at the same time. The strong desiccant solution coming from the energy storage tank flows into the osmotic membrane to dehumidify the indoor air. After the air dehumidification, the desiccant solution becomes weak and is delivered to the solution regenerator by the solution return pump. Before entering into the solution regenerator, the weak desiccant solution is heated by the solar heat.

In the cold winter season, the indoor air does not require dehumidification. Under such conditions, the valve (C) and (D) are opened, while the valve (A) and (B) are closed. The indoor circulating fan, the solution circulating pump, and the solution return pump are all running, and the ultrasound-atomizing regenerator stops working. The desiccant solution in the energy storage tank is used for storage of the solar heat during daytime, and release the heat to the indoor air through the osmotic membrane during nighttime.

The wind turbine in this system runs all the time throughout the year. It generates electric power from the wind energy and supply electricity for the power equipments in the system including the ultrasonic generator, the solution pumps, and the fans.

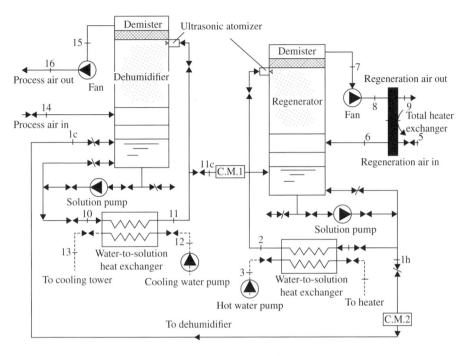

Figure 6.7 Schematic diagram for a liquid dehumidification system based on ultrasonic atomization technology

The ultrasonic atomization may be used to develop the air dehumidifier as well. As shown in Figure 6.7, the air dehumidifier and the desiccant regenerator in the dehumidification system are all designed with the technology of ultrasonic atomizing. Since the air entrained with desiccant droplets is not allowed to mix with the processed air, the high-performance demister must be used in the air dehumidifier.

6.3   Future Work

6.3.1   Development of Ultrasonic Transducer

The regeneration of desiccant is a blandly new domain for the ultrasonic applications. So, the ultrasonic transducer should be specially developed in the future work for this purpose in order to achieve a better effect. For the applications of power ultrasound, the crucial problems to be solved are related to the efficient ultrasonic generator and the transmission of ultrasonic energy. In the solid-gas systems, the challenges confronted mainly include the impedance mismatch between the ultrasonic transducer and air (which makes it difficult for the sound waves to propagate into the solid medium), and the high acoustic energy damping in the porous medium. These difficulties may be overcome by adequately designing the ultrasonic application system, for example, the transducers developed for solid desiccant regeneration are featured by good impedance matching with air, large amplitude of vibration, and high power capacity as well as extensive radiation area. Meanwhile, the acoustic frequency and power of transducer should be cautiously designed according to the acoustic characteristics (including the acoustic impedance, attenuation, and absorption) and mass load of desiccant used in the system.

Another key problem is the heat dissipation of the ultrasonic transducers during their work. As we know, waste heat always occurs during the energy conversion (here, from the electric energy to the ultrasonic energy). The temperature of the ultrasonic transducer will be very high if the heat dissipation cannot be released in time, and this will greatly reduce the ultrasonic transducer's working efficiency and result in the breakdown of the equipment. In order to ensure the high working efficiency of the ultrasonic transducer, on the one hand, we should develop a cooling system particularly for the high-intensity ultrasonic transducer; and on the other hand, equally important, we should develop a new piezoelectric ceramic material which is featured by large piezoelectric constant, low dielectric loss, and small temperature coefficient so as to reduce heat dissipation.

6.3.2   Development of Desiccant Materials Adaptive to Ultrasound-Assisted Regeneration

In the present work, we only use the currently existing desiccant materials for the study of ultrasound-assisted regeneration. Normally, theses desiccant materials have been developed mainly focusing on their dehumidification and regeneration features, and hence, they may be not very suitable for ultrasonic regeneration. In future work, the newly developed desiccant materials ought to take into account both their acoustic characteristics and their dehumidification or regeneration features. For the solid desiccants, the higher rigidity of the material is favorable for sound energy transmission; and for the liquid desiccants, its viscosity should be as low as possible for a better effect of ultrasonic atomizing.

6.3.3   Development of Demister

For liquid desiccant regeneration with ultrasonic atomizing, a demister must be required to avoid the escape of liquid droplets with the air flow. Currently, the most commonly used demisters in the liquid desiccant system are of the baffle type. But, it is only suitable for removing droplets with the diameters above 100 µm. The atomized droplets by the ultrasound are often below 100 µm, and some are even lower than 30 µm. So, the high-efficiency demisters particularly for the size range of atomized particle by ultrasound are to be developed. The influencing factors on the performance of a demister mainly include the velocity of airflow, the operating pressure, liquid and gas viscosity, liquid surface tension, and characteristics of device geometry. These factors need to be considered when developing a demister in order to achieve a higher separation efficiency.

6.3.4   Environmental Impact

The possibility of a new technology application depends, to a large extent, on its environmental impact and economic benefit. For the environmental impact, the possible harm to human body done by high-intensity ultrasonic radiation must be seriously assessed. Exposure to ultrasound can be either through direct contact, a coupling medium, or the air (airborne ultrasound). Limits for exposure from each mode should be treated separately. In general, direct contact exposure to high intensities of liquid-borne ultrasound is not allowed. So, the limits for human exposure to the airborne ultrasound are emphasized. In Grigoreva's work, the authors thought airborne ultrasound is considerably less hazardous to man in comparison with audible sound, and proposed 120 dB to be adopted as an acceptable limit for the acoustic pressure for airborne ultrasound [1]. After two years later, Acton [1] proposed a criterion below which auditory damage and/or subjective effects were unlikely to occur as a result of human exposure to airborne noise from industrial ultrasonic sources over a working day. He based his criterion on the belief that it is the high audible frequencies present in the noise from ultrasonic machines, and not the ultrasonic frequencies themselves, that are responsible for producing subjective effects. He extended this criterion to produce a tentative estimate for an extension to damage risk criteria, giving levels of 110 dB in the one-third octave bands centered on 20, 25, and 31.5 kHz. In his other paper published in 1974 [2], Acton announced that additional data obtained for industrial exposures confirmed that the levels set in the proposed criterion were at approximately the right level, and that there did not seem to be any necessity to amend them. The International Radiation Protection Association (IRPA) [3] has drafted the first international limits for human exposure to airborne acoustic energy having one-third octave bands with mid frequencies from 8 to 50 kHz, which is shown in Figure 6.8 [4]. In Figure 6.8, the line

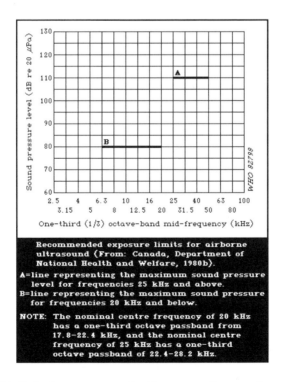

Figure 6.8   Limits for human exposure to airborne acoustic energy issued by IRPA [4]

A represents the maximum sound pressure level for frequencies 25 kHz and above; the line B represents the maximum sound pressure level for frequencies 20 kHz and below. The nominal center frequency of 20 kHz has a one-third octave passband from 17.8 to 22.4 kHz, and the nominal center frequency of 25 kHz has a one-third octave passband of 22.4–28.2 kHz.

Measures should be taken to avoid the potential hazards caused by the high-intensity ultrasound, for example, persons exposed to the high levels of noise associated with the ultrasonic equipment should be protected either by wearing devices like earmuffs, or by acoustic barriers constructed around the equipment to reduce the noise levels. Moreover, an economic analysis for the new regeneration technology should be carefully made before it is applied to an actual desiccant system.

References

[1] Grigoreva, V.M. (1966) Effect of ultrasonic vibrations on personnel working with ultrasonic equipment. *Soviet Physics–Acoustics*, **11** (3), 426–427.
[2] Acton, W.I. (1968) A criterion for the prediction of auditory and subjective effects due to airborne noise from ultrasonic sources. *Annals of Occupational Hygiene*, **11** (4), 227–234.
[3] Acton, W.I. (1974) The effects of industrial airborne ultrasound on humans. *Ultrasonics*, **12** (5), 124–127.
[4] IRPA (1981) Draft: Guidelines on Limits of Human Exposure to Airborne Acoustic Energy Having One-Third Octave Bands with Mid Frequencies from 8 to 50 kHz. International Radiation Protection Association, International Non-Ionizing Radiation Committee (IRPA/INIRC), November 1981.
[5] World Health Organization (1982) International Programme on Chemical – Environmental Health Criteria 22: Ultrasound, Geneva, Switzerland.

Appendix A

Basic Equations for Properties of Common Liquid Desiccants

A.1 Lithium Chloride (LiCl)

Lithium chloride, a kind of desiccant used in air-conditioning systems, is a chemical compound with the formula LiCl. The salt is a typical ionic compound, although the small size of the Li^+ ion gives rise to properties not seen for other alkali metal chlorides, such as extraordinary solubility in polar solvents (83 g/100 ml of water at 20 °C) and its hygroscopic properties. Molar mass: 42.394 g/mol; Density: 2.068 g/cm^3; Melting point: 605 °C; Boiling point: 1382 °C.

A.1.1 Solubility Boundary

The solubility boundaries of these two salt solutions are defined by several lines. For salt concentrations lower than that at the eutectic point, the solubility line defines the conditions at which ice crystals start to form. This is the ice line. For higher concentrations, the solubility boundary defines the conditions at which salt hydrates or anhydrous salt crystallize from the solution. This is the crystallization line.

For LiCl–H_2O solutions, the crystallization line, A-B-C-D-E in Figure A.1 [1], also defines the transition points separating the ranges of formation of the various hydrates, as indicated in the figure. Equations for each range can be adjusted to experimental data from the literature and have the general form for the crystallization line [1]:

$$\theta = \sum_{i=0}^{2} A_i \xi^i,$$ (A.1)

where $\theta = T/T_{c,H_2O}$; ξ is the mass fraction of the salt in the solution; T is temperature in K; and T_{c,H_2O} is critical temperature of water in K. For the ice line, the equation is slightly different:

$$\theta = A_0 + A_1\xi + A_2\xi^{2.5},$$ (A.2)

where the parameters A_i are included in Table A.1, for each range of the boundary.

Ultrasonic Technology for Desiccant Regeneration, First Edition. Ye Yao and Shiqing Liu.
© 2014 Shanghai Jiao Tong University Press. All rights reserved. Published 2014 by John Wiley & Sons Singapore Pte Ltd.

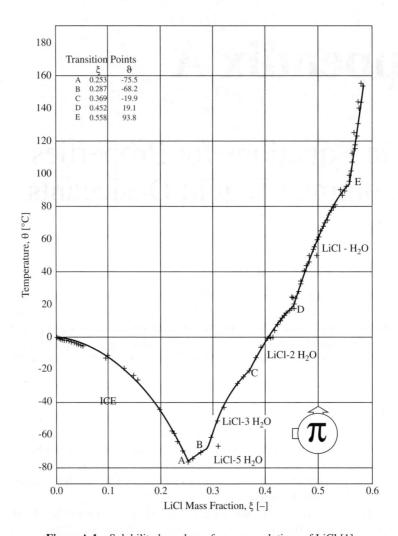

Figure A.1   Solubility boundary of aqueous solutions of LiCl [1]

Table A.1   Parameters of the equations describing the solubility
boundary of LiCl-H_2O solutions

Boundary	A_0	A_1	A_2
Ice line	0.422088	−0.090410	−2.936350
LiCl-5H_2O	−0.005340	2.015890	−3.114590
LiCl-3H_2O	−0.560360	4.723080	−5.811050
LiCl-2H_2O	−0.315220	2.882480	−2.624330
LiCl-H_2O	−1.312310	6.177670	−5.034790
LiCl	−1.356800	3.448540	0

A.1.2 Vapor Pressure (Pa)

The equilibrium pressure of saturated water vapor above aqueous solutions of the salts considered here is perhaps their best studied property. Despite the amount of experimental data, spanning almost a century of research, no consistent formulation for the accurate prediction of this property has yet been published. Most attempts made are limited to short ranges either on concentration or temperature. Formulations derived from basic principles have not done better than the empirical ones and are confined to very dilute solutions. The formulations in the following reproduce quite accurately the known experimental data and default to congruent values at the boundaries.

The formula is established in terms of the relative vapor pressure, π (relative to water at the same temperature), which is the function of the mass fraction and temperatures. The general equation can be written as below [1]:

$$\pi = \frac{p_{s,\text{LiCl}}(\xi, T)}{p_{\text{H}_2\text{O}}(T)} = \pi_{25} f(\xi, \theta), \tag{A.3}$$

where $f(\xi, \theta) = A + B\theta$; $A = 2 - \left[1 + \left(\frac{\xi}{\pi_o}\right)^{\pi_1}\right]^{\pi_2}$; $B = \left[1 + \left(\frac{\xi}{\pi_3}\right)^{\pi_4}\right]^{\pi_5} - 1$, $\pi_{25} = 1 - \left[1 + \left(\frac{\xi}{\pi_6}\right)^{\pi_7}\right]^{\pi_8} - \pi_9 \exp\left(-\frac{(\xi - 0.1)^2}{0.005}\right)$; and the parameters are given for the LiCl-H$_2$O solution in Table A.2.

A.1.3 Specific Thermal Capacity (J/(kg·°C))

The equations proposed for the calculation of the specific thermal capacity of aqueous solutions of lithium are as below [1]:

$$c_{s,\text{LiCl}} = c_{\text{H}_2\text{O}} \times [1 - f_1(T) f_2(\xi)], \tag{A.4}$$

$$c_{\text{H}_2\text{O}} = A + B\theta^{0.02} + C\theta^{0.04} + D\theta^{0.06} + E\theta^{1.8} + F\theta^8, \tag{A.5}$$

$$f_1(T) = a\theta^{0.02} + b\theta^{0.04} + c\theta^{0.06}, \tag{A.6}$$

$$f_2(\xi) = d\xi + e\xi^2 + f\xi^3 \quad \text{for } \xi \le 0.31, \tag{A.7}$$

$$f_2(\xi) = g + h\xi \quad \text{for } \xi > 0.31, \tag{A.8}$$

where $\theta = T/228 - 1$; T is temperature in K; ξ is the mass fraction of the salt in the solution; and the parameters are listed in Table A.3.

Table A.2 Parameters for the vapor pressure equations of LiCl-H$_2$O solution

π_0	π_1	π_2	π_3	π_4	π_5	π_6	π_7	π_8	π_9
0.28	4.30	0.60	0.21	5.10	0.49	0.362	−4.75	−0.40	0.03

Table A.3   Parameters for the specific thermal capacity of the aqueous solutions of lithium

A	B	C	D	E	F		
88.7891	−120.1958	−16.9264	52.4654	0.10826	0.46988		
a	b	c	d	e	f	g	h
58.5225	−105.6343	47.7948	1.4398	−1.24317	−0.1207	0.12825	0.62934

Table A.4   Parameters in the density equations of the aqueous solutions of lithium

i	B_i	ρ_i
0	1.993771843	1.0
1	1.0985211604	0.540966
2	−0.5094492996	−0.303792
3	−1.761912427	0.100791
4	−44.9005480267	−
5	−723 692.2618632	−

A.1.4   Density (kg/m³)

The densities of aqueous solutions of lithium can be calculated by the following equations [1]:

$$\rho_{s,LiCl} = \rho_{H_2O}(\theta) \times \left(\rho_0 + \rho_1\bar{\xi} + \rho_2\bar{\xi}^2 + \rho_3\bar{\xi}^3\right), \tag{A.9}$$

$$\rho_{H_2O} = 322 \times \left(1 + B_0\theta^{1/3} + B_1\theta^{2/3} + B_2\theta^{5/3} + B_3\theta^{16/3} + B_4\theta^{43/3} + B_5\theta^{110/3}\right), \tag{A.10}$$

$$\bar{\xi} = \frac{\xi}{1-\xi}, \tag{A.11}$$

where $\theta = 1 - T/647.3$; T is temperature in K; ξ is the mass fraction of the salt in the solution; and the parameters are listed in Table A.4.

A.1.5   Dynamic Viscosity (MPa·s)

The dynamic viscosity of aqueous solutions of the lithium can be calculated with the following equations [1]:

$$\mu_{s,LiCl} = \mu_{H_2O} \times \exp\left(\eta_1\bar{\xi}^{3.6} + \eta_2\bar{\xi} + \eta_3\bar{\xi}/\theta + \eta_4\bar{\xi}^2\right), \tag{A.12}$$

$$\bar{\xi} = \xi/(1-\xi)^{1/0.6}, \tag{A.13}$$

$$\mu_{H_2O} = 5.5071 \times 10^{-5}\vartheta_1\vartheta_2, \tag{A.14}$$

$$\vartheta_1 = \theta^{0.5}/(1 + a/\theta + b/\theta^2 + c/\theta^3), \tag{A.15}$$

$$\vartheta_2 = \exp(\sigma_{H_2O}(\phi_1 + \phi_2 + \phi_3)), \tag{A.16}$$

$$\sigma_{H_2O} = 1 + d(1-\theta)^{1/3} + e(1-\theta)^{2/3} + f(1-\theta)^{5/3} + g(1-\theta)^{16/3},$$
$$+ h(1-\theta)^{43/3} + i(1-\theta)^{110/3} \tag{A.17}$$

$$\phi_1 = a_o + a_1\beta + a_2\beta^2 + a_3\beta^3 + a_4\beta^4 + a_5\alpha + a_6\alpha\beta, \tag{A.18}$$

$$\phi_2 = b_o\alpha\beta^2 + b_1\alpha\beta^3 + b_2\alpha\beta^5 + b_3\beta\alpha^2 + b_4\beta^2\alpha^2 + b_5\beta^3\alpha^2 + b_6\beta\alpha^3, \tag{A.19}$$

$$\phi_3 = c_o\alpha^3\beta^3 + c_1\alpha^3\beta^4 + c_2\alpha^3\beta^6 + c_3\alpha^4 + c_4\alpha^5, \tag{A.20}$$

where $\theta = T/647.226$; $\alpha = 1/\theta - 1$; $\beta = \sigma_{H_2O} - 1$; and the parameters are listed in Table A.5.

A.2 Calcium Chloride (CaCl$_2$)

Calcium chloride, CaCl$_2$, is a salt of calcium and chlorine. It behaves as a typical ionic halide, and is solid at room temperature. Calcium chloride can serve as a source of calcium ions in an aqueous solution, as calcium chloride is soluble in water. Calcium chloride can be produced directly from limestone, but large amounts are also produced as a byproduct of the Solvay process. Molar mass: 110.98 g/mol; density: 2.15 g/cm^3; melting point: 772 °C; and boiling point: 1395 °C.

A.2.1 Solubility Boundary

For CaCl$_2$–H$_2$O solutions, the crystallization line is more complex, particularly due to the formation of various tetrahydrates. Some of these are metastable. In the boundary represented in Figure A.2 [1], we decided to show only those generally reported as stable in the literature, that is, the α and β tetrahydrates. The equations describing the solubility boundary of CaCl$_2$–H$_2$O solutions have the same form of LiCl-H$_2$O (Equations (A.1) and (A.2)), and the parameters are given in Table A.6 [1].

A.2.2 Vapor Pressure (Pa)

The formulas for calculating the vapor pressure of CaCl$_2$ solution are the same form of LiCl-H$_2$O (Equation (A.3)), and the parameters are given in Table A.7 [1].

Table A.5 Parameters in the dynamic viscosity equations of the aqueous solutions of lithium

a	b	c	d	e	f	g
0.978197	0.579829	−0.202354	1.9937718	1.09852116	−0.5094493	−1.76912427
h	i	a_0	a_1	a_2	a_3	a_4
−44.90055	−723692.26186	0.5132047	0.2151778	−0.2818107	0.1778064	−0.0417661
a_5	a_6	b_0	b_1	b_2	b_3	b_4
0.3205656	0.7317883	−1.070786	0.460504	−0.01578386	1.241044	−1.263184
b_5	b_6	c_0	c_1	c_2	c_3	c_4
0.234038	1.476783	−0.492418	0.1600435	0.003629481	−0.7782567	0.1885447
η_1	η_2	η_3	η_4			
0.090481	1.390262	0.675875	−0.583517			

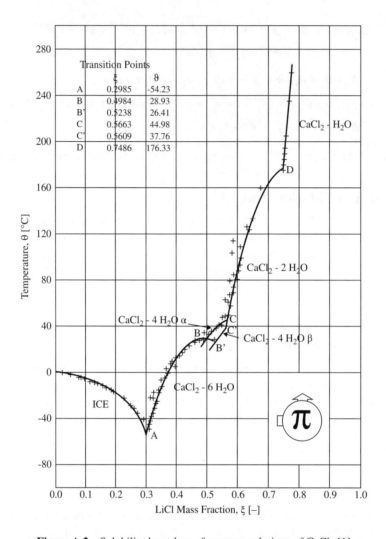

Figure A.2 Solubility boundary of aqueous solutions of CaCl₂ [1]

Table A.6 Parameters of equations describing the solubility boundary of CaCl₂−H₂O solutions

Boundary	A_0	A_1	A_2	A3
Ice line	0.422088	−0.066933	−0.282395	−355.514247
CaCl₂-6H₂O	−0.378950	3.456900	−3.531310	0.0
CaCl₂-4H₂O α	−0.519970	3.400970	−2.851290	0.0
CaCl₂-4H₂O β	−1.149044	5.509111	−4.642544	0.0
CaCl₂-2H₂O	−2.385836	8.084829	−5.303476	0.0
CaCl₂-H₂O	−2.807560	4.678250	0.0	0.0

Table A.7 Parameters for the vapor pressure equations of CaCl$_2$ solution

π_0	π_1	π_2	π_3	π_4	π_5	π_6	π_7	π_8	π_9
0.31	3.698	0.60	0.231	4.584	0.49	0.478	−5.20	−0.40	0.018

A.2.3 Specific Thermal Capacity (J/(kg·°C))

The equations proposed for the calculation of the specific thermal capacity of aqueous solutions of lithium (see Equations (A.4)–(A.8)) can be applied to that of aqueous solutions of CaCl$_2$ solution. The corresponding parameters are listed in Table A.8 [1].

A.2.4 Density (kg/m³)

Equations (A.9)–(A.11) with the parameters listed in Table A.9 are used for the density calculation of CaCl$_2$ solution [1].

A.2.5 Dynamic Viscosity (MPa·s)

Equations (A.12)–(A.20) with the parameters listed in Table A.10 are used for the dynamic viscosity calculation of CaCl$_2$ solution [1].

A.3 Lithium Bromide (LiBr)

LiBr, another kind of desiccant used in the air-conditioning systems, is prepared by treatment of lithium carbonate with hydrobromic acid. The salt forms several crystalline hydrates,

Table A.8 Parameters for the specific thermal capacity of CaCl$_2$ solution

A	B	C	D	E	F		
88.7891	−120.1958	−16.9264	52.4654	0.10826	0.46988		
a	b	c	d	e	f	g	h
58.5225	−105.6343	47.7948	1.63799	−1.69002	1.05124	0	0

Table A.9 Parameters in the density equations of CaCl$_2$ solution

i	B_i	ρ_i
0	1.993771843	1.0
1	1.0985211604	0.836014
2	−0.5094492996	−0.436300
3	−1.761912427	0.105642
4	−44.9005480267	–
5	−723 692.2618632	–

Table A.10   Parameters in the dynamic viscosity equations of $CaCl_2$ solution

a	b	c	d	e	f	g
0.978197	0.579829	−0.202354	1.9937718	1.09852116	−0.5094493	−1.76912427
h	i	a_0	a_1	a_2	a_3	a_4
−44.90055	−723 692.26186	0.5132047	0.2151778	−0.2818107	0.1778064	−0.0417661
a_5	a_6	b_0	b_1	b_2	b_3	b_4
0.3205656	0.7317883	−1.070786	0.460504	−0.01578386	1.241044	−1.263184
b_5	b_6	c_0	c_1	c_2	c_3	c_4
0.234038	1.476783	−0.492418	0.1600435	0.003629481	−0.7782567	0.1885447
η_1	η_2	η_3	η_4			
−0.169310	0.817350	0.574230	0.398750			

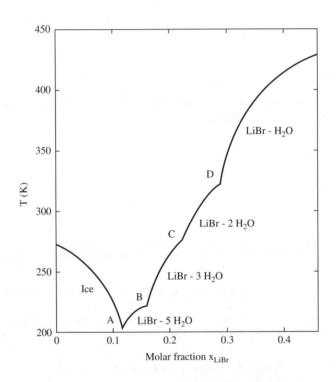

Figure A.3   Solubility boundary of aqueous solutions of LiBr [2]

unlike the other alkali metal bromides. The anhydrous salt forms cubic crystals similar to common salt (sodium chloride). Lithium hydroxide and hydrobromic acid (aqueous solution of hydrogen bromide) will precipitate lithium bromide in the presence of water, that is, $LiOH + HBr \rightarrow LiBr + H_2O$. Molar mass: 86.845 g/mol; density: 3.464 g/cm^3; melting point: 552 °C; boiling point: 1265 °C; and solubility: soluble in water, methanol, ethanol, ether slightly soluble in pyridine.

Table A.11 Parameters of the equations describing the solubility boundary of LiBr solutions (Equation (A.21)) [2]

i	LiBr-5H$_2$O^a			LiBr-3H$_2$O^b		
	a$_i$	m$_i$	n$_i$	a$_i$	m$_i$	n$_i$
1	2.61161×10^1	1	1	2.47039×10^1	1	1
2	2.38994×10^4	1	3	4.65459×10^3	1	3

i	LiBr-2H$_2$O^c			LiBr-H$_2$O^d		
	a$_i$	m$_i$	n$_i$	a$_i$	m$_i$	n$_i$
1	1.62375×10^1	1	1	1.00743×10^1	1	1
2	2.47098×10^3	1	3	3.94593×10^3	1	4

i	Ice^e		
	a$_i$	m$_i$	n$_i$
1	1.33842×10^1	1	1
2	-4.39293×10^1	2	1
3	4.02577×10^3	3	1
4	-5.52364×10^4	4	1
5	3.28383×10^5	5	1

^{a}$T_L = 202.8$ K, $T_R = 222.4$ K, $x_L = 0.1175$, $x_R = 0.1604$.
^{b}$T_L = 222.4$ K, $T_R = 277.1$ K, $x_L = 0.1604$, $x_R = 0.2213$.
^{c}$T_L = 277.1$ K, $T_R = 322.2$ K, $x_L = 0.2213$, $x_R = 0.2869$.
^{d}$T_L = 322.2$ K, $T_R = 429.15$ K, $x_L = 0.2869$, $x_R = 0.4613$.
^{e}$T_L = 273.16$ K, $T_R = 202.8$ K, $x_L = 0.0000$, $x_R = 0.1175$.

A.3.1 Solubility Boundary

The solubility boundary of aqueous solutions of LiBr is shown in Figure A.3 [2]. The crystallization line A-B-C-D in Figure A.3 defines the transition points separating the ranges of formation of the various hydrates.

The crystallization line of aqueous LiBr solution can be expressed by the $T-x$ relation as below [2]:

$$T(x_{s,LiBr}) = T_L + \frac{T_R - T_L}{x_R - x_L}(x_{s,LiBr} - x_L) + T_t \sum_{i=1}^{N} a_i(x - x_L)^{m_i}(x_R - x_{s,LiBr})^{n_i}, \quad (A.21)$$

$$x_{s,LiBr}(T) = x_L + \frac{x_R - x_L}{T_R - T_L}(T - T_L) + \sum_{i=1}^{N} b_i \left(\frac{T - T_L}{T_t}\right)^{m_i} \left(\frac{T_R - T}{T_t}\right)^{n_i}, \quad (A.22)$$

$$x_{s,LiBr} = \frac{\xi_{s,LiBr}/M_{LiBr}}{\xi_{s,LiBr}/M_{LiBr} + (1 - \xi_{s,LiBr})/M_{H_2O}}, \quad (A.23)$$

where $x_{s,LiBr}$ and $\xi_{s,LiBr}$ are molar fraction and mass fraction of salt LiBr in the solution, respectively and M_{LiBr} and M_{H_2O} are molar mass (kg mol^{-1}) of LiBr and water, respectively; the subscripts L and R denote here the left and right endpoints of the interval, respectively.

Table A.12 Parameters of the equations describing the solubility boundary of LiBr solutions (Equation (A.22)) [2]

i	LiBr-5H$_2$O^a			LiBr-3H$_2$O^b		
	bi	mi	ni	bi	mi	ni
1	−6.17446	1	1	−7.17618E−1	1	1
2	−1.4677E+3	3	1	−1.02551E+1	3	1

i	LiBr-2H$_2$O^c			LiBr-H$_2$O^d		
	bi	mi	ni	Bi	mi	ni
1	−1.06305	1	1	−9.25082E−1	1	1
2	−1.90921E+1	3	1	−7.22341	3	1

i	Icing area^e		
	bi	mi	ni
1	1.22335	1	1
2	−1.67781	1	2
3	−2.65346E+2	1	4
4	−1.93594E+3	1	5
5	−5.16209E+3	1	6

^{a}T$_L$ = 202.8 K, T$_R$ = 222.4 K, x$_L$ = 0.1175, x$_R$ = 0.1604.
^{b}T$_L$ = 222.4 K, T$_R$ = 277.11, x$_L$ = 0.1604, x$_R$ = 0.2213.
^{c}T$_L$ = 277.1 K, T$_R$ = 322.2 K, x$_L$ = 0.2213, x$_R$ = 0.2869.
^{d}T$_L$ = 322.2 K, T$_R$ = 429.151, x$_L$ = 0.2869, x$_R$ = 0.4613.
^{e}T$_L$ = 273.16 K, T$_R$ = 202.8 K, x$_L$ = 0.0000, x$_R$ = 0.1175.

To introduce a dimensionless temperature variable and to make the coefficients a_i and b_i dimensionless, the water triple point temperature $T_t = 273.16$ K was arbitrarily selected as the reference temperature value. The parameters in Equations (A.21) and (A.22) are given in Tables A.11 and A.12 [2], respectively.

A.4 Vapor Pressure (Pa)

The new correlations of the thermodynamic properties of aqueous lithium bromide have been constructed by Z. Yuan based on a Gibbs free energy fundamental function [3]. The equations mainly include as follows:

$$\left(\frac{\partial g}{\partial \xi}\right)_{T,p} = \mu_{\text{LiBr}} - \mu_{\text{w}}, \tag{A.24}$$

$$\mu_{\text{Libr}}(\xi, T, p) = g + (1 - \xi)\left(\frac{\partial g}{\partial \xi}\right)_{T,p}, \tag{A.25}$$

$$\mu_{\text{w}}(\xi, T, p) = g - \xi\left(\frac{\partial g}{\partial \xi}\right)_{T,p}, \tag{A.26}$$

$$g(\xi, T, p) = (A_o + A_1\xi + A_2\xi^2 + A_3\xi^3 + A_4\xi^{1.1}) + T(B_o + B_1\xi + B_2\xi^2 + B_3\xi^3 + B_4\xi^{1.1})$$
$$+ T^2(C_o + C_1\xi + C_2\xi^2 + C_3\xi^3 + C_4\xi^{1.1}) + T^3(D_o + D_1\xi + D_2\xi^2 + D_4\xi^{1.1})$$
$$+ T^4(E_o + E_1\xi) + \frac{F_o + F\xi}{T - T_o} + p(V_o + V_1\xi + V_2\xi^2 + V_3T + V_4\xi T + V_5\xi^2 T$$
$$+ V_6T^2 + V_7\xi T^2) + \ln(T)(L_o + L_1\xi + L_2\xi^2 + L_3\xi^3 + L_4\xi^{1.1})$$
$$+ T\ln(T)(M_o + M_1\xi + M_2\xi^2 + M_3\xi^3 + M_4\xi^{1.1}), \tag{A.27}$$

where μ_{Libr} and μ_w are chemical Potential of LiBr and water in solution (in J/g), respectively; g is the Gibbs function; ξ is mass fraction of LiBr in the solution; T is temperature of solution in K; and p is ambient pressure in kPa.

The coefficients of the function were obtained by using a multi property curve fit to a carefully reconciled set of the best available vapor pressure (see Table A.13 [3]). The correlations have been proved to have good accuracy over the full range of liquid concentration from pure water up to the crystallization line and from 5 to 250 °C.

A.5 Specific Thermal Capacity (J/(kg·°C))

The equation proposed for the calculation of the specific thermal capacity of aqueous solutions of LiBr is as below:

$$c_{s,LiBr} = -2T(C_o + C_1\xi + C_2\xi^2 + C_3\xi^3 + C_4\xi^{1.1}) - 6T^2(D_o + D_1\xi + D_2\xi^2 + D_4\xi^{1.1})$$
$$- 12T^3(E_o + E_1\xi) - 2\frac{(F_o + F_1\xi)}{(T - T_o)^3} - 2pT(V_6 + V_7\xi)$$
$$+ \frac{1}{T}(L_o + L_1\xi + L_2\xi^2 + L_3\xi^3 + L_4\xi^{1.1}) - (M_o + M_1\xi + M_2\xi^2 + M_3\xi^3 + M_4\xi^{1.1}), \tag{A.28}$$

where ξ is mass fraction of LiBr in the solution; T is temperature of solution in kelvin; and the coefficients are given in Table A.14 [3].

A.6 Density (kg/m³)

Density of LiBr solution can be calculated by the following equation:

$$\rho = (V_o + V_1\xi + V_2\xi^2 + V_3T + V_4\xi T + V_5\xi^2 T + V_6T^2 + V_7\xi T^2)^{-1}, \tag{A.29}$$

where ξ is mass fraction of LiBr in the solution; T is temperature of solution in K; and the coefficients are given in Table A.14.

A.7 Dynamic Viscosity (Pa s)

Dynamic viscosity of LiBr solution can be expressed by Equation (A.30) based on the data from Diguilio et al [4].

$$\ln \mu_{s,LiBr} = \sum_{i=1}^{4}\sum_{j=1}^{2} a_{ij}x^{j-1}T^{i-1} \tag{A.30}$$

Table A.13 Parameters of the Gibbs function (Equation (A.27)) for calculating the vapor pressure of LiBr solution [3]

	0	1	2	3	4
$A_i, i = 0 \ldots 4$	5.506219979E+3	5.213228937E+2	7.774930356	−4.575233382E−2	−5.792935726E+2
$B_i, i = 0 \ldots 4$	1.452749674E+2	−4.984840771E−1	8.836919180E−2	−4.870995781E−8	−2.905161205
$C_i, i = 0 \ldots 4$	2.648364473E−2	−2.311042091E−3	7.559736620E−6	−3.763934193E−8	1.176240649E−3
$D_i, i = 0,1, 2, 4$	−8.526516950E−6	1.320154794E−6	2.791995438E−11	NA	−8.511514931E−7
$E_i, i = 0,1$	−3.840447174E−11	2.625469387E−11	NA	NA	NA
$F_i, i = 0, 1$	−5.159906276E+1	1.114573398	NA	NA	NA
$L_i, i = 0 \ldots 4$	−2.183429482E+3	−1.266985094E+2	−2.364551372	1.389414858E−2	1.583405426E+2
$M_i, i = 0 \ldots 4$	−2.267095847E+1	2.983764494E−1	−1.259393234E−2	6.849632068E−5	2.76986853E−1
$V_i, i = 0 \ldots 4$	1.176741611E−3	−1.002511661E−5	−1.695735875E−8	−1.497186905E−6	2.538176345E−8
$V_i, i = 5 \ldots 7$	5.815811591E−11	3.057997846E−9	−5.129589007E−11	NA	NA

Summary statistics.
Average error: Vapor pressure = 2.97%.
$R^2 = 0.99993$.
Specific heat = 0.398%.
Number of data points = 1764.
Specific volume = 0.177%.

Table A.14 Parameters of the specific thermal capacity (Equation (A.28)) and density equation (Equation (A.29)) of LiBr solutions [3]

i	0	1	2	3	4
C_i	2.648364473E−2	−2.311041091E−3	7.55973662E−6	−3.763934193E−8	1.176240649E−3
D_i	−8.52651695E−6	1.32015479E−6	2.7919954388E−11	NA	−8.511514931E−7
E_i	−3.840447174E−11	2.625469387E−11	NA	NA	NA
F_i	−5.159906276E+1	1.114573398	NA	NA	NA
V_{0-4}	1.176741611E−3	−1.002511661E−5	−1.695735975E−8	−1.497186905E−6	2.538176345E−8
V_{5-7}	5.815811591E−11	3.057997846E−9	−5.129589007E−11	NA	NA
L_i	−2.183429482E+3	−1.26698509E+2	−2.364551372	1.389414858E−2	1.583405426E+2
M_i	−2.267095847E+1	2.983764494E−1	−1.259393234E−2	6.849632068E−5	2.767986853E−1

Table A.15 a_{ij} in Equation (A.30)

i \ j	1	2	3	4
1	15.434	−1.796	−454.0	−1645.0
2	−1.497E−01	8.581E−02	=3.187	−11.190
3	3.211E−04	−4.050E−04	−6.116E−03	2.286E−02
4	−2.398E−07	6.025E−07	2.69E−06	−1.336E−05

where, x is LiBr mole fraction, T is temperature of LiBr solution in K, a_{ij} is coefficients as shown in Table A.15.

References

[1] Conde, M.R. (2004) Properties of aqueous solutions of lithium and calcium chlorides: formulations for use in air conditioning equipment design. *International Journal of Thermal Sciences*, **43** (4), 367–382.
[2] Patek, J. and Klomfar, J. (2006) Solid-liquid phase equilibrium in the systems of LiBr–H2O and LiCl–H2O. *Fluid Phase Equilibria*, **250** (1-2), 138–149.
[3] Yuan, Z. and Herold, K.E. (2005) Thermodynamic properties of aqueous lithium bromide using a multiproperty free energy correlation. *International Journal of HVAC&R Research*, **11** (3), 377–393.
[4] DiGuilio, R.M., Lee, R.J., Jeter, S.M., Teja, A.S. (1990) Properties of Lithium Bromide-Water solutions at High Temperatures and Concentrations – I. *Thermal Conductivity, ASHRAE Transactions*, **96**(Part 1): 702–708.

Index